HAZARDOUS DUTY

One of America's Most Decorated Soldiers Reports from the Front with the Truth About the U.S. Military Today

COLONEL DAVID H. HACKWORTH
(U.S. Army, Ret.)
WITH TOM MATHEWS

AVON BOOKS NEW YORK

AVON BOOKS
A division of
The Hearst Corporation
1350 Avenue of the Americas
New York, New York 10019

Front cover inset photograph by General Sladin Bradley
Front cover photograph by Leif Skoogfros
Published by arrangement with the author
Visit our website at **http://AvonBooks.com**
Library of Congress Catalog Card Number: 96-19169
ISBN: 0-380-72742-0

The William Morrow edition contains the following Library of Congress Cataloging in Publication Data:
Hackworth, David H.
Hazardous duty / David H. Hackworth, with Tom Mathews.
p. cm.
Includes index.
1. Hackworth, David H. 2. United States. Army—Officers—Biography. 3. Soldiers—United States—Biography. I. Mathews, Tom. II. Title.
U53.H25A3 1996 96-19169
355'.0092—dc20 CIP
[B]

First Avon Books Trade Printing: September 1997

Printed in the U.S.A.

QPM 10 9 8 7 6 5 4 3 2 1

Dedication

For Eilhys—chief editor, and inspiration, soulmate

And to the brave men and women on "hazardous duty" who have put country over career to help me tell this story. Without such courageous soldiers for the truth, no nation can remain free.

CONTENTS

HAZARDOUS DUTY

PROLOGUE

After Vietnam, I needed a place to lick my wounds, to think, to start the healing. I needed to save myself. All the things that had made me love the Army and my country were mangled. By the time I left I wanted to nuke the Pentagon, Congress, and the White House. I felt like lining up LBJ, Richard Nixon, Maxwell Taylor, and William Westmoreland against a stone wall and shooting them. General Westmoreland, of course, felt more or less the same way about me.

I moved to Australia in 1972, to the Uki Valley near the base of Mount Warning in Queensland. A creek ran through the valley and there was a dirt track that dead-ended in a box canyon running up the tree-covered slopes of Mount Chowan. The spot still looked the way nature created it millions of years ago—rugged forest, unchanged by man or his machines. Under the double canopy of trees it was quiet, cool, musky, like the good side of Vietnam's rugged highlands. In the dark shadows of that rain forest, I finally found peace and I went there as often as I could. As the sun was going down, I'd sit on a high rocky perch commanding the valley and shut out everything except the noises of the jungle and the last brilliant light of the sunset.

Then one day I met a farmer on the road who said the land at the back of the Uki Valley was for sale. I chased down the owner, we agreed on a price, and in thirty days this pocket of rain forest was mine. It was autumn and the weather was perfect. I bought a small tent and some gear and camped out. The evenings were warm and the sky cloudless, the stars glittering like diamonds in the zodiac far above me. I set up camp by a creek that emptied into a pond with a small sandy beach shaded by tall trees and tropical ferns. I cooked over an open fire, "inhaled" a little dope, slept under the stars, and dreamed of Eilhys.

During my first months by the pond, I got rid of everything that reminded me of the Army. First I sent my five-thousand-volume library of military books to the Australian Infantry School. Then I methodically burned, buried, or gave away all the trophies and souvenirs I had been keeping for my I-love-me wall. I got rid of all things Army except one set of fatigues, a pair of jump boots, my Ranger patrol cap, and a footlocker full of notebooks, maps, and documents recording a soldier's life.

That left my decorations.

One day a schoolteacher and his class of twelve-year-olds were camped by the creek. I liked them. They were out for adventure and I had given them the run of the place. When the kids learned I had been a Yankee soldier, they wanted to hear war stories and see my medals. I didn't much feel like telling the stories, but I could see the medals blew their minds. Several of the kids asked if they might have one. I thought I'd go them one better. I organized an "Awards Ceremony" and told them all to turn out at first light.

When the sun came up the next morning, I walked down the dirt track by the creek through the fog. Long yellow beams of light filtered down on the encampment where the pretend soldiers stood in formation: three rows of ten, all soldier straight, like rows of corn in a soft mist, a platoon of troops standing tall at reveille formation.

I looked and for just a second I might as well have been in Italy, Korea, or Vietnam. I had to remind myself that I was not back in the Army, that I was just playing a game, that I meant to give the kids a kick.

It had rained the night before. The only noises were the rush of the creek and the sucking of my boots lifting out of the mud. I was bareheaded. No steel pot. I wore my camouflage fatigues, though I had ripped off all patches, badges, and insignias of rank.

One of the kids stood in front of the formation. He was th commander. When I walked up, he saluted and sounded off.

"Sir," he said. "The soldiers to be decorated are present."

Down the ranks of honor I went, awarding each hero several medals: a Distinguished Service Cross and a Purple Heart to one boy, a Silver Star and Distinguished Flying Cross to another, a Cross of Gallantry and Bronze Star for valor to a third, and so on through the rows. I told each warrior how proud his country was of his sacrifice and heroism. Solemnly, I shook each valiant hand.

After I gave the last medal to the last brave soldier, I hesitated for some reason. As I turned the Silver Star I had just pinned to his chest around in my fingertips, I saw an inscription that read: SGT. DAVID HACKWORTH, KOREA: FEB. 1951.

My baptism by fire and blood in hell's inferno.

Suddenly I could feel the freezing cold and the hot flow of adrenaline surging through my body. For just an instant I was back on a frozen battlefield among the dead, wounded, and dying. I could hear the roar of olive-drab Sherman tanks with their fronts painted like dragon's mouths. Red bloodstains spread across the pure hospital white of the ice-covered rice paddies where my 3rd Platoon of Company G had been clobbered. Broken bodies lay smashed on the frozen ground.

Lieutenant Phil Gilchrist's platoon was charging a snow-covered knoll to relieve pressure on what was left of my platoon. We'd jumped off that morning with forty guys. Now, four hours later, we were down to eight. The North Koreans were fighting to hold their ground. As Gilchrist's ridge exploded, a platoon-size enemy force rushed from a dike to his rear. I waded into them. It was hand-to-hand combat, rifles and submachine guns barking and chattering, bayonets slashing and sticking, grenades exploding, men flying to pieces like rag dolls caught in a lawn mower.

I could smell the cordite and see the ugly black smoke of explosions and the red flashes of fire. Up on the ridge, Gilchrist could see that he was being flanked. I could hear him give the command to withdraw. Then Sergeant Phillips said to him, "Don't worry about the gooks on the dike, Lieutenant, Hack is keeping them down."

Gilchrist scrambled down the hill a ways to see what was going on. "Sure 'nuff, Hack," he wrote me in a letter. "There you were—shot in the head—with your blood frozen on your face before it could clot, keeping the enemy on the far side of the dike from raising up and firing at us. Hey, Hack, did I ever tell you that the whole 1st Platoon was grateful. That was above and beyond the 'call of duty.' "

The Silver Star lay there on my fingertips. Then, as quickly as the vision had appeared, it vanished. I saluted the boy soldiers. Then I turned and walked up the road through the fog with no more medals, hoping for no more memories of war, of lost and shattered brothers, the killing and the waste and the futility of it all. When I got back, I took off my fatigues and burned them.

The images recede now, softening as the years blur into the past. At the time, I thought I was closing out my life as a soldier, an adventure that began when I was a little boy myself in Santa Monica, California—the day the last Japanese bomb was exploding on Pearl Harbor. In the weeks that followed, soldiers were everywhere in my neighborhood, digging in antiaircraft guns, erecting barbed wire along the beach, putting up barrage balloons, flashing great searchlights through the night skies looking for Japanese airplanes.

Panic struck, army camps went up almost overnight. The one nearest my house on Ocean Park Boulevard between Fifth and Sixth streets spread across one square city block that had been a park. A fence went up. Behind the barbed wire, tents on wooden platforms stood row on row. Little streets ran between them and there was a parade ground where soldiers drilled and bugles played.

I was ten years old and I already loved the Army. I was brought up by my grandmother. Her brother, my Uncle Roy, a vet shot up and gassed in World War I, was a war hero and also not quite all there. He'd been awarded the Distinguished Service Cross by Black Jack Pershing himself. He carried it with him everywhere and let me play with it a time or two. Maybe that's when I got stuck with the glory bayonet. As an eight-year-old I knew I needed to get one of those. Roy lived out at the Veterans Home, but in his day he had whipped the kaiser. He planted in my mind my first love of the U.S. Army. Movies filled out the romance: *Dawn Patrol* and *The Fighting 69th* with Pat O'Brien as the chaplain and Jimmy Cagney as the eight ball who falls on a grenade to save the lives of his comrades. Heroism, sacrifice, duty, honor, country. I couldn't get enough of them. They filled my imagination and swelled my scrawny little chest.

Within the park, the Army was billeting an infantry battalion from the I Armored Corps, General George Patton's boys. They trained out in the Mojave Desert and rotated back into Santa Monica. All that desert dust and the military's obsession with polished boots gave me and my buddies the chance to do something patriotic for the soldiers and to make some spending money. We found wood and made shoeshine boxes, then wangled polish and rags and set up for business at the front gate.

When the soldiers came out, three or four little boys were always there to nail them, and we soon became rich. Then others around the neighborhood noticed how much ice cream we were scarfing down and they got wise. The bigger kids made boxes, too, and started elbowing out peewees like me. That was how I learned to think like a guerrilla: *If only I could get inside the camp, I could get to the boots before the boots reached the big boys at the front gate.*

One day I slipped through the barbed wire at the back of the camp and darted into the first-squad tent. Each of the tents had about ten soldiers. They welcomed me and very soon I had more boots than I could polish. Then the first sergeant caught me. He thought what I was doing looked a lot more like espionage than enterprise, so he banned me from the camp. But I kept sneaking back in, and the soldiers hid me. Their average age was about twenty-six, and many had their own kids. They missed them, so they adopted me as the unofficial company mascot.

The unofficial company tailor made me a little uniform. In those days the troops were still wearing World War I gear, so I had little leggings, little boots, and a Smoky the Bear campaign hat. Then they hauled out an M1903 Springfield rifle, the World War I weapon, and taught me the manual of arms. So there I was all decked out and squared away and polishing shoes. I also learned the first lesson of intelligence: When you have a secret keep your mouth shut.

Then the First Sergeant caught me again, this time in uniform—death by firing squad for a spy. He was really hot until the guys showed him how well I could do the manual of arms. After I got out there presenting and shouldering arms like a madman, he made me the authorized company mascot. In those days, the First Sergeant was God. So after that I no longer had to dodge through the camp or hide in the tents.

In that camp I fell in love with soldiers. I loved their life, their closeness, their cheerfulness. They were like a family. They had high purpose; they were unified, dedicated. To me, they represented everything that was good. A sense of great adventure surrounded them. At age ten I knew my destiny. Nothing would be better than to be a soldier. At fourteen, afraid I would miss World War II, I lied about my age and went to the South Pacific with the U.S. Merchant Marine. At fifteen, I used my false Merchant papers to enlist in the U.S. Army, winding up on the Morgan Line around Trieste, trying to keep the Yugoslavs and the Italians from wasting each other. Now, at sixty-five, I find myself spending a lot of time back where it all began for me—the rugged mountains of Bosnia, Croatia, and Serbia.

As I look back at the fifty years that have passed since my first trip to Yugoslavia, I see that eventually the sunshine Army takes every warrior on a forced march into reality. At the movies you smell popcorn, all buttery and warm. On the battlefield it's cordite, acrid, exotic, sour as death. Once the first bullet has whistled over your head, once you have seen the guy next to you take a slug, all the romance and adventure and glory vanish. Blood changes the game. No one knows more about life and death than a combat soldier.

Sounding off against stupid generals and venal politicians finished me as a professional soldier. But I still had loyal friends in the Army, and there was one last chance that I might be redeemed. In 1974, when Hank "the Gunfighter" Emerson was Commanding General of the 2nd Infantry Division in Korea, General Creighton Abrams, the Army Chief of Staff, came out to Asia and stayed with him for several days. Both men were romping, stomping war fighters. At the end of the visit they spent a few minutes alone together over a drink. Hank was short an assistant division commander. He had an idea.

"I really want a hotshot general as soon as possible," he said. "This is a troubled division and I'm trying to shape it up."

"You can have any officer in the Army," said Abrams.

"Fine. I'd like to recall Dave Hackworth. He's living out in Australia."

Abrams had the surprised look of a buck just hit between the horns. Moving quickly, Emerson told him that he felt that the Army had mishandled me after I sounded off.

"What do you mean?" Abrams said.

"They treated Hack like some sort of a leper. They were going to court-martial him. It should have been evident to everyone that he was valuable to the Army, that he was someone who had truly distinguished himself in combat."

Abrams took a long drink. He was silent for quite a while, three or four minutes. That was his way. He never shot from the hip. He took another pull on his drink, looked down, looked up, got into his hard-thinking mood.

"Hack was a tremendous battalion commander," he said, then he fell silent. He puffed on his cigar, looked up at the ceiling, looked out the window, thinking. The silence went on for nearly five minutes. Finally, he said, "God, you know, I understand what you are saying, Hank. I know Hackworth was a tremendous combat leader. It's a good idea, but given what happened, I just can't do it. The conditions today just wouldn't allow it, wouldn't allow me to recall that boy."

When Hank told me the story later, I thanked him; then I said, "Oh, Hank, you're full of shit. You think that ever could have happened?"

"Well, I guess you're right, Hack," he said. "But we all need guys who are truth tellers."

What he said that day has always stayed with me. It must have been ten years later, toward January 1989, when Hank and I were invited to the U.S. Army Pacific Command in Hawaii to brief the brass at Schofield Barracks about our experiences during the Vietnam War. It was the first military base I had been on since 1971. While I was out in Australia, I didn't want to know anything about the Army. I just zeroed it out of my memory bank.

On the day of the briefing I rented a snappy white convertible and advanced on the front gate. When I pulled up, an MP was standing at rigid attention, tall and starched, in shiny boots and helmet, the way only an MP can look.

"Hi," I said to him. "I'm supposed to go to a conference here."

"What's your name?" he said.

"Hackworth."

He pulled out a clipboard. When he came to my name, it must have said Colonel Hackworth, because he came back to a rigid brace and snapped an incredible salute.

My hand started to go up. Then it started to shake. It rose from the seat of the car to my knee and then it stopped and wouldn't move another inch. My whole mind and body were saying, *You're not into the saluting trip. You're no longer part of this game.*

I dropped my hand. The MP didn't know what to do, because you're not supposed to drop a salute until it is returned. Finally, he dropped the salute, hesitated, then, completely confused, he snapped another one.

And the same thing happened to me all over again. The shaking hand, the frozen arm, paralysis.

I gave up. No way that hand was moving.

"Hey, everything's cool, man," I said. "What building is it? Which way do I go?" And I drove on in.

The experience was unnerving. I had a friend who was a shrink for the CIA and I asked him what had happened to me.

"Your whole system was rebelling against the military," he said. "You had dropped out. For you it was over, completely over."

But had I? At Schofield Barracks it had made me feel good that the Commander, U.S. Army Pacific, had wanted his officers to learn from Vietnam. But what really blew me out was how much the Commander and others who had been in the military's ticket-punching, bureaucratic dancing school since Vietnam had forgotten. Our institutional memory was shot. During that trip I met a lot of old friends from Vietnam: Colonel Mike Sierra, who had been a brand-new lieutenant in my parachute brigade; Colonel Dave Crittenden, a two-fisted stud out of the Hardcore Battalion; and feisty Herb Garcia, who had been my battalion Executive Officer in the 101st Airborne. All of them felt the same dismay I did. Nobody at the high command level could dial into the past and its invaluable lessons. The Army acronym for this is CRS—Can't Remember Shit. It's a disease—the higher up the ladder you go, the worse it gets until it fully blooms as terminal military Alzheimer's.

Several months later in the spring of 1989, I was invited to speak at the Army War College about my new book, *About Face*. At the end of my talk I criticized the new Beretta 9-millimeter pistol, the replacement for the old .45-caliber sidearm.

Steve Silvasy, who was then a brigadier general, snapped back, "Wait a minute, Hack. Hold on. That Beretta is a damn good weapon. I've put thousands of rounds through mine. Never had a problem. Hack, this time you're barking up the wrong tree."

I thought, *How could all the guys I've talked to be wrong?*

Even my daughter Leslie, a crack shot and a Beretta owner, had complained about the weapon's slide.

I'd known Steve for years. He was a great warrior in my parachute brigade in Vietnam. He was a young leader with strong convictions, not afraid to express them to the boss. An old salty airborne sergeant told me Steve became so pissed off about the high jam rate of the M-16 rifle that he slammed it down in front of his battalion commander and said, "Why don't we have a decent fighting weapon like the AK-47 instead of this piece of junk?"

All this rushed through my mind as Steve was vigorously defending the Beretta in front of a roomful of students and faculty members.

"Hey, Steve, I might have bad info. Maybe I'm wrong." I moved to another subject.

But a few weeks later, a friend who worked in the Army Chief of Staff's office bootlegged me a copy of a letter General Carl Vuono had sent to the chairman of Beretta. The letter said that the slide problem had not been corrected and it put the company on notice that the U.S. Army planned to cancel the contract.

With great glee, I sent Steve a copy of the Chief's letter, along with a note: "Guess I wasn't barking up the wrong tree. I'm concerned about your early stages of CRS."

In just twenty years, Steve had gone from a romping, stomping, hands-on warrior who was one of us, to a guy who had become one of them, someone who had forgotten where he came from, someone who couldn't remember shit. As the troops say, "And so it goes."

After my lecture at the Army War College broke up, a full colonel, an Army aviator, bopped up.

"Hack," he said. "I was the brigade aviation officer when you commanded the Hardcore down in the Delta."

I thought the guy's face looked familiar. Smart-looking dude. Rows of ribbons ablaze on his chest. The top decoration was the Distinguished Flying Cross. Then, BINGO, it all came back to me.

February 1969. I had a yo-yo brigade commanding officer, Colonel Ira A. Hunt, who was into instant glory—his glory. It was clear he intended to make general on the backs of the kids in the rifle companies and didn't give a damn how many of them he killed to prove that he was Patton reincarnated. He didn't know shit from Shinola about war or war-fighting. He was an engineer, one very ambitious ticket puncher. The rest of the infantry battalion commanders and I spent most of our time spiking his harebrained ideas, ideas that got our units in bad trouble. He was more dangerous than a platoon of Viet Cong sappers who had gotten through the wire with satchel charges.

One day I had a major fight with this colonel about the way he was playing squad leader with my warriors.

"You don't give orders to my subordinate units," I told him. "I'll be damned if I'm going to take your shit."

"Let me remind you, Lieutenant Colonel Hackworth, that I am your brigade commander and you are being insubordinate," he snapped.

"You can pull rank all you want, Colonel," I blasted back. "Because no matter what you are, you are not a soldier."

Besides being the ultimate Perfumed Prince, Ira Hunt was a screamer. His face got as red as a matador's cape as he stormed off to his command post screaming, "I could relieve you, Hackworth. I could relieve you."

A few minutes later I was ordered to report to Brigade Headquarters. I knew that sorry son of a bitch was going to fire me in front of his witnesses and on his turf. But when I got to his CP, I found him pacing up and down in front of a bank of radios. A long-range recon patrol was in a world of hurt. Ira Augustus Hunt, in his typical imperial style, had flung this small recon element into a hairy situation. The patrol's chopper had been shot down on the landing zone. The LRRPs had no artillery or gunship support because the colonel had neglected to work up a contingency plan that would keep reserves and firepower locked and cocked. Now they were in the boonies surrounded by Viet Cong, about to be snuffed out. Between the chopper crew and the LRRPs, they'd taken sixteen casualties out of eighteen guys.

The colonel was wetting his pants. This would be his Little Bighorn. There would be no way he could cover up the disaster even though he and the division commander, Major General Julian Ewell, were so close the grunts were beginning to wonder whether they were an item. Ewell couldn't do a column left without snapping off Hunt's well-entrenched nose.

"Colonel Hackworth, do everything you can to extract them," he told me. He was speaking softly now. I was no longer the plebe being hammered by the sadistic upperclassman, now the guy needed me to pull his nuts off the barbed wire.

"All right," I said. "But if I get those men out, I never want you to fuck with my battalion again."

I headed for the brigade helicopter pad. Chum Roberts, my artillery liaison officer, and I climbed into a waiting light observation helicopter (LOH) and zoomed off to the Badlands. By the time we reached the LRRPs, it was getting dark. We spotted the downed chopper and we could see flashes from the firefight. When I got on the LRRPs' radio frequency to find out what was going on down there, all I could hear was "Mayday, Mayday, Mayday."

No one answered when I tried to radio back. From the muzzle flashes, I could see they were surrounded. If we didn't get to them fast, they were dead. For a few minutes we circled, waiting for a rifle company I had saddled up to come to the rescue. The reinforcements were choppering in, but they needed more time. This was the single most frustrating moment of my life. I was right there above the fight, but I couldn't talk to anyone on the ground. I had no way to influence the action. I felt like a handcuffed surgeon standing over a patient who was bleeding to death.

Charley Company was still twenty minutes out. Once they arrived I could insert them on the flank of the firefight and they could shoot their way into the besieged LRRPs' perimeter. But suddenly the LOH pilot said he was running out of fuel, we should break off. The choices were lousy. If we left before the reinforcements came in, the LRRPs would be fucked. At night there was no way Charley Company could find them, and by morning it would be too late.

I told the pilot to stay on station.

"Look," he shouted. "We're going to run out of fuel and crash. We've gotta break off."

There was only one thing to do.

"Get your shit together," I yelled to Chum. "We're going down there."

I told the pilot to land, but being sound of mind, he refused. So I thumped my .38 against the side of his flight helmet and said, "Land or you're dead."

He took us straight down like an elevator out of control.

The sky lit up and slugs zipped all around us. As we hovered over the downed chopper, our guys on the ground started grabbing the LOH's skids. It felt as if they were going to pull us out of the sky. Chum and I had to kick off the panicky hands so we could unass the bird.

We jumped out and landed in waist-deep water and the bird whirled back to base with its fuel light blinking red and the pilot sitting in two feet of shit.

Then Chum got on the radio and called in wall-to-wall artillery while I got the LRRPs and crew to form a good 360 perimeter. Between Chum's blistering artillery and the renewed fire from the LRRP, the Viet Cong lost their advantage. Charley Company, led by Captain Gordon DeRoos, checked in. We vectored his flight to our position and inserted one bird at a time in a little LZ that we managed to clear next to the downed chopper. As each squad of infantry came in, they pushed the perimeter outward while we evacuated the wounded on the outgoing birds. Soon all of DeRoos's boys were in and all the LRRPs and crew were out.

Then Chum and I jumped on the last slick and called it a night.

Colonel Ira Hunt never fucked with me or my battalion again while that sorry bastard commanded our brigade. He went on to make two stars before he was found out and retired a few years later. Somebody finally got his number but not in time to prevent him from allowing a lot of good men to be killed or wounded.

The colonel aviator sitting across a cup of coffee with me at the War College was the same pilot who had reluctantly dropped Chum and me off at the LRRP rat fuck twenty years before. He was awarded the Distinguished Flying Cross. I told him I often wondered what I would have done if he had refused to land after I laid that pistol aside his temple.

"I've thought about that myself," he said. "But you were one wild guy back then, and I wasn't game to find out."

What the armed forces and the country need is to stop brooding about the past and to think things through in a positive, constructive way. I don't want to bash the military. But as an institution it is not above criticism. It desperately needs honest critics who know what they are talking about, not ideologues or fools or people who have been co-opted, but tough-minded patriots who will push for a lean, mean, invincible, and affordable defense force. Maybe I can help.

The Vietnam interview I gave ABC's *Issues and Answers* at Cao Lanh in 1971 was the defining moment of my life. After I attacked the military's conduct of the war, I was not welcome within the U.S. Army as a force for reform. Vietnam left me full of bitter anger and disappointment, mainly with the military but also with my country and how it treated the warriors who fought there. After moving to Australia, I began to recover. I carved a farm out of the rain forest, then ran and eventually sold a top restaurant in Brisbane. But for almost twenty years I still held that anger close, poisoning my life. For me, writing *About Face* was healing, a catharsis. Not until I wrote the book was I able to start coming to terms with my own rage. I didn't realize until I finished that I no longer felt angry. I no longer wanted to slit William Westmoreland's throat, I just wanted to expose him for what he was: a reckless butcher. Then, one day in 1989, while I was sitting with my old Army mate Tim Grattan at his place by the lake south of Whitefish, Montana, I suddenly realized something: All the rage that had caused me to leave my homeland for self-imposed exile in Australia was gone. I looked around and saw that America was where I wanted to live. At that moment I started thinking, *Here is where I want to park my bones when I depart from this planet.* When I found myself saying that, I knew I was ready to come home. I bought a chunk of Montana and started planning my next venture: World Headquarters in Whitefish.

I ended *About Face* with the words "This is just the beginning."

At the time, I didn't know how much groundpounding I would have to do to keep that promise. Since *About Face* came out, I have spent seven years chasing grunts on the training fields and in battle, this time with a pen and notebook instead of an ammo belt and an M-16. As a civilian, I have waded into the middle of every major

American military operation overseas: Desert Storm, Somalia, Haiti, Korea, and now Bosnia.

Most of these shows didn't deal with war as I knew it. They have dealt instead with what the Pentagon likes to call "Operations Other Than War," aka Peacekeeping/Peacemaking/Peace Enforcing and humanitarian missions. Often I wonder why we persist in calling these military interventions "peace" missions when there is so little peace to keep. We are lurching into ratholes like Mogadishu, Port-au-Prince, and Tuzla because we have reached the point where every time CNN shows something horrible on television we think it's our moral duty to strap on our body armor and go out and fix it.

Principle—that is the word we keep hearing. In practice, of course, we are spending billions of dollars we can't afford on hare-brained military operations overseas. Instead, we should be devoting our limited resources to doing something about the violence and rising body count in our own cities and preparing for the real threats to our national security.

What we are seeing is a new face of war.

Two new faces, to be precise: one high-tech, shiny, and all bristling with smart bombs, the other low and dirty as a sucker punch.

The Gulf War showed us what high tech can do. On the first day of the air war, Baghdad was a modern city. By the third day there was no electricity, no running water, no regular transportation. Technology took out technology. If we had kept it up long enough, we could have blown Saddam Hussein and his thugs back to the days of Ali Baba and the Forty Thieves.

The low-tech version that snared us in Somalia and Haiti—and now has us by the short and curlies in Bosnia—caught us unprepared, even though we should never have let it take us by surprise. During the Cold War, Uncle Sam and Ivan had everybody lock-stepping behind them, left face, right face, forward march, walk walk walk, run run run. For forty years the little guys followed their superpower leader and did what they were told. If they didn't, they got the hell beat out of them.

Now Ivan has collapsed and while we Americans are trying to figure out who we are, the Ratko Mladics and the Mohammed Farah Aidids of the world are saying to themselves, *Hey, time to get a little power, time to get a little turf, time to even up some old scores.*

That's where we stand right now. We're looking at the little guys with their AK-47s, their satchel charges, and their anger. Flying the flag of revolution, they want to turn back the clock. For the next twenty years, they're the ones we're going to be fighting. To beat them, we need to strike a better balance. Right now we are paying far too much for high-tech whiz-bang weapons and nowhere near enough for the basics: better boots, rifles, and body armor.

We are not the only ones looking into the new face of war. The Soviet Union has fallen apart. The same thing could happen to China. With all its multiethnic divisions, China could turn into one gigantic Yugoslavia, twenty times bigger, twenty times worse. If fixes are not made in a hurry at home, even the United States could find itself balkanized.

It wasn't supposed to be that way when we won the Cold War. Remember how hopefully President George Bush talked about the New World Order. What we have gotten instead is the New World Disorder. For the moment we don't have to worry about nuclear war, but shit happens, and we don't know whether to jump into Rwanda, rush up to Somalia, invade Haiti, or bomb Bosnia. Within the new global madhouse, wars are nasty and brutish—and nowhere short.

I'd like to march you through some of these operations so you can see for yourself what our warriors are facing and where your tax dollars are being spent. Right now we are gearing up to fight the wrong kind of war. High-tech war, the highly deceptive, surface marvels of Desert Storm, has mesmerized us with illusions of invincibility, of bloodless wars with almost no casualties. This view of war as an arcade game is totally wrongheaded. We shouldn't be toying with it the way we do. It entices us into spending vast sums of money creating the wrong weapons systems to fight the wrong enemy. What if we become so dependent on this high-tech stuff that a few very smart nineteen-year-old hackers can shut us off? All that great gear is like the toy dinosaur with a worn-out battery. It goes *wheeeeeeeeeee* and falls over.

At the high-tech level, war has become so destructive that it's obsolete. Why didn't we fight Russia? We didn't fight the Russians because we knew damn well that if we blew up Moscow they were going to blow up Washington. The firepower on both sides was so awesome that both knew they couldn't win. It will nonetheless go down in history that both went broke just getting ready to fight. The Soviets developed cardiac arrest and bellied up. Now we are clutching at our own heart. The difference is, we had a credit card and they didn't. So we ran up a debt of about $5 trillion, for which we're now paying megabucks each year in carrying costs. Hurtling forward with more high-tech war toys is going to cost us even more.

We have all the satellites and aircraft carriers we need, all the tank divisions, bombers, and fighter squadrons. We've gotten drunk on weapons systems. Swilling down the propaganda offered by the Pentagon, the politicians, and the defense contractors, we are forgetting the first thing about war, which is the warrior. Our warriors today are wearing and carrying basically what their fathers packed in Vietnam: the same M-16 rifle, same jungle boot. I saw our grunts in Desert Storm putting tape over the boot holes where water is supposed to drain—this time to keep out the sand. Their M-60 machine gun is flat worn out. I stood at the Berlin Wall nearly forty years ago with the same weapon. You fire a few rounds and the thing jams. Jammed weapons mean dead soldiers. Our radios are still the old backpack monsters, heavy, with a line-of-sight transmission that can be blocked by the nearest hill. In the Gulf and Somalia the grunts were stuck with them while reporters talked to their editors in New York over cellular phones that weighed a few ounces.

The reason the grunts who will be doing our fighting get the lowest priority is that the companies that make the boots, the battle uniforms, the radios, and the rifles are not in the same league with the people who make the missiles, satellites, aircraft, and carriers. It's the high-tech guys who really write the heavy PAC checks. As a result, our troops go into battle with the wrong gear and get the shit shot out of them by an enemy who's smart enough to equip his warriors with the stuff that counts.

Why is this happening? Several generations of self-serving politicians and generals have produced defense gridlock in Washington. Do you remember that fine scene in the movie *Patton* where one of Patton's units is on the road, but it has bogged down and no one seems to know why? General Patton is wondering what the hell is going on. He drives to the head of the column himself and finds a guy with a mule blocking his advance; so he pulls out his pistol and shoots the mule in the head. "Get that son of a bitch out of the way," he says. "Let's get this goddamn thing going." Then he stands in the middle of the road directing traffic until the column gets moving and he can return to the attack.

That's the kind of leadership we need, not only from our generals but from our political leaders. Today, the President never moves an inch without a brigade of damage control counselors and housebroken generals around him. Today, once you get beyond the rank of lieutenant colonel, very few real war fighters are left. Bird colonels undergo a personality change. They can see power blinking all around them. They know they've made it a long way and if they can just keep their noses clean and look like recruiting posters, a big office with a big window is waiting for them down at the end of the hall. They are basking in the possibility of a star. They can see the brightness right before their eyes: one star, two stars, three stars, four stars—the whole Milky Way. This mentality corrupts good soldiers. They become politically correct. They adopt corporate strategies and drop martial values. This blinds them to the sacred interests they swore to uphold when they took their oaths as ensigns and lieutenants: duty, honor, and country. When your career shines so brightly, it's easy to forget what your job is really about. All that counts is grabbing that star and moving on up into the galaxy.

The Perfumed Princes thrive in peacetime because bureaucracies hate troublemakers. In peacetime, the system wants order: Don't make waves, don't churn things up—that becomes the byword. War fighters and truth tellers challenge the bureaucracy at the very moment it's most happy with the status quo. They get bad efficiency reports. In today's military, with all the downsizing, one bad report, a single dent on a non-politically correct officer's fender, is enough to destroy a career.

So we are breeding out risk takers, giving them bad fitness reports, eliminating them as fast as we can. Warriors can only be warriors. It's in their nature. Our young warriors are just as good as they were fifty years ago, in some ways better. The young soldier today is a lot smarter than I ever was. Our squad leaders, lieutenants, captains, and majors are as sound as those who won World War II. But show me a general today who is called Old Blood and Guts. Show me an admiral named Bull. Thank God we have one named Snuffy directing our operations in Bosnia. But he's among the last of the breed. Instead of Army leaders like Iron Mike Michaelis or Marines like Victor "Brute" Krulak we are top-heavy with modern corporate managers, not war fighters but highly political generals with graduate degrees hanging out their rear pockets. If you put a profile of our top-ranking military leadership up against a profile of corporate leadership, you won't find much difference between them because they spend so much time studying and copying each other.

Since *About Face* was published in 1989, a lot of soldiers who are truth tellers by nature have gotten in touch with me. Four hundred people contributed to *About Face*, and I can no longer count everyone who helped with this new book—officers, NCOs, and grunts from the Army and Marines, the Air Force and Navy, people who talked to me on battlefields while they were ducking incoming, people who talked in Washington, where they were doing the Pentagon dance. Sometimes it's by phone: "Hack, you'd better look into this."

Sometimes it's by letter: "Dear Hack, you're not going to believe this one." As each day passes, more sign on as soldiers for the truth. Over the past seven years some of these wonderful people have gone from first lieutenant to major or from major to colonel, though once they come within sight of the stars, all but the very best of them begin to clam up.

Before I write a story or speak my mind these days, I check with these soldiers to see what they think. Many are far more competent than I am in technical areas of the military, many are in positions to see what is really happening. They are not yessir types. They tell me when I'm full of it, but they're also willing to tell me when I'm right. All of them realize their careers would be destroyed if their superiors knew they were communicating with me. We've established a bond of trust. Right up front I tell them, "Believe me, whatever you say to me will not be traced to you. I will do everything within my power to keep this from whipping back on you."

They know I'm not going to compromise them. Some have spent weeks working with me on a story. Unpaid. They collect their information and when I ask where it came from they say things like "Sure, I went to DIA," "I called a pal at CIA," "I phoned NATO's J-3 shop. Here's the poop." Behind every story lies a wealth of phone calls, faxes, letters, cassettes—blood, sweat, and extremely heavy risk. Each guy is motivated for different reasons. I try my best to weed out tips that appear to be self-serving or for revenge. Most of my sources just want to stop the goofy things that happen in the military and in Washington. They want to stop us from buying a bad piece of equipment or to keep us out of the wrong fights, or to make sure we go into battle with the right gear, the right training, the right leadership, the right mission.

World Headquarters in Whitefish, where I live much of the year, is a million miles away from the centers of power in Washington and New York that so corrupt our top military leadership and the media. People ask why I do it. A few months ago my daughter Leslie asked, "Dad, when are you going to stop and smell the roses?" But I get that from everyone. "Hack, when are you going to take a break?" Or "Hack, you're running out of time." I hope to hell they're wrong. I'm not ready for the Old Soldiers Home quite yet. Something keeps driving me back to the barracks and battlefields. And I know what it is.

All my life I've seen us get it wrong. In Korea I saw soldiers die because of hypocrisy and deceit coming from the top in Washington and bad battlefield leadership from General Douglas MacArthur to platoon leaders. After that war I swore I would never again stand quiet while incompetent, uncaring politicians and generals put American soldiers in harm's way. I want to look after the grunt, to make sure our soldiers on our battlefields are well trained, well equipped, well led—and not sent to hot spots on missions that don't make sense. That's what I'm all about and that's what this book is about. As long as I'm around, I'm going to sound off, to persuade the American people to understand, to wake up and stop the insanity we get ourselves into every time the military industrial complex starts licking its chops over another war. There's just no excuse for the way we screw up, killing and maiming so many good young men and women, and wasting so many taxpayer dollars.

I hope this book helps us not to repeat the mistakes of the past. I hope it makes us pay more attention to the ordinary soldier, our greatest strength. I hope it keeps us from being mesmerized by high-tech wizardry that's worth far more to engineers in Seattle or Los Angeles and to business tycoons across America than it is to the grunt in battle. I hope it keeps us from losing our shirt to the military industrial complex. I hope it will help us to see that spending billions on weapons that don't work or are redundant doesn't make sense, that we should be spending our money to prepare for where our real fights are likely to be.

Right now, this is not happening. So as I said at the beginning, if you really want to know what's going down in places like Kuwait City, Mogadishu, Port-au-Prince, or Tuzla, let me share the stuff in my reporter's notebook that didn't get told.

Everyone at Baumholder is waiting for the Perfumed Princes. They are due on the base in Germany at noon and they are bringing with them the President of the United States. Bill Clinton wants to give a little pep talk to the troops he is sending to Bosnia. Since the speech will come at twelve-thirty, the orders of the day are to fall out at nine-thirty. That way the lucky grunts at Baumholder can stand in the cold and fog for three hours—and wait.

Perfumed Princes wear stars, grunts don't. To measure the distance between them you have to do your calculations in light-years. The Princes get promotions and villas, the grunts get the dirty end of the stick. After fifty years around battlefields I now see this all too clearly.

At eight o'clock that morning, the men of the 3rd Platoon of Company B, 4th Battalion, 12th Infantry, 1st Armored Division (Old Ironsides), come piling into the dayroom. The temperature outside is in the low twenties. They wear winter uniforms and new Army-issue versions of the old poncho and blanket "Baumholder coats" we improvised when I trained here nearly forty years ago.

An Army minder is there to do the introductions and police the meeting. A woman.

"He's a reporter," she tells the grunts. "If you don't want to read what you say in the press, don't say it. You might be embarrassed. You don't have to say anything."

I stand up.

"Yep, that's right, you guys. Just remember: name, rank, and serial number." I hold my hands up over my head as if I'm surrendering.

It totally defuses them. The minder blushes, the grunts lean forward.

"Hey, how many guys in this room?" I begin. I can count twenty-three. A Bradley platoon is supposed to have thirty-one. "Where are the other eight guys?"

They look at each other uncomfortably. The minder frowns. Two years training for Bosnia and the platoon is heading out missing a quarter of its fighting strength.

"How many of you have been here for a year? Raise your hand."

Eight hands shoot up.

"Eight, only eight," I say. "The rest of you are new guys? And the outfit is supposed to have been getting ready for this mission for two years?"

It doesn't make a whole bunch of sense to me.

"What about your winter gear?"

No Gore-Tex. They are still wearing heavy sweaty stuff from the seventies that doesn't breathe.

"How's your body armor?"

They've been issued old Kevlar vests whose shelf life has run out. In a firefight they'd be wrapped in the equivalent of cardboard, not stuff that stops slugs. The brass, of course, all wear the new Ranger body armor, as do the elite troops such as Special Forces. Priority always goes to the top brass and the connected, those who are somebody's boys. So here we are spending hundreds of billions of dollars on high-tech whiz-bang stuff, for fat defense contracts, for all that logrolling in the Pentagon and Congress, for all those sweet plums of the military industrial complex—toys and riches for the Perfumed Princes—while these guys are heading out for duty, honor, and country still suited and booted like their grandfathers.

And so it goes until we fall out to wait for the President, about five thousand troops standing in the wind. The day is freezing cold. The gray fog has grounded the choppers and the President has to drive in from Ramstein Air Force Base. For three hours the division cools its heels, real warriors lined up like toy soldiers for the photo opportunity that comes when the President and the Princes sweep in and take their places on the platform at the head of their public relations theater.

After the twenty-one-gun salute and the forced "Hoo-ahs" from the troops, the President offers his remarks, the same pablum he offered back home about saving the poor people of the Balkans from their own savagery. It isn't a Commander-in-Chief-to-soldiers talk. The speechwriters have gotten it all wrong. But the President looks pleased in his pretend-warrior jacket. The troops look at each other and shrug.

When the charade is over, we reassemble in the dayroom. Earlier that morning the grunts were worried about not having proper equipment, about mines, about not having enough troops to do the job right. Now what worries them is leadership.

"Did the President address your concerns, your anxieties?" I ask them.

The answer is no—in thunder.

These are not toy soldiers. They are warriors and they don't like being used as props by politicians and generals playing to television cameras. The low point had come when the President said if any of them got killed, he would assume full responsibility. The rhetoric sounded like the cheap lines of a big man on campus playing for brownie points, not something inspiring from a concerned Commander in Chief.

"Let me get this right," said one NCO. "He's going to assume the responsibility. But we are the ones who are going to die."

At that moment it hit me. All my life I thought I loved the Army. Now, listening to this mechanized infantry platoon, I began thinking, *No, I never loved the Army. What I always loved were the troops. And I love them still. I love them because they are what it's all about, soldiers on hazardous duty sharing the dangers of war, forged strong and pure and true. There is no bullshit about them. Because they do the dying. They are the absolute truth. Soldiers for the truth.*

CHAPTER 1

GOATS AND GUNFIGHTERS: THE GULF, 1990

The A-10 Warthogs were blistering an Iraqi position on a ridgeline about two kilometers in front of me—bombing, strafing, rocketing. Earlier that morning Captain Ed Carter had ordered the air strike. Green Berets were shooting laser beams at the targets. Riding the beams in, the Warthogs were scorching the Iraqis with Maverick missiles and iron bombs, then hosing them down with 30-millimeter nose guns. Four hits, two misses. Fifty thousand dollars a pop. Good day for the military industrial complex, bad day for Saddam Hussein's boys.

"That's it for now," said the air observer. "Let's take a break for lunch."

I was sitting on the top of a two-hundred-foot-high water tower watching the air show, shooting the breeze with Green Beret Sergeant Tom Jobe and scanning the Iraqi armor and fighting positions. An eerie scene: giant sand berms riddled with blackened craters stretched away to the horizon. After taking in the terrain, I decided to climb down and stretch my legs. I set off with a Special Forces sergeant to walk the perimeter, picking my way through a camp littered with shrapnel. I couldn't get any farther forward without stumbling into an Iraqi position.

About ten minutes into our stroll, the sergeant suddenly yelled, "Get down. Incoming!"

He hit the dirt in front of me digging faster than a dog scratching for a bone.

An incoming artillery round makes a distinct zipping sound and then a nasty CLUNK and BANG. I hadn't heard anything like that. *Has this guy lost it?* I wondered. *What's his problem?*

Then, out of nowhere—BOOOOOOOOM!

I looked up, and off to my right a Warthog was nosing up into the sky after dropping a five-hundred-pound bomb close enough to give an Irishman a shave. I looked again and a second Warthog was diving at me with exactly the same idea in mind.

Then I got down real fast.

When I lifted my nose out of the dirt, black smoke was rolling across the perimeter. I took off for the water tower like a targeted deer thinking, *Turn off this bad shit before someone like me gets squashed.* As I grabbed the ladder and started for the top, the battalion commander, an Egyptian commando, burst from his command post and started yelling at me. "Stop it, Colonel. Stop it," he cried. "The Americans are bombing us."

I stopped and turned around.

"I'm no longer in the Army," I yelled. "I'm with *Newsweek*." Then I scooted up the ladder.

The friendly fire was a little present from the United States Air Force and *Newsweek* magazine. Only a few weeks earlier—in December 1990, to be exact—I had been sitting in an office on Madison Avenue looking at an editor who reminded me of a general. I had first met Maynard Parker in Vietnam years before when I was hunting the North Vietnamese Communists and he was an aggressive young reporter chasing down stories. Now, a million light-years from the Central Highlands, Parker was editor of *Newsweek*.

"Hack," he said, "I want to hire you as a special correspondent. Go to the Gulf for a few weeks. Give me a story. Are we really ready to fight over there?"

Saddam Hussein had rolled into Kuwait and parked just across the border from our favorite gas station, Saudi Arabia. President George Bush and General Colin Powell were organizing an enormous army in the desert. No one knew what would happen next. The President was embarrassed and furious because Saddam Hussein had misread him, misinterpreting a year of coddling as a green light to seize Kuwait. Reporters from all over the world had hotfooted it to the Gulf only to be isolated like a dangerous virus by the Pentagon's Thought Control Police. I suppose the media war had made "General Parker" think of me.

The previous July, at Eilhys's insistence, I had sent him a story for *Newsweek*'s "My Turn" column. The Cold War was over and everyone was getting dizzy over peace dividends and deep defense cuts. I had seen the same thing happen after other wars. I was worried we would blow our budget on billion-dollar toys to keep the defense establishment happy and shortchange ourselves on what we really needed to confront the new face of war.

All this bothered me so much I wrote a piece called "Save the Boys, Not the Toys." I faxed it over to Maynard, who called me straightaway.

"Great piece, Hack," he said. "It's going to run next week."

The story led me into a new career and a long-term relationship with *Newsweek.* No sooner had Maynard bought it than Saddam Hussein decided to prove my point. In early August of 1990, he invaded Kuwait.

Now, a few weeks before Christmas, Maynard and I were looking at each other across a desk on the twelfth floor of the Newsweek Building. I was thrilled. All my life I had felt like an uneducated dumbbell and here was a leading newsmagazine offering me a very special job. Maynard was saying, "Now, let's see. We can do about X dollars. How about X dollars a week?"

I was so stunned I couldn't even talk. Me, a war reporter? *Forget money*, I was thinking, *Hell, I ought to pay you.*

My silence seemed to puzzle him.

"Well, okay," he said. "Double that."

My God, I'm thinking. *This guy's talking about big bucks. I'd go tomorrow and pay my own expenses.*

"General" Parker gave me a hard look.

"Well, I think I can go X more. But that's the most I can do."

Pulling myself together, I finally broke radio silence.

"Look, Maynard," I said. "That's fantastic."

I would have gone for zilch.

A few days later I climbed on a plane with my *Newsweek* credentials and set off for the Middle East. All the way to Saudi Arabia I felt more jittery than I ever felt going into combat. Fighting was my profession. I knew what to do when the mortars came crumping in and bullets started snapping over my head, but I had never done any reporting. Over the Atlantic I sat in the dark thinking, *You are fresh meat, man. Welcome to journalism, motherfucker.*

The plane touched down in Dhahran at two-thirty in the morning. Before doing anything else, I checked in with General Norman Schwarzkopf's Joint Information Bureau. The JIB was Thought Control Central. The night I arrived it was dark but a colonel was there on duty. My assignment from *Newsweek* was to do a cover story on the combat readiness of the allied coalition. Since I had never been a reporter before, I approached the mission the same way I would a military estimate. I was thinking, *Gotta see a tank unit, an artillery unit, and an infantry unit. Gotta see the choppers, a fighter squadron, the close air support guys, the bombers. Can't miss logistics, the engineers, and the docs.* I needed to see everything: the 24th Division, the 101st Airborne, the 1st Cav support units, the headquarters staff weenies, a cross section of our fighting and support strength.

The colonel gave me some application forms. I filled in eleven of the little slips and headed for the hotel thinking I had really hit the deck running. I had no idea how naive I was. The Army and the Saudis had approved my credentials, but they weren't happy to see me. Because of my background, they knew they couldn't bullshit me. If I got a look into any of their locked closets, I was sure to notice any dirty laundry. And they could no longer order me to shut up.

So they stonewalled me.

I started going down to the JIB three, four, and five times a day. "What's happening?" I said. "What about my requests?"

"We're working on it," they told me. "The unit is thinking about it. These things take time. You're not in the pool. Other guys have been here a lot longer than you. You've got to get in line."

Colonel Bill Mulvey, Chief of Thought Police, was the officer in command of the JIB. Slick, full of excuses, he was the kind of guy who would spin his own mother if he thought it would help him with the boss. Every time I went to see him, he said, "Hack, we're giving you top priority. I personally got a call in to the general. Don't worry about a thing. It won't be long."

"Thanks, Bill," I said, taking the colonel at his word as an officer. Then some honest sergeant took me aside and said, "Man, is he ever sticking it to you." Colonel Mulvey represented everything that is sick with military public relations: manipulation and hype over the truth, hypocrisy over the officer's code. How long would these flacks last if they had to follow the West Point honor code—don't lie, don't steal, don't cheat? For sure, most military flack shops would be empty within a week.

Under Colonel Mulvey, the JIB was screwing everyone, of course. There was no reason to take it personally. I was still so green I assumed the three Ps would get me through: Politeness, Patience, and Persistence. Not at this finishing school.

Finally, one of the honest young officers under the colonel's command took pity on me. One day we were talking in private out of sight of his boss.

"Look, Hack," he said. "Off the record—don't get me in trouble on this—but don't waste your time hanging around here. You ain't going nowhere, man. They're going to put you on ice."

Those were the facts of life. Before I gave up, I filled out a whole fistful of official requests—twenty-one in all. All denied. I still have the copies. I guess I feel perversely attached to them, like a writer to his first rejection slips.

If anything, the official JIB briefings were worse than the Five O'clock Follies in Saigon. During the Vietnam War any reporter with guts could grab a chopper and get out where the white phosphorus lit up the boondocks and you could get a good look around. But Colonel Mulvey and the JIB had everyone in this war zone pinned down. After a week or so I got sick of looking at them. My only hope was to break out of the daily trot between the hotel and the JIB. So I hustled a four-wheel drive and started probing operations. *Screw them,* I thought. Their stonewalling just brought out the guerrilla in me.

On my first patrol I ran into one roadblock after another. The military had corked the city tighter than a moonshiner's jug. The only way you could get to the front was to penetrate the checkpoints, but anyone with a press pass was immediately intercepted and returned to the kennel.

It constantly surprised me how many guys were eager to point their weapons at me. I thought we were on the same side. Jet-lagged, totally pissed off, I started thinking, *Here I am, a shavetail reporter on my first assignment, and I'm fucking up. I'm never going to be able to get the story.*

Precisely at that moment, someone in heaven sent up a flare and my guardian angel saw it.

Hack, you poor dear, she said. *I'm going to send you some help. One of your lieutenants in Vietnam. He's a full colonel now.*

About five o'clock that afternoon, the phone started ringing. I picked it up.

"Hack, it's Bobby."

"Bobby who?"

"Bobby Harkins. Dak To, 327th Airborne, 1966, Bravo Company, sir. Remember me?"

"Bobby! God yes, Ben Willis, Bill Carpenter and Hank 'the Gunfighter.' Holy shit, how are you?" I said, feeling like I had just drawn a royal flush in a no-limit poker game. "I really want to talk to you."

"Right. Me too. I'm taking you to dinner. I'll pick you up in a few minutes."

"Roger on that," I said. "I'll be out front of the hotel."

Grabbing a notebook, I headed for the street. I was standing there assuming Bobby would take me to a restaurant where we could have a good catch-up when a Humvee drove up and out he jumped. He wasn't thinking hors d'oeuvres or maître d's.

"I'm taking you to my outfit," he said.

Vrooooom, vroooooooom. We roared across town laughing and telling stories and catching up on a lot of lost years. The Humvee dropped us outside the mess hall. When we walked inside, I found myself back in the Army, surrounded by paratroopers chewing on the worst chow in town. The trays were loaded with MREs, the Pentagon's replacement for the K-rations of yesteryear. Officially, the acronym stood for "meals ready to eat." The grunts called them "meals refused by Ethiopians."

For the first time since I hit the desert, I felt right at home.

Bobby Harkins was commander of the Dragon Brigade, one of the most important units in General Gary Luck's XVIII Airborne Corps. It provided combat support and service, logistics and communications to some of our best warriors. If I could get a fix on what was happening with his outfit, I knew, I could figure out the screwiest American deployment I had ever seen.

After dinner, Bobby said, "Come with me, Hack." Up a ramp, through a door, past a security guard, and into an improvised auditorium we went. I walked into a big room and found myself looking at thirty or so officers and NCOs. Bobby had turned out every key man in his command.

"Tell this guy whatever he wants to know," he said, pointing to me. "He was my battalion commander in 'Nam and he's a straight shooter."

The shit slingers from the JIB were nowhere in sight. If Mulvey had known where I was, he'd have stroked out. I was surrounded by soldiers for the truth. It was like walking into the vault at Fort Knox with a shopping cart and permission to help myself to all the gold. I pulled out my notebook as guys started standing up and talking.

"Sir, I'm an intelligence officer," said the first man. "I'm responsible for monitoring Iraqi radio traffic. We read all their mail and we know exactly what they are doing."

"Sir, I'm a logistics officer," said the second guy. "Here is our status. We couldn't fight if we wanted to. We don't have enough ammo, we are short on trucks and tanks. We don't have enough TOW missiles. And those we have we can't trust."

"Sir, I'm an engineer. I've examined the enemy's barrier plan. Right down the line. Here's how it looks."

And on it went.

I was scribbling as fast as I could. By the time I left that room maybe four hours later, I knew the enemy's order of battle and our state of readiness—or I should say lack of readiness.

Or weakness.

We were nowhere near ready to fight. This was after Christmas. *So what's the big rush?* I thought. Back home a lot of Democrats were reluctant to go to war, but the Republicans were gung ho. Congress and the White House were directing a lot of heat at the Secretary of Defense and the Chairman of the Joint Chiefs of Staff, and only Lieutenant General Calvin Waller of the Army and General Alfred M. Gray, the commandant of the Marine Corps, had the guts to get out in the open and say, "We're a long way from being ready to fight this war."

After I got back to the hotel that night, I looked over my notes and did some calculations. It had been nearly five months since Saddam Hussein left the reservation. For years the Pentagon had been telling us that our military forces were ready to fight anytime anywhere. According to our standard battle doctrine we were supposed to be able to deploy 10 stateside divisions to Europe in 10 days. After nearly 150 days we had nowhere near that strength in Saudi Arabia even though we were moving onto a battlefield that needed almost no advance preparations. Everything was already there: modern airports, big storage depots, first-rate communications. The Saudis pumped gas to the world. We would never run out of petrol for our tanks, trucks, or airplanes.

So why weren't we ready?

Throughout the Cold War we had been forcing trillions down the Pentagon's gullet like a farmer fattening a goose—around $13.8 trillion in 1996 dollars. To pay for the Bush-Reagan arms buildup, we had taken out a second mortgage on the country. What had the American people gotten for all that money and sacrifice? A tin-pot dictator like Saddam Hussein felt safe waltzing into Kuwait. Nearly five months later he was still giving us the finger. We had been promised military forces that were lean, mean, fast, and invincible. Where were they when Saddam Hussein gobbled up Kuwait? What would have happened if a real enemy like the Russians had come galloping through the Fulda Gap? Had we bought and paid for a reliable military defense—or was someone just selling us a bill of goods?

After Bobby opened the magic door for me, word got around I was in-country. Every day for the next few weeks, a dozen or so soldiers would knock at my door, privates, corporals and sergeants, captains, majors and lieutenant colonels, all asking to talk. We swapped war stories until they saw they could trust me. Then it would be, "Hey, we're not ready yet. We don't have half of our vehicles and no chemical detection gear." Or, "Hey, they are bullshitting the troops."

They knew they were going to meet the enemy, and they were not afraid, but they wanted to fight the Iraqis at the right time. They didn't like politicians who wanted to zap Iraq every time their polls went soft, or generals who just saluted and said "We're good to go" when we damn well weren't. Maybe that's why Bobby decided to call me. Maybe it was just a cub reporter's good luck. Either way, with Mulvey and his Thought Police all over everyone, Bobby and his men put their necks on the line to get the truth to the people back home.

I didn't rush the story on to New York. I couldn't help thinking, *What if it was an organized leak? What if that night with the troops had all been a setup? What if brave Bobby had become twisted like Mulvey?* I didn't want to go ahead until I could confirm the story with my own eyes.

The next morning I snuck past the Saudi roadblocks, slipped up the coast, and stumbled into a Marine unit firing antitank missiles from the tops of Humvees. I thought, *My God, these gunners are exposed. In a hot firefight they would last only a few minutes.* The TOW missile system, deployed on top of a Humvee, was a good way to get killed. The gunner's head was sticking out of the top of the vehicle about seven feet off the ground and there was nothing to protect him. He was a sitting duck. A real widowmaker. A scientist or a Perfumed Prince had put this thing together, not a warrior.

The vehicles were neatly lined up behind a little sandbag firing line. I asked a sergeant why they weren't deployed tactically. He groused, "Safety comes first in today's Marines." The layout was like the rifle range at Camp Pendleton. *Not a way to train for combat,* I thought, remembering the old saw, A soldier does in combat what he's learned in training.

Then the shoot began. Eight missiles were fired: boom, boom, BOOM, BOOM, BOOM. One hit the target. The rest corked off in wild directions, or sputtered, shook, or just malfunctioned. This outfit would have been in a world of hurts if their task had been to stop a platoon of balls-to-the-wall Iraqi tanks coming at them.

That night, Mike Stone, the Secretary of the Army, a friend I had made through *About Face*, called. He said, "Hack, what do you consider our biggest shortcoming out there? What do we need urgently?"

Perhaps it was the dismal show I'd seen that morning that gave me an idea—the NCOs had blamed it on old missiles that had been in storage too long—or maybe it was my experience forty years before in Korea when grunts tried to stop Communist T-34 tanks with another lousy antitank system—the 2.36-inch World War II bazooka. That had been like throwing ice cream cones at tanks to stop them. Whatever it was, I said, "Get the French Milan. It's the best grunt AT missile in the world. It can be fired from the prone and it has a hell of a punch. The French Foreign Legion has it. I've seen it work. It's hot shit."

Stone said, "I'll look into it and see what I can do."

"Hurry," I said and told him about that morning's disaster.

I'd met Stone the year before. He'd invited me to his Pentagon office. I liked him. He seemed a straight shooter, though he'd bristled when I told him there were too many Perfumed Princes in the top ranks.

"Hack, you're dead wrong," he said.

I didn't stick with the subject. I knew when to retreat.

Then he told me that Carl E. Vuono, the Army Chief of Staff, was retiring and he was looking for a new guy. I said I didn't like Vuono and I knew he didn't like me. I didn't tell him I thought Vuono was the Army's head ticket puncher and number one Perfumed Prince.

"I'm glad he's leaving," Stone said. "He's an airhead. Who would you pick, Hack?"

"Bill Carpenter. He's the best war fighter you've got."

"Christ, the man doesn't talk. I sat next to him at lunch out in Korea recently and for an hour he didn't say one word except 'Pass the salt and pepper.' "

"He's not a talker," I said. "But what a hell of a troop leader and a great fighter. Besides, you don't need a salesman."

Stone looked down, thought for a while. Finally, he said, "Hack, the Chief of Staff must be the number one salesman in the Army. Given the cutthroat environment inside the Beltway today, if a service chief doesn't perform at least as well as Willy Loman on a bad day, that service is in trouble."

General Gordon Sullivan, a Pentagon veteran, became the new Chief a few months later. Several years after that, Carpenter retired as a lieutenant general. He refused to be assigned east of the Mississippi, which meant the Pentagon was out. He could rise no higher than three stars.

Now, out of the past another name came to me. The last time I had gone to the Army's National Training Center at Fort Irwin I had met a very bright young captain named Marty Stanton. We clicked. He was gruff and straight. He was also extremely bright, not a hairy-chested guy who wants to slug it out with you the first time he meets you, but a total professional. I began to wonder where I could find him. At one point he had told me he was going to Saudi Arabia to serve as an adviser to the Saudis. *Maybe he's right here on this battlefield,* I thought. So I did a little poking around and after cutting through a lot of red tape, in early January I was able to track him down and call him.

The duty officer of his unit answered the phone. No way did I want to give my real name. The JIB had taught me what happened when I used the name Hackworth.

"Hi," I said. "This is Mike O'Mara. I'm a Spec four here in Dhahran and I'm Major Stanton's cousin. Is there any chance I can talk to him?"

The duty officer hesitated, then said, "I'm really sorry to have to tell you this. . . ."

"What?"

"I'm surprised that your family hasn't told you. . . ."

Oh shit, Marty's dead, I thought.

"He's a prisoner of war. He was captured on August fourth."

"Oh God," I said.

"That's the bad news. The good news is that he's just been released with the rest of Saddam's hostages. He's on the way back. The intelligence types are debriefing him now, but he'll be coming back to headquarters tomorrow. Give me your number and I'll have him call you."

"No, no, uh, I'd rather call him."

The officer gave me Marty's home number. A few days later I called.

"How you doing, Marty? This is Hack. I hear you were up north. Wanna tell me about it?"

"Hey, Dave! Sure. It's a weird story."

We arranged to meet the next day. Marty was on his way back to his unit, which was deployed up at the front, but he made time for lunch. It turned out that on August 2, the day the Iraqis invaded Kuwait, he had been in Kuwait City as a tourist.

The morning of August 2, at 4 A.M., he woke up, looked out his window—and saw Iraqi infantry disembarking from buses. He called the embassy in Riyadh to report and was instructed to keep observing and reporting. He did this for three days until the Iraqis rounded up the hotel guests.

The Iraqis hauled him out of the hotel and interned him, moving him four or five times to different sites in Iraq over the months that followed. With his soldier's eye, he had seen things that would have been lost on civilians. He saw that Saddam Hussein's top unit, the much touted Republican Guard, was a third-rate mob. The tank crews were sleeping in the shade. They never pulled maintenance and treated their tanks like camels. The units he had seen were nowhere near as proficient as American forces.

"We'll knock the snot out of them in one minute," he told me.

The Iraqis had been very careless with their prisoner. Based on what Marty told me, the Iraqis were not ten feet tall, not professional. He felt sure the first time the Iraqis found themselves in the rocket's red glare they would cut and run.

This was not the story Americans were getting from the White House, the Pentagon, or CNN. To hear them tell it, we were about to step into the ring with Mike Tyson on a bad hair day.

I say this ruefully because in the beginning I was as wrong as everyone else.

Before leaving for the Gulf, I relied heavily on the Pentagon and my contacts within the intelligence community to draw up my assessments, a major mistake. I had plenty of good private sources, of course; but even some of my deeper throats turned out to be shallow.

At one point a friend gave me a study done by the Army War College predicting a bloody fight in the desert. According to this analysis, the United States was veering toward a war with battle-tested troops who had almost eight years of hard combat experience against Iran under their pistol belts. The Iraqi Army would be fighting on familiar battlefields. We would be the away team with an army so green it hadn't seen any significant combat since Vietnam.

The report looked reliable to me. The Army War College is one of the places where the Pentagon is supposed to get its best military thinking. After reading the study, I thought we were looking at a minimum of six months hard combat with perhaps fifty thousand casualties. No walk in the sun.

Now here was Marty back from the enemy camp, having clearly seen a bush-league force where everyone else was talking Arab Terminators.

I should have known better. I was still such a raw hand as a reporter that I'd walked right into a trap I would never have fallen for as a soldier: I believed the Pentagon's Poets of Propaganda and our vaunted intelligence agencies.

I wonder what we pay all those people for? They didn't warn us about Saddam's designs on Kuwait until it was too late. Then in their panic, they just nodded while President Bush built him up as the Middle East's Adolf Hitler. But, what can you expect of people who failed to predict the collapse of Lenin's Evil Empire—even though it came on their watch and even though they had been paid more than anyone in human history to keep an eye on Moscow? How would it look if the Iraqi Army was not the Bavarian Doberman George Bush made it out to be, but just a mangy piss-weak mongrel?

For me, the lesson was mortifying. From then on, it protected me from the worst distortions of the Great Media Mind Fuck: *If you can't see it with your own eyes, if you can't smell it and touch it and hear it and feel it, then you don't know what the hell you're talking about.*

Somehow—and fast—I had to beat feet up to the front.

Mark Peters is the only guy I ever met who called me yellow and walked away. He stood on the far side of six feet and even with cameras flying from his lean, fit frame, he moved like an antelope. He had the busted nose of a guy who never shied away from close combat, on the job or off. In an earlier life he had pulled a hitch in the romping, stomping Rhodesian Scouts, before that country turned into Zimbabwe. So he knew angles of fire as well as camera angles. He didn't give a shit about the first, but he cared passionately about the second. He was a smart, boisterous, two-fisted guy, one of *Newsweek*'s best. The minute I met him we clicked.

"I've heard of you," I said. "I'd like to work with you."

"Fine," he said. "I've heard of you, too."

The perfect partner for a guerrilla.

In the desert, top photographers were the true hard core, the guts of an outlaw band called the Unilaterals. No one moved like they did, no one knew as much about the battlefields. Many were freelance shooters, good men and a few good women. Mark knew most of the others from scrapes in Africa and other continents. Most didn't give a rat's ass about Colonel Mulvey, and all thought regulations were made to be broken.

I loved the Unilaterals. They didn't suck up and they didn't waste time hoping Colonel Mulvey would bless them with a spot on one of his press pools. Their temperaments didn't allow it, nor did the reality of their work. Reporters could always sit in their hotels and get enough bits and pieces from the JIB feeds, *Stars and Stripes*, and each other to patch together a butt-covering dispatch. But the photographers couldn't fake shots in hotel rooms or the JIB. They had to move out and take chances. They were brave and they were generous. "Hey, I know a good place to fish," they'd say. "Go over there beyond the dunes. See Captain Tahler. Use my name."

They had discovered ways of penetrating the blockade around Dhahran, so I knew it could be done. In guerrilla war all's fair. All I had to do was outguerrilla the JIB.

Mark had a good fix on the roads out of town. The next thing we needed was a disguise. We got some help from a truth teller in the XVIII Airborne Corps who had been slipping me information and sneaking me into his units. An airborne sergeant named Marty Martello, the all-volunteer Army's equivalent of a shrewd Sergeant Bilko, scrounged us helmets, boots, and outer garments, everything but rifles. He also knocked off a decent coat for me that was long and very warm, perfect for cold nights out where the camels roam.

I took a camouflage jacket that blended with the desert to a seamstress, who sewed in secret pockets. Up in the top one was the green Army identification card I carry as a retired colonel. I could pull it out and get by most military personnel. Then I bumped into a friend from Vietnam who had been a Green Beret in my outfit. He was working as a pilot for one of the oil companies. One night he invited me over to his place for some homemade moonshine and I told him the problem I was having with the roadblocks.

"No problem," he said.

He wangled some blank ID cards from the oil company. From then on we could pass as instant roustabouts. I didn't want to fake military credentials. That was a supreme no-no. But I still knew how to look like and act like a colonel whenever it helped. All that was left was our "orders." They took the form of a fax from "General Clifton" to Colonel Hackworth.

The fax read like this: "Colonel Hackworth, you are authorized to conduct covert missions behind enemy lines. You will report to no one but high command intelligence and to me. To all who read this, give Colonel Hackworth every consideration and anything he needs. Anyone who fails to do so will be reported directly to me. Major General Anthony Clifton."

"General" Clifton was really Tony Clifton, *Newsweek*'s Gulf bureau chief, an Australian who had covered Vietnam and two dozen other hot spots over the years. A good man. Tony drafted the fax. To make it look more authentic he sent it to me through the hotel teletype machine. He also got us a four-wheel vehicle. Mark and I sprayed it desert brown and painted it with the Desert Shield reverse chevrons. From then on, whenever we hit the road, it was just Colonel Hackworth and his driver, loyal "Sergeant Major" Peters, out for a little drive. With this rig we were able to get up to the front and ramble wherever we pleased while the bulk of the regular press corps was still held hostage in town.

Sure, I can hear the Army spluttering over the ruse, and I can hear the hotel-bound media guys sitting by the pool going "Tsk, tsk." But the U.S. military declared war on freedom of information during the Gulf. A lot of reporters grumped but the grumping accomplished very little. A reporter's duty is to bring the truth home. If his own government forces him to take guerrilla action, he should do what must be done to get the truth to the people.

Before Mark and I headed for the desert on our frequent "raids," we went to the supermarket and loaded up with fruit cocktail, other goodies, and a basic supply of canned food. We picked up all the newspapers we could find and as many cans of Coke as we could carry. That way when we went by a tank or artillery or infantry unit, we could bring the news forward, even though it was a week old. Just by giving a paper to each platoon and a few cans of fruit cocktail, we were able to build a lot of goodwill. We packed sleeping bags, a portable radio to pick up the BBC, which was accurate in its coverage, and we always carried four or five extra cans of petrol. We also had flashlights and defective gas masks issued by *Newsweek*, which put us right on par with the troops. Both masks had big holes in their sides. If Saddam Hussein had thought of it, he could have blown us away with a good whiff of cheap perfume.

If we sinned, we sinned good. With the JIB playing Info-Cop, we ran the blockade like Colombian drug runners. Sometimes we hit a roadblock and got snookered, but most of the time we made it. For anyone I considered really stupid or naive, I could flash my phony orders from General Clifton. Other times I could use my oil company ID. I had them all tucked in my secret pockets. The act got smoother and smoother with each trip.

Many of the roadblocks were permanent and we got to know exactly where they were. I carried a compass and we cut across the desert as if we were in a yacht race, shooting azimuths and praying my congenitally flawed sense of direction would improve.

"Okay," I said to my "sergeant major" behind the wheel. "We're eleven miles from Al Kazaam. Let's cut across here for a couple of miles, go around the village, and miss the checkpoint." Once past the obstacle, we pulled off the sand barrens and hauled ass down the road.

Every now and then the Americans or the Saudis surprised us with temporary roadblocks. When that happened, the best technique was simply to smile and salute. Because I'm an older guy, it was easy for a guard to take me for a colonel or even a general. When a roadblock loomed up, Mark and I flipped on our helmets, drove right in and I threw a snappy salute. According to military drill, the soldier couldn't quit saluting until I dropped my hand.

So I held my salute, which left the guard standing there with a dumb expression on his face, his weapon at present arms, no longer pointing at us.

"Drive on, Sergeant Major," I said, affecting a British accent. "We don't have time to waste here."

"Good day, soldier," I added, dropping my hand, and we roared down the road, laughing all the way to the front.

The desert was an endless expanse of rocky ground and dunes, burning hot by day and freezing cold at night. In a single day the thermometer dropped sixty degrees between high noon and three in the morning. Sandstorms caused brownouts where you could lose sight of men and tanks ten yards away. Dust as fine as talcum powder clogged everything: engines, aircraft, artillery, rifles. The desert floor was alive with scorpions and vipers and prickly camel thorns. Heat mirages deceived you: trees, water, enemy tanks, or anything else your overheated brain conjured up. "The whirling sands created electrical disturbances that shorted out electronic gear and made the compasses go crazy. The stinging winds left sores and blotches scattered across your face. Under traditional bedouin tribal law, a man could be pardoned for killing his wife when the weather got really bad."

I didn't see many Saudis anywhere near the front line—they didn't do much fighting themselves and they hired local tribesmen to perform their hard work. If you stopped beside an Arab infantry platoon or encountered a recon vehicle, they would invite you to sit down and drink their coffee and eat dates. The bedouin soldiers were very gracious, humble people. If you kept your hair cut high and tight, they allowed you to pass. The secret was that salute and a friendly smile.

After a while we got to know everyone in the sector. Mark and I breezed along busting through the roadblocks. I always brought bags of candy for the Saudis. They'd wave and shout, "Hi, Colonel." The rumor was that I was an intelligence officer collecting information, a rumor I probably started myself.

At first, my own side wasn't that friendly. The mind control boys had fanned out and done their job. Early on Marty Martello helped me get into a battalion of the 37th Engineers from the XVIII Airborne Corps. In combat the first guy that goes in is the engineer. He clears the minefields, he checks the roads, he's the one who prepares the battlefield.

No one, of course, had told anybody up the chain of command we were on the way. We just arrived. They stuck us in a tent with the security detail, and Mark and I both had a cot, pretty good conditions. From our base, we could range out during the daytime and roam around the battlefield.

My first day there, I wanted to go out for my daily four-mile morning walk. So I stowed my gear. Mark was talking to the grunts in the tent. I went out to stretch my legs, thinking I'd inspect the perimeter. I always do this wherever I go, just walking until I get in my four miles, keeping in shape for Eilhys and her forced marches waiting for me in Central Park.

As I was moving along, this young buck sergeant, great looking, a paratrooper about twenty-one years old, threw his M-16 on me and said, "You. What are *you* doing here?"

His voice was nasty and his face was getting red. The mind control brigade was getting to our kids and giving them Gestapo manners.

"I'm a reporter with this battalion," I said. "The colonel has invited us to stay five days. It's all quite legitimate."

"You can't be out here," he snapped. "You can't be in the battalion area without an escort."

He was still pointing his weapon at me. I have an Irishman's temper and guns pointed in my general neighborhood always trigger it.

"Oh, go fuck yourself," I said. "You can take that weapon and shove it. If you want to shoot it, you go ahead and shoot it. And if you're not quick about it, I'm gonna shove it up your ass."

The guy didn't know what to say. And I was just getting warmed up. Here I was, an American, and he was treating me like an enemy paratrooper who had dropped into the middle of their camp.

I turned around, left him gaping, and finished my walk. When I got back to the tent, Mark came out with a serious look on his face.

"Holy shit, Hack, now you've done it."

"How?"

"They're going to kick us out. They say you abused a sergeant."

"What's the story?"

"The sergeant major's been here and the colonel wants to see us."

When laid-back Mark got worried, I knew we were in deep shit.

They marched us to Lieutenant Colonel Robert Holcombe, commander of the 37th Combat Engineer Battalion, a professional from El Paso, Texas.

"Did you do it?" he asked me.

"Yeah," I said. "The goddamn guy pointed a rifle at me. Lucky I didn't stick it up his ass." All the time I was thinking, *You went to all this trouble to get here and now you've compromised a friend. The sergeant was just doing his duty, following Schwarzkopf's chickenshit orders. You're the one who screwed it up.*

So I said to the colonel, "The kid didn't have to wave his weapon around that way playing Nazi sergeant. But I'll admit I screwed up. I lost my cool. I should have said, 'Right, Sergeant, let's go back to the head shed and get this straightened out.' "

Colonel Holcombe thought our visit was a little more approved than it actually was and he accepted the apology. After our little heart-to-heart, he went up the chain of command asking, "What about these press guys?"

"What press guys?"

When he said, "Hackworth and Peters," they really freaked. But he was a good man, a lot smarter than the moron running the JIB, and he let us stay. The effect of his decision was better coverage for the military.

The longer I stayed the more I realized what great young fighters we were deploying in the desert. The problem wasn't with the grunts or the NCOs. My own mentor, Sergeant Steve Prazenka, had taught me long ago, there are no bad outfits, only bad officers. The problems I saw came not from the bottom but from the top.

Mark and I were living in a security tent with rotating teams that watchdogged the perimeter. The more time we spent with these soldiers, the more I saw what incredibly bright people were volunteering for the Army. One night I sat talking to two kids who sounded like West Point captains. They were discussing the sociological implications of the war, the political aspects, the international consequences.

"What's your job?" I asked one of them.

"I'm a mechanic, PFC," he said.

"What about you?" I asked the second man.

"Specialist fourth-class," he said. "I'm a dozer operator."

What an incredible Army, I thought. *The soldiers have never been so bright.*

Mark's balls were bigger than his brains. He lived on the edge. He was so brave that he didn't always think out the consequences of what he was doing; he'd just attack. This meant that as reporter and photographer we were often working at cross-purposes. For the reporter, the essence of a story is not only what you see, but what people are thinking, subterranean currents, the causes behind the effects. For the photographer, the image is everything.

Out in the desert, when Mark spotted an enemy position he immediately wanted to advance.

"No way," I would tell him. "We'll stand back at least five hundred meters."

The five-hundred-meter rule, tried and true, had kept me alive on more battlefields than I want to count and I wasn't going to start breaking it now for Mark's scrapbook. The effective range of an AK-47 is five hundred meters. Stay farther back and you improve your odds on living. But Mark thought my math was bullshit. This always sparked us off. We drove across the desert having roaring arguments.

"You're a coward. An old man. Yellow belly. No guts."

"You're too rash. You're going to get wasted."

One day we were driving up a main road and we stumbled on an Egyptian unit. Its tanks were dug in and its main guns were pointed toward the Iraqi fighting positions a couple of sand dunes away.

"We are not going up there to take a photograph," I said. We hunkered down about six hundred yards in front of the Iraqis, arguing like two rug merchants. Finally, I said, "Mark, here are the keys. You take the vehicle. I'm walking back down the road."

I had seen a Special Forces camp about ten kilometers back.

"If you make it, I'll meet you there. I'm not going forward to get killed."

"You're a coward," he said. "Give me those keys."

I tossed them over and started getting my stuff out of the vehicle. We'd been through this act before and he was calling my bluff. I turned to him and pointed to a bunker down the road.

"You see that bunker?" I said. "Inside that bunker I can see two guys. Iraqi soldiers. They are bent down right now looking over the berm, looking at us. And Omar is saying, 'Do you see what I see Abdul? It's two crazy Americans. Here come two more fools. We'll capture them and give them to Saddam Hussein and he'll give us the People's Hero Award.' "

Mark thought about it and for once he decided not to go for broke. Only a few weeks back, Bob Simon, a first-rate correspondent from CBS, and his crew had been scooped up while trolling in the desert. I threw my stuff back in the vehicle and we drove on looking for a little less trouble.

For me, it was a bit scary moving among so many Egyptian and Syrian units. They wore the Soviet helmet, the uniform, the shoulder boards. They toted AK-47s and had T-72 tanks. Whenever I was with them, my gut was always churning. And then it hit me! All the Soviet gear was blowing me out. I had spent my whole life from age fifteen, dodging that stuff. As a boy soldier, I had been carefully programmed: *When they start coming, you get your butt down and start shooting.* Now they were all around me. I had been going around feeling like a Yank wandering through a Soviet division on the wrong side of the Berlin Wall.

Up at the front during Desert Shield, I was looking at an army that had been rebuilt since its destruction in Vietnam, a spirited, well-trained, professional army capable of burying the ghost of that lost war. "You can't issue an order without someone asking why," Lieutenant John DeJarnette from Sedalia, Missouri, told me one day when I stopped to talk to him. These were thinking soldiers—the best kind—and disciplined. I never saw a soldier or Marine out of uniform. I could wander into a position at midnight and the grunt would be there with his gas mask at his side, his vest on, and his gear all up close, rifle at the ready. Back in the old Army you might see his rifle and gear spread from asshole to appetite. Small details, but they showed that this all-volunteer Army had completely recovered from its Vietnam hangover.

I saw other differences from the brown shoe Army, and some of them worried me. Within one unit I visited, typical of the new Army, 67 percent of the men were married. In my day we wrote letters home; these guys sent and received videos. On the tapes I saw wives talking and the children playing and waving. I scribbled a note: *What a big hurt if casualties come.* We now have an Army of dads and moms. In the old Army nobody was married because you couldn't afford a wife or family. The traditional attitude was, "When the Army wants you to have a wife, they'll issue you one." These new soldiers were all one big family. How would they handle it when one of their brothers was killed? Would it draw them together and harden their fighting spirit or would the unit fall apart? Until the fighting started, there was no way to be sure.

Everywhere I looked, I saw women soldiers. The flacks were saying what a great job they were doing. Out in the field, unit commanders and NCOs were telling a different story, always off the record because they were so frightened of bucking the forces of political correctness. Marty Martello told me that in his unit they had to conduct routine "fuck patrols" to keep male and female soldiers who had paired up from going at it in the desert. For the ten men and ten women who were getting it on in a given unit, everything was awesome. But among the eighty guys who were flying solo, it was a serious morale killer. No one should have been surprised when it turned out that we developed a Desert Storm baby boom. One day while I was visiting a hospital, a nurse told me that she was seeing women soldiers getting pregnant just to get the hell out of the desert and that it was not uncommon for some of them to swap urine samples to get that ticket home.

I sensed another problem as well. For many of these professionals the Army was merely a job, with all the strengths and weaknesses of any other. It was not a calling like the old Army. Their promotions, their future, depended on how they played along. Under those conditions, it's hard to stand up for what you believe and develop a healthy attitude about sounding off.

That's one of the things I love about draftees. They may be a bit short on the professional side, but they are not afraid to bitch. The military has taken a chunk of their life away from them, taken them out of the university, away from their families, and they are angry.

If they see an injustice, if they see something wrong, they are quick to blow the whistle.

A few of the soldiers I met on that trip shared my concern. One good man said to me, "We worry about some of the guys. They joined up to get schooling, not to fight."

Even if some of them had enlisted for the GI Bill, I felt certain they would all fight and fight well. Everywhere I went they were eager, leaning forward. I was seeing some crack units.

"Good to go, sir," they said. "Good to go."

During the weeks leading up to the air war, Mark and I perfected our routine. We would go out for five days, then head back for two days to file and do radio and TV interviews for *Newsweek*. Then we would refit for the next trip.

Back at the hotel, I got all the tactical maps and tacked them up in my room. I put on layers of acetate and drew arrows along the most likely lines of attack, analyzing the Iraqi positions, war-gaming what I would do if I were General Schwarzkopf. Stormin' Norman and I went to the same schools, we studied the same drills, so by early January I had a pretty good idea of how the war was going to unfold. I marked my maps with an arrow coming around the flank from the west. That way we would not assault the Iraqi trenches headlong as we did against the Germans during World War I. It was only common sense to go in through the back door.

I wrote a story for *Newsweek* and hit dead center. Stormin' Norman thought I had given away his Hail Mary battle plan. He was furious. I heard rumors he was going to lift my press credentials and boot me out of the theater. The truth is that General Schwarzkopf's plan was not exactly original. The end run is as old as war and everyone in the desert knew of it except that military genius, Saddam Hussein.

The timing for the air war was top secret. I stumbled on it completely by accident. One evening I got in an elevator and noticed a guy whose face looked familiar. I couldn't place him, but we started talking and it came out that he was the technical representative for one of the big defense contractors, the firm that made the Apache helicopter. He stepped out of the elevator and disappeared, but I saw him again later in the dining room.

"Do I know you?" I said. It turned out we had met in Vietnam. He was a retired lieutenant colonel, a former 'Nam gunship pilot.

That was the evening of January 16. After dinner I said to him, "I wonder when things are going to happen?"

"Tonight's the night." He leaned over and whispered, "It's gonna happen tonight."

He knew what he was talking about. His job tied him into the whole military communications web. He was one of the eight thousand civilian technicians scattered around the war zone nursing along new weapons systems with teething problems.

When a critical part was needed for an Apache, the tech obtained it directly from the United States—by Special Express. Within twenty-four hours the part was installed in the broken chopper and the bird was flying again. Before SpecEx joined the war effort, a chopper could be down for weeks. The system was great for defense contractors, who could crow after the war about how reliable their wonder weapons were.

A lot of systems were being hand massaged this way. If General Schwarzkopf hadn't had the techs and SpecEx, he would have had a hell of a time sustaining the war. While I'm into capitalism and the private sector, I can't help wondering whether this kind of privatization is really the best way to run a large-scale military operation.

That particular night, the Apache tech rep dumped a professional and moral dilemma right at my foot. What was I going to do with his tip? I had never been in this position and I felt a sharp conflict over my role. During the 1960s I had written pieces from Vietnam for military journals and later, in the 1970s, some newspaper pieces in Australia after I left the Army, but they had drawn on knowledge I already had. I still saw myself more as a professional military observer than a shoe-leather journalist. Whatever I was, I knew if I got a scoop that threatened American lives or national security, I couldn't use it. I would have to wait. The First Amendment does not give reporters the right to put the lives of soldiers or the future of the country in danger.

Later that night I got a call from Maynard Parker. Back in New York, "The General" was feeling restless.

"Hack, what do you hear?" he said.

"I think it's going to be pretty soon," I told him. "If I were you, I would send some reporters to the hospitals in-country. If the hospital lights are on and they're calling in all the medics, that would be a pretty good sign the attack's coming soon."

"Good idea," Maynard said.

I left it at that.

The night the air war began was totally weird; no other word fits. The Air Force cut loose about two-thirty, on the morning of

January 17. The sirens were blaring. Everyone was running around. A *Newsweek* phototechnician, David Berkwitz, banged on my door, yelling, "We're at war! We're at war!"

"Why are you waking me up?" I said. "We're a weekly. If we're at war, we can find out in the morning."

But he was pissing-in-his-pants nervous and needed somebody to hold his hand, so I figured what the hell and got dressed and went downstairs. I stepped off the elevator into pandemonium. Sirens wailing, tracers lighting up the black sky. Jittery gunners were shooting at stars, mistaking them for incoming SCUD missiles and enemy airplanes. Total panic.

I went into the dining room and fixed some cereal; then, bowl in hand, I strolled back to watch the media frenzy. A French team from Television 7 was running all over togged out in contamination suits. They had on gas masks with long hoses and peaked caps and rubber booties far bigger than normal human feet.

Ducks. They look like ducks, I thought.

These Martians were scurrying back and forth, in front of the elevator, and down the steps to the air raid shelter in the cellar. Most of the hotel staff were in the shelter. But these guys were newsmen, a camera team. BOOM, BOOM, BOOM. You'd hear the antiaircraft fire and explosions. Then quiet. The extraterrestrial ducks would run up, grab their cameras, and take a few pictures. Then it would be BOOM, BOOM, BOOM. Another round. And the ducks would do an about face, follow their leader, and waddle back all in a line to the cellar.

It reminded me of December 7, 1941, in Santa Monica, when I was a little kid and World War II jumped off: everyone looking up at the sky, everyone panicking, fire from the antiaircraft batteries firing at clouds, the citizens seeing Japanese soldiers behind every orange tree, rumors that the Japanese had invaded Oregon and were shelling Santa Barbara. The next day the only crater I could find was left by an antiaircraft shell that plummeted to earth right in front of my grandmother's church. I thought what a shame it hadn't blown up the church and Dr. Langford, my overzealous Sunday school teacher. But it was only a dud.

Basically the Gulf War was over within the first six minutes of the air attack—except for the *Today Show.* Bryant Gumbel asked me for an interview. When I got there, a colonel named Harry Summers, one of those well-fed military experts the networks love so much, was telling Americans we were looking at a one-year war with heavy casualties. Colonel Summers was probably cribbing from the same War College scenario that had deceived me. At that moment the air attacks were decapitating Iraq's command and control system, shutting the place down as a working military machine. Bryant Gumbel asked me what I thought.

"It's a six-minute war," I said.

Their jaws dropped.

"It's all over," I said.

No one believed it.

Afterward I thought a lot about that moment. There was something peculiar about the media frenzy, something that went deeper than the excitement and stress of covering an important story. It was almost as if frightened reporters who knew nothing about military realities wanted to inflate the war to inflate themselves. I don't know how they looked on the screens back home. To me, many of them looked as bloated as the balloons in the Macy's Thanksgiving parade, but not so funny—or harmless.

I had no intention of covering the war as another media phony. The first chance Mark and I got, we threw our rations and fart sacks into the four-wheel drive and took off for the front.

Our objective was to hook up with a Special Forces unit. We had a name Mark had picked up from one of the Unilaterals, a beautiful French photographer named Isabelle. Isabelle was so brave and headstrong that four Iraqi soldiers had surrendered to her out in the field. That drove Mark crazy. He was never quite sure whether it was more important to seduce her or find eight soldiers who would surrender to him—twice four to keep male honor intact. Isabelle had steered us to an outfit headed by a Green Beret named Ed Carter.

Once we reached the front, we drove around trying to find him. We wandered into one camp only to discover we had rambled into an Egyptian parachute unit. The Egyptians were more alert than the Saudis, professional warriors. The minute we drove up, they grabbed us. But they didn't know what to make of us. We were escorted into their command post. Ordinarily, it would have been run by a battalion commander, a lieutenant colonel, but that day the division commander, a general, also happened to be visiting.

I introduced myself as Colonel Hackworth and told him we were doing a recon.

The battalion commander was very friendly. He was planning an operation and asked me for my advice on a tactical problem. I was leaning over his map explaining how I would organize the troops and direct the air strikes—using all the tricks of the trade—when I heard his aide whisper something, and the colonel looked down at my feet.

My pants weren't bloused. Not only that, but I was wearing my Montana walking shoes, a heavy boot but not by any stretch of the imagination a GI boot. As telltale, I had on blue top socks and they were hanging down. The colonel glanced at me.

"You are not in the military," he said.

"I didn't say I was in the military."

"Yes, you did. You said you were Colonel Hackworth."

"I am Colonel Hackworth—here's my identification." And I handed over my military ID card.

"But we know you are a member of the press."

Somehow, while I was having so much fun helping him with his tactical problem, word had reached him we were on the loose in his sector. He nailed us. For the first time I felt a little pucker of anxiety because the Egyptians could be really mean hombres. The colonel was angry we had not identified ourselves straight off as reporters. But what really made him furious was that right in front of his general he had asked for military advice—and I had offered it. At that delicate moment, a squared-away Special Forces officer walked into the tent. I guess the furious Egyptians had sent for their American adviser. Luckily for me, Captain Carter, a Vietnam vet, had read *About Face.* He cut out the sting, assured the Egyptians we weren't spies. Then he took us over to his own camp and kept us there. We made it our base.

Carter was actually a warrant officer, which explained why he was a forty-five-year-old captain. The Egyptians were so into hierarchy they would only accept advice from commissioned officers. So for the duration, the Army had given him a brevet commission. His job was to run a Special Forces A Team with ten first rate troops, all senior NCOs, experts in weapons, communications, demolitions, close air support, and killing.

The well dug in camp was far forward; it had taken a lot of incoming rounds and the ground was littered with twisted shrapnel. I spent a lot of time on top of the water tower with a kid wrapped in a poncho against the stiff wind. He was from an intelligence outfit and his job was to monitor Iraqi radio traffic. He was writing everything down, feeding the reports to the Special Forces and to his higher command. I could talk to him and get the word on what the Iraqis were up to.

The kid spoke Arabic, as did many of the members of Captain Carter's team, a unit from the 2nd Battalion, 5th Special Forces Group. They were fine soldiers, expertly trained, who got off on calling Desert Storm a warrior's Super Bowl. This breed was different from my generation of Green Berets, snake eaters who never missed a barroom fight. The earlier generation were only into sex and adventure. These guys had master's degrees in Arabic literature and understood the bedouins as well as they understood the Iraqis. Not only were they mentally sharp, but they also knew how to fight—and it was wonderful being with them.

The third day in camp, the Warthogs almost got me at lunch break. When the first five-hundred-pound bomb fell I started thinking, *Man, I've always been a burr under the saddle of the Air Force and they're really going to get even here. They're going to eliminate me.* I scooted back up the ladder and shouted at the kid, who was already on the radio shutting off the attack. Not long afterward, Mark came back from trolling near the Iraqi fighting positions.

"You missed your Pulitzer Prize," I said. "You could have been right here when American planes were bombing their own guys, clicking pictures of those Warthogs diving down, the bombs exploding. You missed it, kid. Isabelle would have had it." I loved to slam it to the "sergeant major."

The Egyptians blamed Captain Carter for the five-hundred-pound bombs that had fallen, blasting the shit out of their perimeter.

"The Air Force bombs our ass and I have to eat sheep's brains to make it right," he groaned, but that night he went over for dinner and by the next morning the coalition was good to go.

The close call came about three weeks into the air war. It turned out the Warthogs had been tasked to hit an Iraqi bunker with an antenna and flying a flag. Our camp with the tall water tower and flag had looked a lot like the Iraqi positions two kilometers away. The pilots had the right ball park, but being hotshots, they hadn't checked in with our air controllers, so no one had laser-painted the enemy. When they saw our complex and didn't notice the flag was Egyptian, down they dove and blew the hell out of what they saw, nearly blowing us all away.

I wrote down their call signs and nailed them in *Newsweek*. About a year later I ran into their squadron commander at Myrtle Beach when the squadron was being disbanded, and we had a big fight. He wasn't pleased with what I had written about his outfit's performance that day. I told him tough shit. He was a prickly character, an observation seconded by his young pilots who told me he had spent a long time in the Hanoi Hilton. But I intended to get the message through to him and everyone else. Friendly fire isn't friendly. It can kill you just as dead as anything from the enemy. And his boys had damn near done that to us.

Whenever American senior officers came around, Captain Carter hid Mark and me. Other grunts out in the desert did the same thing. They wanted their story told. It angered them that the brass was keeping the media penned back at the JIB. That meant there could be very little serious reportage about the men and women down on the ground, the risks they were taking, the conditions they were enduring, how they felt. The Thought Police denied this information to the people back home. It seldom happened in World War II or Korea. It seldom happened in Vietnam. But it always happened in Desert Storm, and the grunts resented it as much as I did.

The day we left the camp for the last time, a platoon of the 1st Cav Division came sweeping along the side of the road in tactical formation. Mark was hanging out the back window shooting pictures. One of the JIB rules was that you couldn't photograph a deployed American outfit, and Mark was clicking away with a long lens that looked more like a cannon than a camera. The platoon sergeant spotted it.

"Stop that vehicle," he roared and he ordered his men to run onto the road with their weapons pointed.

Once again, I knew we were in trouble. But I also knew the rules of engagement. They were very simple. An American soldier couldn't shoot unless he was shot at.

"Hit the pedal," I yelled at Peter Sharp, a British tag-along reporter who was driving that day. "Let's get the fuck outta here."

All we had to do was speed up to 80 miles an hour—we were doing about 35—and we would have been out of sight by the time the platoon got on the road. But Peter chickened out. He stopped.

The Americans ordered us out of the vehicle and got very mean and abusive.

"Go fuck yourself," I said to the sergeant when he got around to me. "I'm not getting out of the vehicle, so do whatever you want."

"Fine," he said, "I'm going to arrest you."

Once again our own side was going to lock us up as if we were Iraqi spies simply because Mark had taken a few shots out of his window.

Just then, out of the camp sailed a Humvee with members of Captain Carter's team.

"Will you tell this asshole we are good guys, not spies?" I called over to them.

"Sergeant, why don't you just cool it and get out of here?" said the leader of the A Team, a very awesome Green Beret master sergeant. "These are our friends. They are working with us. They are on special duty here. Piss off."

That's how we got away. Captain Carter and his men knew we would never compromise their plans or endanger the unit. They could trust us. So they let us stay. They treated us as team members, fed us, billeted us, just the way it used to be in places like Anzio, Inchon, or Khe Sanh.

Note to Colonel Mulvey and the JIB: It can be done.

Very early on, Mark and I decided to sneak up to the border between Saudi Arabia and Kuwait. It was a dumb thing to do because we wound up in no-man's-land with the friendly lines far behind us on the south side of Kuwait City.

The first time we approached the border, we came up on a zone with a few villages. I was driving because Mark wanted to take some video shots. He was freelancing for ABC. The network couldn't get its own crews up there, so it had given Mark a video camera and offered him big dough for good footage. He was on top of the vehicle filming away and I was thinking, *This is getting too close.*

"Oh, don't worry about it," Mark said. "Creep up a little bit more."

My gut was saying, *Fuck this. Look at those houses. Dangerous people could be coming out of them in about one minute and then Bang Bang You're Dead.*

But Mark was concentrating on his camerawork and as usual was totally oblivious to the danger. Just at that moment an Iraqi soldier darted across the road from behind the house. He had an AK-47 at high port.

"Hang on," I yelled to Mark. Gunning the four-wheel into a U-turn, I hightailed it out of there. We roared back two kilometers, then stopped. There were no more Iraqis in sight, but it was the wrong place to be out for a Sunday drive. We spent a few minutes catching our breath and trying to figure out what to do next. I wanted to go back up to the front but on foot so we could snoop and poop, use the terrain to our advantage and have a good look into the Iraqi positions.

While we were contemplating what to do, I saw a patrol moving on the other side of the road. They were wearing green berets. From a distance of five hundred meters we couldn't tell who they were, but they looked friendly and they were moving toward the border, so we caught up with them. By the time we reached the rear of the patrol, we could see that they were Kuwaitis out on recon. I asked the patrol leader if we could tag along with him. He agreed and we wound up following him across the border into Kuwait. This gave us a good opportunity to see the front line Iraqi fighting positions: their tanks, antiaircraft, artillery. We got so close I could see Iraqi soldiers sunning themselves beside the tanks.

At this point, in mid-January, the air campaign was just starting and we never knew who might come and zap us from the sky, so we stayed with the patrol, Mark taking pictures, me making notes. After several hours we followed them back to their base camp. There we found ourselves with a first-rate outfit, a Kuwaiti special forces unit in direct commo by cellular telephone with the partisan movement inside Kuwait City. The colonel in command had gone to some of my own schools: the Infantry School, the Special Warfare School, and the Antiaircraft School. We told him we were with military intelligence. There wasn't much choice since we were wandering around no-man's-land decked out in our uniforms and helmets.

Later when I went out with the Kuwaiti green berets on a ground patrol, eleven deserting Iraqis surrendered to us. The patrol had set up in a perimeter at the border beyond the customs checkpoint on the other side of Khafji. We were just settling in when a group of Iraqis came ambling down the road. At first they looked like a Kuwaiti patrol. But they had their weapons at sling arms and were waving a white cloth.

"Halt. You're surrounded," the leader of the Kuwaiti patrol shouted out to them.

These sorry-ass sons of bitches didn't touch their weapons. They were no more going to fight than guests at a Quaker wedding. They threw their arms straight up in the air. All were frantically waving surrender leaflets. Our perimeter was around a depression in a sand dune. The lieutenant gathered the prisoners inside to grill them.

They said they were from an armored battalion. They groused that their leaders weren't looking after them, there was not enough water to drink, and not enough food. Each day they were getting no more than three or four small spoonfuls of beans and rice. No meat. Their teeth were falling out from malnutrition. The strength of their unit, its endurance and morale, was melting faster than hell's original snowball.

I looked over their kit and weapons. Everything was in bad condition, filthy. The weapons had not been oiled and rust was already eating at them. A warrior always keeps his magazine topped off. These raggedy-asses were wandering around on half-empty. One of them had been a taxi driver in Baghdad. He could speak a few words of English, all of which he used to curse Saddam Hussein. He said, "We no fight for him. We give up. Everyone is going to give up."

As one of the Kuwaiti green berets did the translating I started to think, *Something is very wrong here. These guys are all supposed to be from the Republican Guard, and the Guard is supposed to be the country's most elite unit. How come the best fighters are defecting?* Everything about these guys said born to lose.

What I was seeing in that sand dune squared with what Marty Stanton had seen after the Iraqis took him prisoner. Listless drag-asses with morale lower than whale shit. But at the JIB and Pentagon Follies, they were still talking about the Republican Guard as if it were Hitler's own Wehrmacht: giants who would rip our hearts out and eat them raw.

The next day I told the Kuwaiti colonel we were newsmen. I didn't want to bullshit him any longer. What the hell, in a way we were fraternity brothers. His response was strong and positive—he was glad the American press had moved so far forward to cover Kuwait's side of the story.

As a country, Kuwait had oil money flowing out its ears. But that didn't make life any easier for the Kuwaiti grunt. In the corner of his headquarters the colonel kept a cage with a canary.

"What's the deal with the bird?" I asked him.

"If the bird keels over, we put on our gas masks," he said. "It's our early warning system."

We were able to eat and patrol with the Kuwaitis. They covered for us. But I didn't want to sleep in their camp because I was afraid we would give ourselves and them away. Intelligence types came cruising through the sector all the time and Marine patrols were also working the same ground. We needed another place to live, something more private but nearby. Up near the border we had driven past an abandoned town. Everyone had fled, leaving their apartments empty. We made our way back and found a building with a 360-degree view and a garage where we could stow the four-wheel drive.

"Pull off the road and hide," I told "Sergeant Major" Peters. "I'm going to duck up there and do a recon."

On the top floor of the building, I discovered a penthouse apartment where we could hole up. Perfect. A villa by the sea.

Then the bottom fell out. As I walked out of our newly liberated quarters, I saw a Marine patrol coming down the road just as Mark stepped forward. I signaled desperately for him to hide, but he didn't see me or them and they nailed him.

Mark has dark skin and his eyes have a rakish slant. He could be an Arab or an Oriental, almost anything. Women love him for it. But the Marines almost shot him. When they first got the drop on him they mistook him for an Iraqi and he came within a whisker of being wasted. He had to throw up his hands and march over to them.

"Shit," I muttered under my breath. The only thing I could do was get out there and try to spring him. As confidently as I could I walked out of my own hiding place with a big Irish grin on my face.

And they put their guns at me.

There were three of them: a lieutenant and two sergeants. They could see quick smart I was no Iraqi.

"What are you guys doing here?" said the officer.

"Lieutenant," I said, "we're from military intelligence." I pulled out my Army ID card. After studying him for a microsecond or two, I decided he looked green enough to swallow my orders from "General" Clifton. So I pulled the fax from my jacket pocket and flashed it. He glanced down, then snapped to attention.

"You're the guys we've been waiting for," he said.

I didn't know what he was talking about but this wasn't exactly the moment to show it.

"Yep," I replied. "Semper fi."

"Well, you're two days late."

"Sometimes on battlefields you don't get where you're supposed to be on time," I said. I was trying mightily to sound as cool as a spook from the ice water Company at Langley.

And it worked. The lieutenant said he and his men were an advance intelligence liaison team. Their job was to sweep the sector, glean whatever intel they could, check in with our own Kuwait Green Beret colonel, then scoot back to Marine headquarters. That's what they had been doing when Mark stepped out and scared the socks off them.

The young lieutenant started talking faster and faster, telling me everything he had picked up.

"We've had one hundred defectors last week in this sector alone," he said.

The deserters were crossing the border even though they knew Saddam Hussein might punish their families. Going AWOL was doubly dangerous. Iraqi engineers had placed minefields in front of the

Iraqi Army to stop deserters as well as the Americans. Saddam had also placed special killer squads of sharpshooters at the front to kill defectors. Even so, nearly three hundred had come across in our zone. It wasn't hard to see why. The air war had been pounding the Iraqis for several weeks. Around Baghdad they put up a fierce air defense, but out in the desert they didn't have all the fancy stuff to protect themselves. I hadn't seen any ground-to-air missiles and heard only a smattering of antiaircraft fire.

Under the pounding of B-52s, the Vietnamese had stood up far better, but they could take advantage of rugged terrain and deep forests. Here, Saddam Hussein's boys were as exposed as eggs on a billiard table. It didn't take anything to fry them. Even so, the Viet Cong would have died to the last man. But the Iraqis didn't have that kind of fire in their bellies.

All this was passing through my mind as the lieutenant stood rattling away intel. Finally, he asked us what we were doing right at that moment. We said we were just getting our bearings. I couldn't very well tell him we had been looking for a pad.

"Well, come on," he said. "Let's go up to the camp and see the colonel and get a cup of coffee."

When we pulled into the Kuwaiti position, you didn't have to be a psychiatrist to see that the colonel was about to freak out. He knew we were reporters, but the Marines were talking as if we were secret agents and they all were working for us.

The colonel was trying to decide whether to laugh or have us shot. I was blinking signals to Mark to be cool and not say anything. Mark was talking up one of the Marine sergeants. All the while, I was trying to flash a sign to the colonel that we were just practicing a little innocent deception on the Marines and all was jake.

It was a very tense moment. We were eating dates and drinking Arab coffee. At one point I was able to pull out of the circle and sit off to the side as if I were thinking and writing notes. Mark stood up with the sergeant and walked out. A while later he came back in and winked.

"Everything's cool," he whispered.

It was my first chance to talk to him.

"Don't say shit to these guys," I whispered back.

"No, no, everything's cool."

"How's it cool?"

"I told them I was a *Newsweek* photographer and I gave this guy a whole bundle of film to keep his mouth shut."

I knew we were dead.

The Americans finished their chat with us and drove off. We stayed behind with the colonel pretending we were there for the night. The minute the Marines disappeared, I explained to the colonel that we had been compromised and had to scoot. It was dark by now, not the kind of situation where you'd want to be moving around very much. The ground assault was still a month away. But you couldn't be sure where you would run into Iraqi patrols. They were all over the place. Our side had little recon cells out too and we didn't want to run into any of them either.

So I grabbed Mark and we jumped in our vehicle and left the security of the Kuwaiti armed camp. We drove with our lights off until we came up to an abandoned hospital. I had seen it during the day and marked it in my memory as a great safe house.

"We're going down to that hospital," I said. "We're close enough to the Kuwaiti camp. If there's an attack, we can sneak out of here and help them fight. At least there will be a fair number of guys able to slug it out if the Iraqis get cute." I also figured a hospital would be full of things to make life interesting.

The gates were locked. So we found a chain and used it to hook the lock to the undercarriage of our four-wheel drive. Then we backed up and the gate swung open. We drove inside and locked back up. Everything looked normal and we could get in and out whenever we wanted.

As soon as we hid our four-wheel drive, we started watching the camp from a window in our new command post. Within a few minutes the vehicle with the Marines came roaring back through the Kuwaiti camp gate.

"They're coming back to arrest our ass," I said to Mark. "The minute your sergeant got alone with the lieutenant, he said, 'Those guys are not operatives, they're not intelligence types. They're a couple of wacko *Newsweek* reporters.' "

That's exactly what had happened, but by then we had flown the coop. The colonel didn't squeal and the Marines never found us.

The next morning I went back to do some damage control. The colonel told me he would still help us, but now he was worried the Marines might come down on him. He wouldn't blow the whistle on us, but we had to minimize the time we spent with him. That was a generous offer considering what we had just put him through. We stayed. The beauty of it was that intel on everything happening in the sector and in Kuwait City fed into the special forces camp. By that time the air war was going full tilt. With each raid the B-52s were smacking the Iraqis so hard you could feel the earth shake.

The colonel was there through it all receiving reports over his cellular phone. The Iraqis were taking a lot of casualties. At one point the colonel leaned over from the phone and said the floors of the hospital in Kuwait City were wet with blood. There were so many casualties the wards were doubling up, two guys to a bed and one guy underneath.

The partisans in Kuwait City were also relaying intel on their terrorist strikes against the Iraqis. Saddam Hussein's thugs were scarfing up everyone they could, arresting and torturing their suspects, trying in vain to suppress the rebels. Most Kuwaitis were holed up with enough food and water to survive for at least a while longer. Their morale was high. A land route led into the city and for a while Mark and I considered going in to link up with the underground. Finally, we decided against it. We might be able to report by cellular phone on what was happening in occupied Kuwait, but we would be cut off from the rest of the conflict. And how would Mark get his pictures out, by FedEx? And if the Iraqis caught us and made us prisoners of war, our effectiveness as reporters would be finished.

So by night we stayed holed up in the vacant hospital. One day we were talking to the colonel when a car rolled up all dust-covered and desert painted like our own. Out stepped two sharp warriors in British helmets and British uniforms. By then Mark and I were beginning to look like a grizzled sergeant major and an old, retread colonel, two guys badly in need of haircuts and shaves who weren't going anywhere in this man's army.

"We've been had," I said as these two young bucks stepped out and walked toward us. "These guys are going to arrest us."

The Brits spoke fluent Arabic. After their grand entrances and a few words with the colonel, they set off to walk the camp perimeter.

"Who are those guys?" I asked the colonel, expecting the worst.

"British reporters," he said, looking steadily at me.

Two seasoned Fleet Street irregulars, as it turned out. After a lot of years chasing stories around Africa and the Middle East, they too knew how to play the game.

The third week into the air war, I went out with a Saudi reconnaissance unit that was running an operation between Khafji and the main Iraqi battle positions along the front. American B-52s were arc-lighting Saddam Hussein's troops with bombs, the Marines were clobbering them with artillery, and there was a lot of fire and smoke in the air. But day after day the story was much the same and it bored my editor in New York. This frustrated me enormously. Mark and I were constantly exposing ourselves to life-and-death moments under fire to get a story and all the time my editor was blowing off our reports. He wanted something sexier, something new and different every time. Now, with more experience, I understand that each story must have a fresh angle. But back then, the pressure was infuriating.

We were really feeling the pressure. One day when we were out in the desert, we got a flat tire. This spooked Mark. Flats were the one aspect of life where he was very conservative and wise.

"Look, we don't want to be caught out here without a spare," he kept saying. "If we lose another tire, we're immobile and we're going to get trapped. This is the time to get out. We've got what we can get. Nothing else is going to come."

At any other moment I would have agreed with him, but this time I felt an old infantry man's twitch, something irrational, impossible to explain.

"No," I said. "We're staying one more day."

"Why?"

"I can smell something. It's going to happen. Trust me. Don't ask me why."

That afternoon we trolled around the Saudi recon screen, talking to the bedouin, and stopping at a Saudi National Guard command post. The American adviser seemed unusually guarded and distracted—we felt as welcome as an ex-husband at his ex-wife's wedding reception. Something was up; he was blowing us off. It wasn't what he said, but what he didn't. My gut kept screaming, *Something's in the wind. Hang around.*

Dusk was approaching. We were standing off toward the side of a defilade, down behind a sand dune where we could peer over into the Iraqi lines. The desert in front of us was antitank missile country. The Saudis had deployed TOW-toting reconnaissance vehicles here and there as a screen in case the Iraqis advanced. Bedouin tribesmen were manning the armored vehicles. The idea was to deceive and delay the enemy, to hide the true locations of the main defensive line. In any attack, the Iraqis would think they had struck the main line when they had only hit the recon screen.

The setup was dangerous. I didn't want to be too close to those vehicles. It made more sense to move off by ourselves in the dunes. That way if a firefight developed, we wouldn't take incoming rounds aimed at the armored vehicles. We were driving a Range Rover. After the big discussion over the flat tire, I remembered seeing an abandoned Rover on the main road from Kuwait, a road littered with luxury cars. During the invasion, the escaping Kuwaitis all drove as far toward Saudia Arabia as their gas tanks allowed, then got out and hauled ass on sandals, leaving behind at least a million dollars' worth of great wheels.

"Let's go over there and knock off two tires," I said. "We can put one on and have an extra spare." That made Mark happy. So we drove down the road, found the wrecked Rover, and started doing some liberating.

As we were taking the tires off, a Saudi patrol came by, a few nasty looking dudes with guns. The Saudis are very particular about stealing—they cut your hand off for openers if they catch you doing it. Over in Khafji, now a deserted city, the stores were full of merchandise and the front doors were swinging open. There were supermarkets, all completely vulnerable. It was like a town that had suddenly been hit by a neutron bomb. All the people were gone and you could go in and help yourself. But the Saudis wouldn't touch anything.

And there we were busily looting.

One of the Saudis pointed his submachine gun at us. Even though he was talking fast in Arabic, it wasn't hard to get his drift. He was going to bust us as thieves. The barrel of his weapon was pointed right at my gut. I reached over very slowly, put my finger in the end of it, and gently shoved it aside.

"Point it over there," I snapped, trying to convey the message that I was a high-ranking officer and he was camel shit. When I moved his gun off to the side, he didn't know how to handle it. He hesitated, giving Mark and me just enough time to go into our act.

"Sergeant, let's get going," I growled. "You're taking too long to change that tire."

With that we yanked off the tires, threw them into our Rover and got the hell out of there.

Earlier we had spotted a deserted motel down by the beach and away from the town. It had beds, suites, hot tubs, the whole lot; a lovely place to stay. A band of footloose British reporters and photographers had liberated it and made it their base.

"I can't stand another night in the desert," Mark said. "Let's go to the motel and crash."

"Nope."

"Goddamn it. It's a lot better than staying out here lying in the sand."

He was right and I was tempted, but I held out because I smelled an attack coming and I didn't want to be caught in a bad position.

So we moved over to the flank again, threw down our sleeping bags, and settled in for the night.

A few hours later the whole sector lit up like the Fourth of July. Elements of two Iraqi armored divisions came pounding down on Khafji right on our flank. The first units penetrated the recon line. Through the black night you could see the flash of cannons and hear the noise of a raging battle as the Iraqis tripped the picket. The forward positions fell back. Both sides were shooting wildly, and friendly artillery started slamming in, some just over our dune.

And then came the main event, the air strikes.

Tracers sliced down from the black sky. You could see rockets flash and bombs explode. Red and yellow flames licked through the darkness, and the smell of cordite drifted up toward us. The ground shook as if we were smack in the middle of an earthquake. When the sun came up the next morning, you could see the smoking carcasses of tanks strewn across the battlefield and several hundred Iraqis standing in the wreckage, their hands reaching for the sky.

I climbed on the enemy tanks. These were the first Iraqi vehicles I had seen up close. For four years as a kid in Italy, I had been a tanker and a recon man. Later as a captain I had commanded a light tank reconnaissance company for two years. My first combat in Korea was with a tank-equipped infantry recon company. I had been a driver, a loader, a gunner, a recon section leader, and a company commander, so I was game to size up the damage to the battered and bruised Iraqi armor.

The tanks were a motor sergeant's nightmare. Their tracks were sagging; everything was lousy. These Iraqis couldn't have fought their way out of a retirement home. I looked in the tank ammo wells and found water in them. The ammo was corroded. The tanks showed no signs of maintenance. The lube points were dry; the oil was grungy-black and low. I checked out the main guns. Any gunner who knows what he is doing calibrates his gunsight so it is exactly parallel with the barrel. It's called boresighting. You line both up on a given point. These sights were all cross-eyed. The gunners were more likely to have blown off the front of their own tanks than to have hit a distant target.

Was this the Nasty Green Giant the Propaganda Poets were talking about all the time? The Iraqi crews were all Ready, Fire, Aim. They hadn't coordinated their attack so that the infantry and armor would hit at the same time. They had gone Punch Two, Punch One—then thrown in the towel. Burned and blood-soaked bodies were thrown across the sand like broken rag dolls. They had lost about two hundred armored vehicles along with five hundred POWs and left the stuffing for a lot of body bags.

The Battle of Khafji was a deadly mismatch. Flying above the battlefield on the night of January 29, 1991, the forward-looking emitters in the bellies of our infrared planes had picked up every armored movement on the ground well before they got near our recon line. Heat radiating from the vehicles also showed up on sensitive night vision screens. The silhouette of a tank or any other kind of vehicle showed up as a greenish light, and then the fighters dove down and snuffed them out. They had a field day.

After the lead Iraqi elements squirted into Khafji, Marine Cobra gunships and Air Force Warthogs shot up the armored columns. As we watched, the planes and choppers battered the Iraqi armor, bombing, strafing, and rocketing them. The explosions lit up the night. Even for that master of miscalculation Saddam Hussein, the way he played Khafji was incredibly stupid. It showed just how bad his army really was and how vulnerable his armor was to air strikes, TOW missiles, and tank main-gun fire. The Iraqis were like ants on concrete. Our air and ground elements were like a giant smashing down with a sledgehammer. There was no place to run, no place to hide.

All of this was fairly obvious. What really blew me out was that Stormin' Norman had shied away from counterattacking. Up to that moment, the Battle of Khafji was the biggest fight of the war. It should have demonstrated to the higher command exactly what it could expect from the Iraqis. When you use two divisions, you are not sending out a recon patrol. What the operation proved conclusively was the Iraqis simply couldn't fight—and Schwarzkopf didn't know how to counterpunch.

To my amazement, General Schwarzkopf displayed very slow reflexes. He remained absolutely tied to his larger battle plan and didn't budge. The hallmark of a great general is that he never boxes himself in that way. I couldn't understand why we just contained the advance, shot from a distance, and didn't drive one down their throat. The next morning was the perfect time to whack them, when they were running, groggy from fighting all night. Two United States tank brigades could have been eating lunch at the Kuwaiti royal palace while sucking in the Republican Guard for the Sunday punch.

Instead, General Schwarzkopf sat in his bunker and bellowed and fretted. Maybe he was tangled in the complication of running a coalition force of so many nations. George Patton and Erwin Rommel would have been on the battlefield, sniffing out their opponent's weaknesses, exploiting them, driving their swords in to the hilt. With the instant commo we have today, General Schwarzkopf could have been forward making command decisions while still keeping in touch with the overall theater situation.

At that stage, the battle plan was to fix the enemy with the Marines, while the Army wheeled around the flank to the west with the Hail Mary terminator. Here's the way it was to work. First: the Marines. Fixing, fixing, fixing. Holding, holding, holding. Sucking a counterattack. Second: the Hail Mary.

On the night of the Battle of Khafji, Schwarzkopf should have launched his tank-supported Marines instead of holding them back for the fixing operation. The minute we breathed hard the Iraqis were going to take to their heels. By then they were so weak we no longer needed everything we had on the ground to smash them. If Schwarzkopf had been more alert, aggressive, and flexible, less of a slave to his plan, he would have counterattacked immediately. A counterattack would have drawn out Saddam Hussein's Republican Guard. As soon as the Republican Guard made its move, he could have swung the Hail Mary from the left. We would have destroyed the Guard and the war would have had a far different ending.

By the time of the Battle of Khafji, General Schwarzkopf had sufficient combat power on the ground to pulverize the Iraqis. Besides the Marines, he had the 1st Cav, the British 1st Armoured, the 24th Mech, the 101st Airborne, with another complete corps, VII Corps, unloading from ships and moving to assembly areas. He could have bagged a bunch of the Iraqi Army at Khafji. Instead, Stormin' Norman sat tight in his bunker and screamed at his staff. On television he had lightning reflexes; on the battlefield he was as sluggish as a World War I Liberty tank.

A faster response would have kept better faith with the basic American battle doctrine—our celebrated Air-Land Battle concept. As the war and staff colleges and Pentagon had worked it out, our air and ground forces were supposed to work together in a highly flexible way. When a threat or opportunity arose, they were supposed to be able to reverse field on a dime to attack in another direction. What Desert Storm proved to me was that in spite of the doctrine, our forces were very stiff in the joints. Control was highly centralized, much as it had been during World War II. General Schwarzkopf's approach was: Here are the orders: you can advance five hundred meters to the next phase line. There you will stop and wait. Freedom to maneuver was given only lip service.

Later on, after G-day, I would hear Marine and Army commanders say, "I went nine hundred meters beyond my phase line and had to withdraw."

Where was all that Air-Land Battle jazz the Pentagon brass and flacks had been pumping out when the time came for a test?

The minute the balloon went up, we blew it. As soon as General Schwarzkopf got the reports from his own people of the approaching Iraqi armor, he should have zipped over to Khafji and eyeballed the battlefield—seeing it, feeling it, hearing it, letting his juices flow, making critical decisions. Patton moved with his lead echelon and he always knew what was happening. At Bastogne, when the Germans came pounding in with their surprise offensive, he picked up his binoculars, said, "My God, look at this," then turned three entire divisions around and attacked in a different direction. "Fine, boys," he said. "Let's go." He was right there on the ground, he had the vision, courage, and flexibility to counterattack, and he stopped the Germans cold. That's how we turned defeat to victory at Bastogne. There was no timidity. If General Schwarzkopf had been in George Patton's boots thinking the way he did during the Battle of Khafji, we would have had to fight a longer, bloodier campaign in Europe. Patton would have been out there in the middle of the blood and gunpowder, not back in an air-conditioned bunker with aides to bring in the chow or aides to hold his place in the line to the john.

All the way back from Khafji, I felt higher than an A-10 deadheading home after putting iron on the target. Mark's photographs were superb and it had taken real guts to get them. At one point he had been out on the battlefield clicking away at corpses when a Saudi public affairs officer rushed up and yelled, "Don't take any pictures of the dead." Mark just kept shooting. So the flack butt-stroked him with his rifle in the back. Mark kind of turned around slowly and I thought, *He's going to snap that poor little asshole in two.* But getting smacked with a rifle didn't faze him. He just gave the guy a kiss-my-ass look, turned around, and went on shooting. Without even touching the Saudi, he coldcocked him.

The Saudis just weren't what they used to be. In the days of their grandfathers and great-grandfathers, they were masters of desert warfare and the hard-hitting, lightning raid. Now, under our guidance, they have turned into a sorry version of an American force. We have made the same mistake with them that we have made elsewhere—in Vietnam, Latin America, and a lot of other places around the world. We keep trying to create mirror images of our doctrine, tactics, and equipment, whether or not they are right for the local rumble. Of course, this is very good for the military industrial complex; it allows us to keep selling billions upon billions of dollars of the latest hardware. Sometimes I couldn't help thinking that the Saudis would have been better off and a lot more effective zipping around the desert in Land Cruisers equipped with Milan missiles, hitting and running, using mass swarm tactics the way they did so brilliantly before they got so filthy rich—and seduced by the West.

We were roaring back from the front. It was getting dark. Mark was driving and I was in the backseat of the four-wheel putting my notes together, organizing everything I had seen and heard. By that time the two of us had been working the front for four weeks or so. "You know," I said, "you've gotta be a Scorpio. But what day were you born?" He said, "Eleventh of November." My birthday. We both flipped. The chemistry had been perfect and now I knew why. Mark approached his work the same way I had when I was a kid with more balls than brains. Getting older had made me less reckless, but with "Sergeant Major" Peters at the wheel I didn't have to worry about missing any action, and with me in the back there was a chance that we might find stories and even get back alive. I knew who he was, who I was, and what it takes to make a perfect team.

In Dhahran reporters were completely isolated. Only in the desert could you feel the steady pulse of the military buildup. Not long afterward I decided to do a recon of our western positions. In early February, out in the middle of nowhere at a place called Rafha, I stumbled on a huge supply depot and air assault strip under construction. Engineers were buzzing around everywhere; airplanes were landing, dropping off supplies and roaring back up into the sky; trucks were dumping mountains of stuff. Stacks of ammunition crates, bladders of fuel, boxes of rations, water, medical supplies—everything a general needed to wage war.

Except warriors.

There were few fighting troops to be seen. This gigantic base was just sitting there in splendid isolation, nowhere near the battlefield. One glance told me I had walked into Schwarzkopf's main supply room for G-day.

When you launch an invasion, you normally send the fighting men with the logistics following close behind. This was just the reverse. The logistics were already in place. On the surface, the approach was dangerous because no one was out there to defend the supply dump. I suppose the thinking was that since we had such good detection capability from the air, we would have plenty of time to move in combat units if the Iraqis got wise and moved on the dump. There may also have been security screens over the horizon somewhere, though I hadn't seen any.

What I was seeing confirmed what I had worked out in the abstract over the tactical maps. Clearly we were planning an end run. Obviously, Stormin' Norman was going to have the Marines fix and hold the enemy, then swing around and get into his soft underbelly where there was no way he could protect himself. The big push would come from the west, nursing off that supply depot.

About a week later I organized a long recon beyond Hafr al Batin. The roads were jammed with convoys moving troops and supplies north and deadheading back empty. Combat units were still moving forward: the British 1st Armoured Division, a brigade of the 2nd Armored Division, the 3rd Armored Cav, the 3rd Armored Division with bridging equipment. Hundreds of gas tankers, hundreds of ammo trucks. Some convoys stretched away for fifty kilometers. Huge supply depots bordered the road, dumps storing half a million gallons of fuel.

I had never seen anything like it. The numbers were staggering, maybe 50,000 to 100,000 vehicles moving half a million men. In Vietnam the American buildup took a much longer time. Here everything was quick-smart right down to trucks carrying sacks of mail, semis with pallets of Pepsi, trucks with portable showers and shitters.

The main supply route was called MSR Dodge. Overhead, the skies were filled with wave after wave of C-130 cargo aircraft and huge choppers ferrying supplies to the front. Fat depots had shot up around Hafr al Batin and beyond where C-130s were wholesaling in supplies and units were retailing them out to the field. I saw asphalt helicopter pads a mile long. Choppers took off from the sand sucking up clouds of dust as they rose and slanted away across the desert.

Along the way I met Sergeant Larry Harrison, an engineer from Oshkosh, Wisconsin. He was working on the final shaping and grading of a new gravel road running parallel to the blacktop highway. Graders were kicking up plumes of desert dust all around him. "If we're not ready now, when will we be?" he said. "This road stretches across the face of the moon."

The sight was truly awesome. MSR Dodge was the only major road to the front, the vital artery, the lifeline of the longest supply line in history—12,000 miles from the United States with the last

100 miles tough as hell. If the weather turned wet or the engineers failed to maintain the roads, we were going to bog down the same way Napoleon got stuck with his cannons at Waterloo. His army traveled on its stomach; ours travels on fuel and gunpowder. Getting the right stuff to the right places was the most critical part of the war.

Most of the drivers were civilian contract workers, crazy third worlders who drove like kamikaze pilots, leaving both sides of the road littered with twisted wrecks. Mark and I drove those roads every week, covering as much as two thousand kilometers a week. Mark wheeled along like he took photographs, aggressive, cursing, swerving in and out. At the same time, nineteen-year-olds driving rigs with seventy-ton Abrams tanks on the back were playing chicken with huge semis loaded with supplies. To drive that road was Medal of Honor shit. It was Dunkirk, Bastogne; it was Saipan. There were more people zapped on MSR Dodge than were killed during the war. To me the amazing thing was that in spite of all the carnage, I never saw a general officer down on the road trying to find out why all these kids were being killed and doing something about it. Where were the generals? They were up in the sky in their choppers flut-flut-flutting to meetings.

The chopper gives our Army great mobility and firepower at a very high price: the top brass loses touch with realities on the ground. As late as 1942, General Ridgway rode a horse while commanding the 82nd Airborne. Three years later, Jim Gavin was still walking as a division commander. In Korea, Iron Mike Michaelis lived on the front, walking ridges and talking to grunts. The chopper has eliminated all this personal contact. It has made generals great time-saving managers while turning them into bad hands-on leaders.

Along MSR Dodge, I met a woman from Faraday, Louisiana, a National Guard staff sergeant who was driving a dusty tank carrier called *The Foxy Lady*. "The roads are good to Hafr al Batin," she said. "After that they're bad. Potholes big enough to hide a camel." Then she climbed behind the wheel and roared off and away on the dust bowl express.

Oddly enough, everywhere I drove I saw camels wandering among all the hardware. Strong winter rains had turned the desert brush green. But now the skies were cloudless and blue. Perfect weather to attack. Morale was satellite high, no serious bitches, everyone full of piss and vinegar, confident, eager to go. Along the road where soldiers were filling sandbags, the tanks had names like *American Bliss* and *The American Dream.* The tankers liked the new

M1A1 Abrams, thought it could outhustle and outshoot the Soviet T-72. "Damn good tank," said Sergeant Harry Tenney, a platoon sergeant from A Troop, 1st Squadron, 3rd Armored Cav, a Vietnam vet who had supported my unit in the Delta. "We are ready as we are going to be," he said. "Good bunch of guys. Trained together, and we'll fight together."

Meanwhile, by mid-February the bombing was turning the Iraqi military machine into a giant scrapyard. The Iraqi Army could not take such a beating and keep its fighting edge. And we had plenty of bombs left. At one air base I saw iron bombs stacked as high as apartment buildings stretching off as far as the eye could see. So many only a computer could count them. Based on what I was seeing, it looked as if we were about two weeks shy of G-day. Soon we would start to see probing actions, patrols, raids, artillery exchanges with our guns moving into place, firing, then pulling back before the Iraqis could pinpoint them with counter battery fire. This would keep Saddam Hussein off balance. Our younger soldiers, guys who had seen no combat, were leaning forward, eager for action. At one stop I ran into Sergeant Michael Gutheridge, a kid from California who spoke for all of them. "I'm nervous but I'm well trained," he said. "When the shooting starts, I'll just grab my balls and go for it." Where did we ever find such good men and women?

As February wore on, it began to puzzle me why President Bush and the Pentagon were sounding so conservative in their public statements. Perhaps they were afraid to disclose how well the war was going, in case things suddenly went wrong. *No more lights at the end of the tunnel*, I thought. *They don't want to get caught again.* So they were painting a black picture. Their caution was understandable, but it didn't reassure the average citizen. Instead it fueled anxiety that Saddam Hussein had a secret agenda, that he would soon make good his threat to unleash the mother of all battles. This was happening exactly at the moment this particular mother was about to die in labor.

A few days later I was out reconnoitering the extreme left flank of General Schwarzkopf's army. There were French troops in the area, the 6th Division, mostly French Legionnaires, very, very good soldiers, augmented by an American parachute brigade from the 82nd Airborne. The French mission was to screen the western flank looking into Iraq. All the roads were pointing dead into southern Iraq, and new ones were still under construction.

I was staying with an engineer battalion engaged in building

Main Supply Route Eagle. To be carving a road one hundred miles out in the desert didn't make sense unless somebody intended to use it, and with that road and the supply depots everywhere I now knew I was looking at a main theater of operations, even though I still couldn't see many troops.

General Schwarzkopf didn't move his units until a few days later, but he had everything well prepared. To me, the fascinating thing was that while the scale of the operation was far larger than anything I had seen in Vietnam, General Schwarzkopf had prepped the battlefield much the same way as the Viet Cong. They liked to move in their supplies, cache them, dig their command posts, conduct their reconnaissance, set up medical facilities, prepare the supply routes, cut through the bush, make high-speed trails—all this before the battle. Just shy of the attack, they would rush in their troops for a short, savage fight. Then they would scoot, because they couldn't afford to stay long. Airpower would eat them alive. I had to hand it to Stormin' Norman. In preparing the battlefield, he had absorbed some of General Vo Nguyen Giap's best lessons. There was an important difference, of course. Once we advanced, the only guy withdrawing would be Saddam Hussein.

Out on the western flank, there had been a lot of incoming Iraqi artillery and mortar fire. At one point I was walking down the road while Mark was out trolling and an armored vehicle drove up and out stepped a French lieutenant from the Legion. Good soldier. Very tough. He pointed his submachine gun at me and told me no reporters were allowed in the sector. Then he arrested me and took me down the road, where he turned me over to an NCO from the 82nd Airborne Division. An American sergeant began interrogating me. I told him I was from the press, not a spy. I was just waiting for my photographer to return. We had been staying with the 37th Engineer Battalion, which he could verify.

The sergeant took me to regimental headquarters and put me under guard. As I sat there sweating out the arrest, I was mulling over what I had seen. The French outfit, a light reconnaissance unit, was using a 90-millimeter all-wheel vehicle, not tracked vehicles. They had to be a screening force, not main players, because they were not heavy with tanks and didn't have a lot of combat muscle. General Schwarzkopf was taking a risk by stacking such a huge logistics operation in a zone with such a light force to defend it. He was leading with his jaw.

But the command post, nestled among rolling sand hills, was very well dug in and camouflaged. I was sitting on a sandbag outside the operations tent and even from that vantage point I had a hard time spotting the fighting positions and other installations. All of them were covered with desert-colored camouflage and blended right into the beige sand dunes. At that moment I was thinking how I would like to be a fly buzzing around in the command tent, eyeballing the maps, checking the intelligence, finding out what the hell was going on in this weird war. Suddenly my daydreaming was interrupted by the tall, rugged-looking paratrooper standing guard over me.

"Hey, Mr. Reporter," he said. "How come I know your face?"

I was writing notes when he started up. I told him I had written a book about my military experience, that maybe he had seen me on TV.

"Goddamn," he said. "You're Colonel Hackworth. You're the hot shit dude who tells it like it is?"

He hurriedly set down his rifle, which up to then he had been pointing at me as if I were Saddam Hussein's first cousin. Now he put it against a sandbag abutment and asked if I would give him my autograph.

"Sure," I said.

He tore a piece of cardboard from a ration box and handed it to me. I signed it. From then on, so far as he was concerned, I was no longer a prisoner of war. Forget the Thought Police, I was just being temporarily detained. Instead of grilling me, he started pumping me for war stories. He wanted to know what combat was like. He wanted to know about Vietnam.

All the time he was talking, I had my ear cocked to what was going on in the tent, the command's tactical operations and intelligence center. The radio was squawking, spitting out info: Iraqi artillery rounds were landing at such and such coordinates, platoon-size forces were moving to the southwest, and on and on. The code name for the road we were on was Eagle, and the engineers "working on the hardball" had taken six rounds of mortar fire.

If I wasn't a fly buzzing around the tactical operations center, I was the next best thing.

Then, all of a sudden, the wind started blowing up.

"Here comes another fucking brownout," said my newly found fan.

I watched as the sandstorm started whipping up the desert, enveloping the command post area. In the wildly whirling dust, everything began to disappear. The paratrooper looked at me for a second. Then he said, "Listen, Colonel, why don't you split? Just head out to the main road and hitch a ride. I'll say you disappeared in the brownout."

He didn't have to tell me twice.

I thanked him, we shook hands, and I started walking. By the time I moved off a dozen yards, I could no longer see him. The tent vanished. A blanket of dust settled over the entire regiment. Everything was gone. I caught a ride as soon as I could and got out of there.

Farther along the flank we hooked up with another unit preparing to jump off on the ground assault. One evening toward the third week in February, the battalion commander's briefers started outlining the battle plan. First, the operations officer said, "we'll be positioning ourselves and moving into our preattack positions tomorrow." Then the intelligence officer ran down the latest intel. Finally, the logistics officer got up and said, "Sir, we've been issued eighty body bags today."

They had completely forgotten I was there. I walked out of the tent feeling shaken. During the briefing there had been so many clues that all I had to do was triangulate the numbers—how long it would take to move forward, how long to breech the enemy minefields—and I could see the ground attack was set for February 24. I stood there thinking, *What a scoop.*

It was dark. Suddenly I heard a voice.

"Colonel Hackworth?"

"Yeah?"

"I'm the sergeant major. The colonel didn't hear anything in there, did he?"

The unit's command sergeant major, a wise old pro, was the only one in the tent who knew what to do. His voice spoke with the same authority as Sergeant Steve Prazenka's when I was a kid soldier in Italy a million years ago. And it too had my total respect.

"No, Sergeant Major," I said. "The colonel didn't hear shit."

The reporters back in Dhahran operated on the lower slopes of irony. They had this sour little joke. Here's what we should do, they would say: Let's send the coordinates of the JIB to Saddam Hussein so he can send down a SCUD and blow Colonel Mulvey and all his pukes away. Then maybe we'll find out what the hell is going on.

From the very beginning of Desert Shield, the Pentagon practiced ruthless thought control. If you played by the official rules, you couldn't move an inch without an escort. There was no way to talk to a soldier unless you had a minder at your side. The first time you tried, some public affairs officer would rush up and say, "Hey, come over here. You gotta have a green suiter with you." The mood infected all but the best men and women. Most privates and corporals would talk but almost everyone else from sergeants up were riddled with paranoia.

One night in the desert I loaned my transistor radio to a nice kid named Waldorf. He and a buddy were listening to the BBC news and we were all talking. Waldorf had just said, "Then we built MSR Eagle, and it's pointed right at the heart of Iraq." This was not exactly classified information. It was like saying, "Then we built Interstate 80 and it goes right across the United States." But a platoon sergeant suddenly took Waldorf away. I was never able to find out what happened to him. Paranoia was all-pervasive.

Still, you couldn't blame everything on the Pentagon or the JIB. I often wondered why the media didn't do a better job explaining the war. I have a theory about this. Going into the Persian Gulf, the U.S. Army hadn't seen any big-deal combat for nearly twenty years. But neither had the American press corps. Within the media there were bright, energetic young reporters, superstars, and time servers. But among them were very few people who really understood combat or the military mind, how you get organized for a war and how you fight one.

When I first arrived in Dhahran back in December, I had the good luck to fall in with David Evans, the defense correspondent for the *Chicago Tribune.* Dave had been a Marine lieutenant in Vietnam; he had spent twenty years in the Corps before checking out as a lieutenant colonel. He had also served in the office of the assistant secretary of defense; so he knew how the system worked from bottom to top. We pooled our military experience, his overview, my grunt contacts, and for a while we made a great team. So few people really knew what was going on that they constantly interviewed us. We had adjoining hotel rooms, so we just left our doors open and people came piling in all the time. We virtually ran a war briefing center.

One day I said, "Dave, for the next war here's what we do. We'll go together and rent a building right across from the hotel where the press corps is staying. We'll put up a sign: Military Consultants. We'll get a meter from a New York taxi and when one of these greenhorns comes in, we'll just flip it on, and charge for giving lessons. We'll say: This is a tank. It has treads on it. It has this big gun in the center which goes BOOM, BOOM, BOOM, and it can go across most terrain. This is a fighter aircraft. It doesn't go on the ground but flies in the sky. This is fire and maneuver. These are the key elements of war: weather, terrain, and the enemy. We'll give basic classes to the press and charge by the hour. Then we will offer insider classes for higher rates. For those we will give really big tips and charge for them. We won't work for anybody, but our own little company. And we'll get rich."

An old warrior can learn a lot with his eyes, even when he is standing at the side of a road. From the vehicles going by, I could make a pretty good guess about our combat readiness. The markings on each vehicle were a code. If you understood it, you could identify the unit. For example, I might see a triangle with some 2s and 6s. That would tell me that I was seeing the 2nd Battalion, 66th Armor, an element from the 2nd Armored Division. If I saw a convoy that had no tanks or armored personnel carriers, but there were decontamination units, that told me we were preparing seriously for biochem war. A decontamination unit is not normally the first thing you move to the front. The first thing you bring is a gun. When I looked at a convoy made up of flatbed trucks carrying cases of Pepsi-Cola and portable shithouses, what did that tell me? It said everything needed for war fighting was up front and good to go. Even the modern all-volunteer Army doesn't move soft drinks first. Never happens.

By the end of three hours, just standing there, I could get a good fix on the order of battle and the state of readiness. At the same moment, the clever young reporter standing next to me, innocent of any military experience, might say, "Quite a few vehicles here. Those were tanks, weren't they? Did you notice how dusty they were? Did you notice how close they were to one another?"

That's how it was—most of the press just wasn't competent to report a war. There were honorable exceptions, of course, in addition to David Evans, fine reporters like Johnny Apple of *The New York Times*, Henry Allen of *The Washington Post*, David Lamb of *The Los Angeles Times*, or Peter Arnett for CNN, a war correspondent of the old school going back beyond Vietnam. Vietnam had also produced two good men for *Newsweek*, Tony Clifton, who ran the show, and Ray Wilkinson, who stayed with the 7th Marines, his old Vietnam outfit before he turned to journalism. He started out a longhair, but every time he came back in from the field he looked more and more like a leatherneck. Just shy of G-day his haircut was high and tight and he looked like a middle-aged Marine recruit. For sure he was all semper fi and good to go.

But we have lost our Walter Cronkites and Ernie Pyles. What

you saw in the Gulf were amateurs or established network and print stars who lived high on the hog in five-star hotels. They seldom to never got into the field unless they were in one of the Pentagon's polluted press pools. And they hated the guys I loved, the Unilaterals, who didn't let the military hogtie them. The Unilaterals regularly got scoops, and this infuriated the others. Out in the desert more than once I saw press people in league with the JIB minders fingering the Unilaterals. They said, "See, over there, three of them, they're going around that corner." And the Unilaterals would be arrested.

As soon as Mark and I got back, all grubby and dirty from our latest foray into the desert, we were surrounded by reporters desperate for a story. They must not have been filling the market back home, because when I got back after each trip, I would find a sheaf of messages waiting for me from New York, mostly requests to be interviewed. Sometimes I would do as many as twenty radio shows or print interviews and four or five TV shows in a day or two before heading back to the front.

One day as I was walking by the hotel desk, someone grabbed me and said, "There is an emergency phone call for you. From Dhahran."

"Who is it?" I asked.

"Oriana Fallaci."

I vaguely knew who she was, but I didn't care what emergencies she had, so I just started to walk away. Sergeant Marty Martello, my dear pal with a heart of gold, was with me.

"Oh no, Hack," he said. "That's Superstar Numero Uno. Italian. One of the most famous journalists of all time. I'm reading one of her books now."

"You're kidding. Don't bullshit the troops."

Marty went and got the book and showed me her interviews with Henry Kissinger and a lot of other high rollers and I started thinking, *I wonder why she wants to talk to me.* But I didn't have time to talk to her.

After that she must have phoned a dozen times. Her messages were like orders. I was to report to her. I was to talk to her. I was to do this. I was to do that. After a while I had a little envelope of maybe twenty messages. So finally, I called her back and said, "Okay, I'll meet you at the desk at seven P.M."

When I got down there, I found myself looking at yesterday's pizza, an overdone hot tomato, very arrogant, very full of herself. As I walked up, she was tearing into the poor Indian receptionist. Her accommodations were terrible. Her messages were being lost. Nothing was right. She was tongue-lashing him like a Marine drill sergeant tearing into a boot back in the old pre-PC days.

My first thought was, *She ain't no Sophia Loren.*

Unfair. I soon discovered that she was far more complicated than that. She had come to the Gulf to interview Schwarzkopf, but he'd blown her off. His advisers warned him that she was too dangerous: she had hurt too many powerful men foolish enough or vain enough to sit for her tape recorder. I guess that's why she seemed so frenzied. They had shut her out at the top and she didn't know any other way to function. She was a superstar, she told me, and big things were expected of her. "If I don't get a great story, this is going to destroy my career. I have no contacts here. You must help me. You must guide me. You must tell me where I need to go, who I am to talk to!"

It was as if she had decided I was to be her personal military adviser. She expected me to fetch the story and drop it at her foot like a wet newspaper. But like a booby trap or a cobra, she intrigued me so I said, "Let's talk and I'll tell you what I know."

I tried to tell her how the air war was unfolding. I was saying, "This is a piece of cake. The air is blowing these guys apart. They are decapitated. They have no command and control. Their logistics are shot. I give them a week at the most." I noticed she was giving me a peculiar look. Many members of the press corps looked at me that way. The look said, *He's a little loopy in the brain.*

Oriana didn't believe me. She wanted instant information. Immediately. But only the information she expected to hear. She was very, very commanding, pushy, powerful. And always smoking. As she was scribbling down what I was saying, she would go to pull out a cigarette and I'd say, "No, stop. Get away from me. Leave. You can't smoke here." And I'd chase her away.

I hate smokers. The stress of competition, not any danger on the battlefield, was what really got to the reporters. It always surprised me how many of them hit the weed. You would go down to the hotel dining room and find it disappearing under an air inversion of tobacco fumes. One night I took aside the maître d' and said, "Look, let's cut the room in half, smokers on one side, nonsmokers on the other." There were others who felt the way I did, so he agreed to a DMZ.

A while later I was talking with David Evans when a French Air Force squadron commander in his flight gear came and sat down in the nonsmokers' section. He looked at me and said, "I suppose we will not be able to sit here." I said, "You won't be able to sit there if you smoke." He was smoking a cigarette as he talked to me. "Why not?" he said, and I replied as patiently as I could, "Because this is a nonsmoking area."

He blew smoke in my face.

"If I want to sit here and smoke, I will smoke here."

"You're going to look very funny with that lighted cigarette shoved up your ass, and that's just what I'm going to do if you come here with your friends and smoke."

He went white, turned on his heel, and walked away. The next morning at breakfast he came over and said, "I wish to apologize for my unofficer-like conduct last night."

"I accept your apology," I said. "As a matter of fact, I didn't think you were a French officer. You acted more like a Nazi storm-trooper."

It was as if I hit him with a telephone pole. He was a big guy, over six feet, and he almost dropped to his knees.

Around the hotels that was about as dangerous as it ever got.

One night I was sitting at a table right on the front line between the smokers and nonsmokers when Oriana came up and started smoking in the DMZ. She wanted to pry a gem or two out of me. "Nope," I said. "I won't talk to you unless you put out that cigarette." I don't think she'd ever been treated that way. She didn't know how to handle somebody who'd tell her to go fuck herself. If she was a superstar, she was burning out fast.

Many of the other media celebrities behaved the same way. Those from the American networks clearly thought they were royalty. They had big staffs, but they seldom got the story right, and like Oriana, each one seemed desperate in his or her own special fashion. There was enormous pressure on them. They were like Babe Ruth, who had to hit a home run every time he came to bat to keep his fans happy. They were all running scared, not of the war but of their jobs.

Call it star syndrome or whatever you want. It reminded me of something I once heard from Eddie Adams, the photographer who took that Pulitzer Prize photo of Saigon's police chief the day he blew out the brains of a Viet Cong prisoner.

"I was a kid when I won the Pulitzer Prize," Eddie told me. "Ever since then everybody expects me to win another. The minute I take a shot the heat is on. I start to think everyone is saying, 'Ha, he's lost it. Boy, he's failed. He's shot. He's a has-been.' "

You have to feel sympathy for what that kind of thinking can do to a journalist. But when I see how these pressures fuse and distort news coverage, another question worries me a lot more: Can the readers and viewers who depend on these men and women really trust them?

All of Saudi Arabia was as dry as an AA meeting. One afternoon I ran into Bud McBroom, my old Special Forces mate, who was flying for Air Oil. He invited me to his pad for a catch-up and a few sips of his moonshine. I left his number at the hotel in case New York came looking for me and went over to check things out. He had his own still and the Islamic laws weren't hurting him at all. The stuff was 180 proof, filtered through hickory charcoal. Smooth. You didn't know how badly you were blistered until you stood up.

The phone rang. On the other end of the line was *Newsweek*'s Karen Wheeler, a skillful publicist, asking if I would do the Michael Jackson radio show in Los Angeles. "Oh sure, why not?" I said. A few minutes later the phone rang again, someone handed me the cellular, and I was talking to Michael Jackson. Live.

"What's happening?" he said.

Just then SCUDs started dropping in and Patriot missiles began sizzling up to meet them.

"We're right in the middle of the nightly SCUD show," I said.

With the phone in my hand, I wandered outside to watch the fireworks, completely disregarding the 180 proof I had taken aboard.

"Holy mackerel, Michael," I said. "You just won't believe this. What an incredible sight. Whoooom. You can see the SCUD explode. Whooom. You see the Patriots go off in the sky. The sky is full of green and yellow tracers." Blah, blah, blah. I was gee whizzing it over the phone not realizing that I was three sheets to the wind.

"Oh God, what great reporting," Jackson was saying. "Right from the battlefield, where we've got our own guy."

It wasn't my proudest moment as a reporter. But far worse, I think, was the way that a few television reporters relayed the action to the people back home. They played it as if they were in London during the Blitz or Peter Arnett in downtown Baghdad. All I can say is that there was little danger that night and covering those strikes took little courage.

One guy who really promoted the SCUD as a horror weapon was Charles Jaco of CNN. He was the one and original SCUD Stud. The SCUD was wildly overrated. It was an area-fire weapon, not a pinpoint threat. Saddam Hussein was like a blind man grabbing a softball and throwing in the general direction of somebody on the off chance of hitting him. It meant a body bag if it landed right on top of you, but the chances of that were a million to one. Even more than the other hypesters Jaco grossly overplayed the danger.

The Patriot missile was also a big joke, fine for late-night entertainment if that was your thing, but I doubt that it ever brought down a single SCUD. It made people, especially the Israelis, feel good, but as a wonder weapon it was a dud. It was great only for the military-industrial complex and cheerleaders, like George Bush, who labeled it the "Scudbuster." But it was really nothing more than an expensive pyrotechnic that lit up the night sky.

I didn't know what was happening with the SCUD frenzy until I got four phone calls in a single day from my family. First my daughter Leslie and my son David Joel weighed in from the United States. Both of them were saying, "Dad, come home." When I asked why, they said, "It's too dangerous over there. Those SCUDs are eating people alive." Right after that Eilhys rang from New York and then my youngest son Ben called in from Australia with the same message. They were all saying, "Come home, come home."

I started thinking, *Wait a minute. The perception around the world is that this piece of junk is as bad as a Minuteman missile. You'd think it was killing thousands of people.*

There was only one explanation: bad reporting. It was happening because certain reporters liked to play war hero. The truth is, they were far from the front, their stage was a platform at a huge five-star hotel. In the background airplanes might be taking off to strike the enemy, the occasional tracer from anxious antiaircraft gunners and Patriot missiles lighting up the sky might suggest to an uninformed viewer that the reporter was at the center of a battle zone, but the same reporters were getting room service every day and never missed a night between clean sheets. They were living the good life while they acted out relatively danger-free fantasies in jungle coats.

On the day the kids and Eilhys all called me, Larry King also invited me to be on his show. As I was preparing what I was going to say, I started thinking, *If I see that guy Charlie Jaco, I'm going to tell him to cool it.*

A little while later I was walking down the ramp to where the cameras were set up. There were tropical trees in the background, the airport off in the distance. It made the TV people look as if they were in the middle of some bad stuff in the desert, but they were really in an oasis by the swimming pool. All of a sudden, there was Charlie Jaco, aka Shaky Jake, himself, coming off the ramp after reporting the news. I was heading in the opposite direction going to get wired up for Larry King when I saw him. He was called Shaky Jake because he was always panicking and either masking up or bolting to the hotel's air raid shelter whenever a jet fighter at the airfield farted.

"Hey, Charlie," I said to him. "Cool it with the SCUD shit. You're scaring everybody back home and you're particularly scaring the people I love."

"Fuck you, asshole," he said.

Not the sort of thing you hear on the evening news.

Certain moments make us regress.

I felt as if I were back on the streets of Ocean Park. I'm fifteen years old. The one thing you never said to anybody was that he was an asshole. So I grabbed Charlie and delivered this incredible punch. It was just reflexes. I didn't think; it was the way I'd grown up. I'm driving the punch on in when his producer jumped forward and took it on the shoulder as Jaco cut a fast retreat.

Then I marched up to the little platform and sat in the chair so the technician could wire me up. As I was sitting there, I heard Jaco's producer on the phone to Atlanta saying, "This guy's a bad man. Absolutely nuts. He just grabbed me and started beating me up and you don't want an insane person like this on the show." And at the other end, Tammy Haddad, Larry King's producer, was insisting, "Hey, put him on or look at sixty minutes of empty tube, because there's not enough time to scrounge up somebody else." When King came on, I was able to get in my commercial about the SCUDs. Over the next few weeks I received dozens of letters from civilians and soldiers in America and the battle zone saying, "Thank you for telling the American people the truth about the SCUDs. My family was scared to death."

While doing the show, I was blown out to see the Gulf War now had a logo and theme song. We had gone from Vietnam horror to mainline entertainment. I thought, *No wonder the networks push war so hard. Desert Storm's a hit show. It means big money.*

When the air war began in mid-January, I thought it would not last much more than thirty days, and I believed the follow-up ground assault would pulverize the Iraqi Army within a week. By the middle of February I began wondering whether a ground attack was even necessary. At the time I was worried about Iraq's biological and chemical weapons. Saddam had them and I knew he could deliver them. Given the vulnerabilities of our detection and protection gear, if he used chemical weapons, we were sure to take enormous casualties. Since then, even though chemical weapons were not openly used, over forty thousand American Gulf War veterans have reported symptoms. As with Agent Orange in Vietnam, the Pentagon initially claimed the symptoms were merely a stress "syndrome."

The Pentagon brass is still into overdrive trying to bury the whole issue. But the evidence suggests that at least four separate factors might have contributed to the symptoms: heavy petroleum smoke and fumes from oil fires; the possibility that our bombs might have hit Iraqi ammo supply dumps that contained chemical weapons; the inoculation of a lot of troops with an untested drug, pyridostigmine, meant to counter chemical weapons.

I was also afraid we would run into a world of hurt from mines, both land and sea. I began thinking, *For once airpower is really delivering, really doing the job. Why should we risk our ground-pounders?* No one else was thinking that way. Over the next week even the blind could see G-day was just over the next sand dune. When the ground assault came, I wanted to be with the grunts. I learned my old Vietnam battalion, the 1st Battalion, 327th Airborne Infantry, was going to take part in a hundred-kilometer helicopter assault deep along the Euphrates River. The mission eventually was to serve as a cork in the bottle blocking the Iraqi retreat. I hoped to move out with them.

First, I tried the official route through the JIB. No soap. Then I considered going to the top, asking friends back home to intervene with General Schwarzkopf or to try the Army Chief of Staff. But I had made too many enemies so I dropped the idea. Meanwhile I kept checking with Major Grigson, the 101st flack. I must have met with him a dozen times. "How's my request?" I asked over and over. Finally, right toward the end, just before Ground Day, the major told me to forget it. As a point man for the Unilaterals, I was relegated to go with the Saudis. All along I had broken every rule and for the JIB, G-day was payback day.

When I was a kid, I had "Death Before Dishonor" tattooed on my arm. That principle applied to toddling along behind with the Saudis while the Marine grunts were up front doing the hard stuff. So I made my way to the Kuwaiti special forces camp that had offered me its hospitality earlier in the war. My thinking was that if I couldn't move out with the Americans, at least I could get as close as possible to the front.

The morning of G-day on February 24, I was in the Kuwaiti command post tracking battle reports from our spearhead units. I couldn't see the whole picture but the Kuwaiti special forces were linked to air units and through American Special Forces to the Army command and it was possible to piece together the allied attack. There were some nasty little firefights, but for most grunts the ground assault was a long, dirty drive through the desert. The Marines were advancing against almost no significant resistance. At the same time, I could see the American forces light-years ahead of their game plan. The Kuwaitis wanted to retake their own capital, and they had moved forward with the Saudis to the outskirts of Kuwait City. The partisans were reporting over the colonel's cellular phone that there wasn't going to be any fight: "We're already in the streets. There's only civilians here, not a rear guard, just a few stragglers from the Iraqi Army."

The Iraqis had developed a bad case of the bug-out blues. Two full days before the ground attack, they had started running like rabbits. This presented an interesting question. Our intelligence operatives had to know what was happening. Did they report it? Did the reports get through? Or by that point was anything that contradicted the wisdom of a ground attack simply brushed aside? Did we even have to have a ground assault? Whether we needed it or not, we got one. I stayed with the Kuwaitis and watched the Iraqis skitter off like cockroaches when the light goes on. There was no real battle. The drive into Kuwait City turned out to be not much more than a long, dark, grubby start-and-stop convoy because the Iraqis had fired the oil wells and the whole way in you were sucking petroleum smoke into your lungs.

It was all over even before the Kuwaiti units hit town, and they arrived a day before the Marines. I sat by the radio listening to the pilots describing an incredible gridlock on the highway leading out of Kuwait City. Again, as in Khafji, the jet jockeys took out the lead, middle, and tail vehicles so nothing could move, then pounded the hell out of the column. That became the notorious Highway of Death—except most Iraqis very wisely got out of those trapped vehicles and ran before we could kill them. Iraq collapsed faster than a soufflé. I thought the job might take a week. It didn't even occur to me we would stop at one hundred hours and let the Republican Guard escape. At that moment it looked like the war was over—until next time.

The drive up to the front had been easy because all the fighting toys were in position. The drive south was tougher. Now the road was full of Kuwaitis going home, semis, trucks, and big tankers coming right at you, busting ass to get to the front. I felt down. My sympathies are always with the grunts and I was still worried about mines and chemical warfare giving us a last-minute bloody nose.

It took me about five hours to get back to Dhahran, where the mood had completely changed. Out in the streets the people were strolling and laughing, families were picnicking in the parks and along the sea—and the gas masks had disappeared. The city knew the war was over. I went back, packed my gear, and notified *Newsweek* I was on my way home.

Just before G-day, "General" Clifton knocked on the door with a message from "General" Parker.

"We'd really like you to stay for a few more weeks," he said with all the charm the lovable Australian could muster.

"I'm leaving, man," I said. "This war is over. I don't do occupations anymore."

When I called to make my airplane reservations, the travel agent said with a little twitter, "Sir, will you be returning to the Kingdom?"

"Not in this lifetime," I told him.

CHAPTER 2

SNOW JOB IN THE DESERT:

THE GULF, 1991

Fort Drum, home of the 10th Mountain Division, is a warrior's post, a good-to-go launching pad for some of our finest go-to-hell fighters. Not long after Desert Storm, I went up there to do some poking around on a story. I knew the commanding general. Steve Arnold, a two-star, had served with me at Fort Campbell when he first got out of West Point. Then he was a brand-new second lieutenant, a real hard charger. Now, thirty years later, he was a boy major general whose energy had paid off. We'd also served together in Vietnam, where I had watched him rise from lieutenant to captain and become one hell of a combat leader. He was a very good man. He did things the right way.

While I was on the post, I didn't go knocking at his door. But I was sure he knew I was there. After three or four days he called and said he wanted to see me. So I went over to his headquarters.

"Sit down," he said, and he brought me a cup of coffee. "I want to tell you about Schwarzkopf."

"Great. What's the story?"

As a brigadier, Arnold had been a top operations officer throughout Desert Shield and Desert Storm. He knew as much about the campaign as any man alive.

"I've never been treated the way that man treated me," he said. "It was worse than anything I saw during the Beast Barracks as a plebe at West Point. He vented all his anger and frustration on the staff."

"Wait a minute, Steve," I said, pointing to the space above my left shirt pocket. "This doesn't say U.S. Army anymore, Steve, it says *Newsweek*. Remember, stud, I'm a reporter."

"I had to get it off my chest."

Anyone who knows the military could see that a few good men thought that General Schwarzkopf's balloon had gotten too big. The myth had swollen until the legend was larger than any real man. Not long after the trip to Fort Drum, I picked up *Parameters*, an official publication of the Army War College, and found an article by General Bruce Palmer, Jr., a deputy to Westmoreland during the Vietnam era. Palmer was retired, highly respected—and highly critical of Stormin' Norman. Some time later, *Naval Institute Proceedings*, a publication with forty-five thousand military readers, many of them senior officers, ran its own critical study of General Schwarzkopf's performance in the Gulf. The piece was written by Jim Burton, a brilliant retired Air Force officer who sacrificed his career to blow the whistle on an early version of the Bradley Fighting Vehicle, which he rightly called a "flaming coffin." Burton was a total truth teller.

General Schwarzkopf was the sort of general who expects a colonel to iron his uniform or stand in the john line for him during long airplane flights. He got his nickname, the Bear, not because he was built like a bear, although he was a mean one, but because of the way, from second lieutenant to four-star general, he ate people alive. Rip, tear, bite; he was a real Montana grizzly. When I first met him in Berlin in 1961, he was a horse holder for Brigadier General Fredrick O. Hartel, a pal of his two-star dad. Back then his fellow officers awarded him a triple-A ID: arrogant, abusive to all subordinates, a total asshole. As he rose in rank, the traits went from lower- to uppercase. In Vietnam he was known by the men in his battalion as "the Nazi colonel."

During and after Desert Storm, fair-minded officers were shocked by the way General Schwarzkopf treated Lieutenant General Fred Franks. On G-day, General Franks commanded VII Corps, the spearhead for the Hail Mary. He was a distinguished soldier, badly shot up in Vietnam with most of a foot ripped off, and VII Corps came from Germany with the best tanks, the M1A1s, and the troops who had been trained under the Air-Land Battle doctrine designed to stop the Soviets in central Europe. In Desert Storm, the plan was for the XVIIIth Airborne to contain the Iraqis so they couldn't escape. The Marines were to pin them down in the east. Then, hey-diddle-diddle, right up the middle, would come a massive instrument of terror: VII Corps—the British 1st Armoured Division, the American 1st and 3rd Armored divisions, and the American 1st Mechanized Infantry Division.

Instead, when the ground assault came, General Franks advanced cautiously. He was no hell-on-wheels kind of guy. By around two o'clock in the morning on G-plus-two, Schwarzkopf was angry at how slowly Franks was advancing. He got on the horn and they had a raging conversation that ended with General Schwarzkopf saying he wanted action and that VII Corps should attack all night if that's what it took to secure a crossroads blocking the Iraqi line of escape. Franks's greatest sin was not getting to the Iraqis' rear in time to bag 'em up. Instead, he ordered VII Corps to turn eastward right into the Guard's heavy divisions, which had set up blocking positions. As one British critique put it later, VII Corps "turned directly into the thick of the hedge."

Franks's VII Corps was green to the desert. Long years in Germany, where there was no area for major maneuvers, hobbled VII Corps. It hadn't operated in big battle formations since World War II. Much of his timidity came from the casualties his units were taking in occasional fratricide incidents. How this massive but untried outfit would have done against the Russians—who would have fought back—is something else to think about. So three hours later, when General Schwarzkopf got up and looked at his situation map, he saw the lines hadn't moved from where they had been when he went to sleep. He expected them to be another thirty kilometers closer to the objective, and according to one of my guys, he went nuts, screaming and yelling. That is his way. In the Army you have cool guys and screamers and Stormin' Norman has always been a screamer. It's what many Perfumed Princes feel entitled to do when they get upset. For a moment he considered relieving General Franks and after the operation he vilified him.

A commander is responsible for everything his unit does and fails to do. That must hold true for General Franks. But General Schwarzkopf was the overall battlefield commander. To cut through the fog of war, he should have been out of his well-bunkered command post and up forward, seeing the battlefield as a Rommel or Patton would have done. Under the circumstances, it seems only fair to take a closer look at General Schwarzkopf's own performance. During Desert Shield and Desert Storm, one of my guys told me afterward, the Bear was often "nervous in the service," testy, quick to fly off the handle. Another senior officer told me General Carl Vuono, the Army Chief of Staff, had to hold General Schwarzkopf's hand regularly, conducting long-distance shrink sessions over the phone.

During the ground assault, General Schwarzkopf made several mistakes. As the battle unfolded, he and his deputy, Lieutenant General Cal Waller, hunkered down and read the SITREPs. They were bunker bound. This led to some crucial errors in judgment. Had they made a practice of getting out of their underground bunker and going forward, they would have realized it was not a good idea to allow Lieutenant General John Yeosock to retake command of the Third Army only three days after major surgery. No man could have regained the enormous strength it takes to run a huge army so soon. A commander in better physical condition could have kicked more ass and exploited the advantage more quickly. If Schwarzkopf had visited Yeosock's command post he could have observed his physical condition firsthand, taken the pulse of battle, and moved to relieve him or take command temporarily himself. After that, he should have choppered to General Franks's command post to find out for himself why the forward elements of VII Corps were moving so slowly and holing up every night, forgetting Patton's basic dictum: Haul ass and bypass.

So who is really responsible for letting the Republican Guard get away?

In January, Secretary of Defense Richard Cheney asked Schwarzkopf how he thought the war was likely to end. Stormin' Norman wasn't sure. He didn't cover all the bases. Later, while briefing his commanders before G-day, he said, "We need to destroy—not attack, not damage, not surround—I want you to *destroy* the Republican Guard. When you're done with them, I don't want them to be an effective fighting force anymore. I don't want them to exist as a military organization . . . I want to pin them with their backs against the sea, then go in and wipe them out."

Schwarzkopf did a lot of good things and he chased the Iraqis out of Kuwait. But he didn't do what he said he was going to do to the Republican Guard. Retired Lieutenant General John H. Cushman, a military analyst, wrote in *Naval Institute Proceedings*, they "got out the back door because General Norman Schwarzkopf—whose concept of operations stressed the need to 'destroy' the Republican Guard—did not plan on the basis that to trap that force would be the best way to destroy it."

We advanced rapidly at first, but in the last twenty-four to thirty-six hours we slowed down. Why? Units kept getting "limit to advance" orders. Headquarters was putting on the brakes. There was an obsession with keeping everyone in line, neat and tidy, just like the phase lines of World War I. It was a fatal flaw. Schwarzkopf worried far too much about exposing a flank. He also depended too heavily on the Air Force to fulfill all its promises. He forgot Patton's war-tested philosophy for when you're on a roll: Let the enemy worry about *his* flanks.

If there was a plan to nail the Republican Guard, it was full of holes. Schwarzkopf miscalculated. He thought the Republican Guard might try to escape toward the west and he did block the western escape routes, but he left a northern vent open. The Republican Guard didn't do what Stormin' Norman expected. When the Republican Guard commander saw what was happening, he issued orders to set up his own blocking forces to secure a withdrawal and then he scooted. Most of the Iraqi armor slipped through Basra by G-plus-one and -two.

Later Schwarzkopf left the impression that airpower was supposed to have blocked the northern vent. But in an unpublished critique, Jim Burton, who has made the definitive study of the ground attack, says, "Since the Army does not openly blame the Air Force, I suspect that the northern vent was not viewed as an escape route until the senior leaders actually saw Iraqi forces streaming north out of the Basra pocket in the final stages of the war."

Who should be held to account? "I have to put most of the blame on Schwarzkopf," Burton told me. "You have to give Schwarzkopf credit for being the first to recognize a rout. But he was unable to get Franks's VII Corps moving quickly. And he was unable to cut off the Republican Guard's line of escape. He saw it happening and he was unable to take charge and do something about it. When they saw the Iraqi armor streaking out, they realized, 'Oh my God.' " But by then, it was too late.

Both Powell and Schwarzkopf are extremely sensitive about this point in their memoirs—and they should be. They both gloss over the fuck-up by fudging the record and arguing that their goal was simply to liberate Kuwait. In fact, the very clearly expressed goal was to annihilate the Republican Guard. They failed to accomplish that goal, leaving Saddam Hussein in place and the Guard in position to raise hell all over again whenever it suits the boss.

Schwarzkopf allowed four and a half Republican Guard divisions—two heavy armored—to escape. The armored divisions later put down the Shiite popular uprising in Basra, snuffing out hope for other rebels across the country and a revolution within Iraq. Through the Voice of America we had encouraged the uprising. Rise up, was our message, we'll be there to help you. Then, just as in Hungary and Czechoslovakia during the Cold War, we broke our promises while Iraqi partisans risked their lives. In doing so we lost our best chance to bring down Saddam Hussein—without even going into Baghdad—and to get new management for Iraq.

There may have been a contingency plan for taking Baghdad. On February 11, I was talking to one of my Deep Throat sources. He said, "Mark my words, Hack. We'll be on the streets of Baghdad drinking coffee by May." Later he brought me a little map marked with attack arrows that pointed right across the Euphrates into the Iraqi capital. In my own view it would not have been necessary or wise to plunge into Baghdad. To seize the city would have required the most costly kind of fighting—house-to-house, street-to-street combat where both sides take heavy casualties. But if the Shiite rebellion had taken fire, it would have spread like flames in a dry forest. With no one to protect him, Saddam Hussein like so many other dictators would have started looking for his trusty little getaway airplane. By the time our troops approached the suburbs of Baghdad, there would have been a white flag waving over every headquarters and hanging out of every window in town. I am certain a few hardcore Iraqis would have stuck by their guns, but I don't think they would have put up much of a fight. We should have told the UN we were just going to step a little beyond the UN mandate because we were like policemen in hot pursuit of an outlaw. We certainly could have justified our action, but we were too weak willed to bring down Saddam Hussein.

I wonder if General Schwarzkopf had been in the field smelling the cordite and seeing for himself that an escape door was wide open, that one hell of a lot of Iraqi armor was getting away, whether he would have knuckled under so easily when with General Powell's blessing, President Bush decided to call the war prematurely. Having to do two jobs—run the war and act as leader, nursemaid, and referee to keep the coalition together—perhaps made him eager to settle for the quick fix instead of holding out for the knockout punch.

To make things worse, the cease-fire permitted the Iraqis to keep their helicopter gunships. Those gunships and the Republican Guard tanks blew away the Shiite rebels, and the rest is history. In 1994 two of these Republican Guard divisions moved up to the Kuwaiti border in attack formation as if they wanted a second round. In responding to the feint, we had to spend well over $1 billion on a second trip to the Gulf. Saddam Hussein can now pull our chain any time he chooses and we will have to deploy troops back to the desert, putting a real hardship on the military and costing the taxpayer an incredible amount of money. All he has to do is crank up his tanks and move forty miles and he can set us off on yet another seven-thousand-mile heart stopper.

When the time came to write his memoirs, General Schwarzkopf said he agreed with the President and the Chairman of the Joint Chiefs on every major issue. I doubt it. Didn't he tell David Frost on television he had had some serious second thoughts? What happened to them when it came time to write his story? After General Schwarzkopf's memoir came out, I was talking to a three-star who is a good friend of mine. I asked him what he thought of the book. He said, "It's the most self-serving tripe I've ever read." General Schwarzkopf applauded before the play was over.

It worries me that the Pentagon will conclude that Desert Storm should be the model for future wars. Any such conclusion would be absolutely incorrect. From beginning to end Desert Shield and Desert Storm were a freak of nature. In Saddam Hussein, we fought a rank amateur who did everything wrong. What other opponent will ever grant us six months to build an overwhelming force right in front of his foxholes? The great tank assault on G-day may turn out to be the twentieth century's version of the Charge of the Light Brigade, with a happier ending, of course, since the disaster was all Iraq's. We may never again see an American tank division roaring across the desert, except in CNN reruns of *The War in the Gulf*, and if we fight smart we never need to experience anything like Desert Shield or Desert Storm again.

Instead of congratulating ourselves we should be doing some very hard thinking about what happened in the Gulf. What the White House, Pentagon, and field command did in settling for a hundred-hour war represents the triumph of media packaging over good military sense. They were looking for a sound bite, something snappy, something to make them look good on the evening news. What they should have been searching for was the right way to finish off Saddam Hussein once and for all, so that the United States would never have to go back to those battlefields. Instead, they left him in power with enough strength to drag us around by the nose.

I don't know a lot about high political strategy, but I do know a lot about war. When you fight a war, you should fight it all the way. After Desert Storm I spent months talking to soldiers who fought in the Gulf: brigade, battalion, and company commanders; sergeants, corporals, and privates; soldiers, sailors, and airmen. They all felt we left the job undone even though we had been on the verge of total victory. We had the muscle and the stature to hold the coalition together long enough to finish Saddam Hussein. We erred in not doing so.

Desert Storm was a hollow triumph. It proved nothing: all that trouble, deploying almost 600,000 troops, spending almost $60 billion, though the Arabs picked up most of the bill. All that sacrifice and for what? We didn't finish the job. George Bush called Saddam Hussein another Hitler. He told the American people Iraq had the fourth biggest army in the world, the Iraqis were close to making their own A-bomb, they had biological and chemical weapons and would use them. Yet Saddam Hussein is still there in Baghdad, very much in charge, having the last laugh at unemployed George Bush and Margaret Thatcher. And he's there with an army that's completely rebuilt.

Recent CIA reports say these forces are far more effective today than they were six years ago. So Saddam Hussein is the guy who learned most from Desert Storm. He got rid of his ash and trash, the divisions that got clobbered and ran. He has a rebuilt Republican Guard back up and running, refitted in spite of the blockade, with plenty of ammo and spare parts, disciplined troops looking good. The vanquished always learn, the victors seldom do.

Why did we pull out when we should have pushed ahead? One of the reasons was President Bush and his advisers felt squeamish about the images they were seeing on TV from the so-called Highway of Death. They worried the United States might wind up looking like a barbarian eager for a great bloodbath. That's understandable. Butchery is un-American. But you must never go to war unless you clearly intend to win. War is like a marriage: It's unconditional. And those TV images were highly deceptive, highly sanitized. There was not a drop of blood shown on the tube. Yes, there were stealth aircraft, smart bombs, bright tracers arching across the sky, and awesome Abrams tanks racing across the desert. Glorious polished war games. Man's biggest turn-on besides sex.

But after Desert Storm, I checked out the Highway of Death, the main road leading from Kuwait City to Baghdad. At Myrtle Beach an A-10 pilot showed me his and his squadron mates' gun camera footage. I'm using the word *road* advisedly. You couldn't see the blacktop for the mess that was on it: a crazy jumble of abandoned civilian cars pressed bumper to bumper against jeeps, trucks, and transports—most of them stuffed with loot—and every now and then a tank. Everything you could imagine.

Except Iraqis. Only a few bodies to count—certainly not the ten thousand some press hysterics were claiming.

My brain shot up a flare: Something else was wrong. There were surprisingly few fighting vehicles. No smashed windshields. No jagged holes. No signs of combat. We're talking gridlock—New York City at rush hour. It suddenly hit me. Ten years and who knows how many trillions of dollars into the Reagan-Bush arms buildup, the United States had sweated blood for seven months to produce nothing better than the mother of all fender benders.

This isn't death's highway, I thought. *It's the world's biggest and most costly traffic jam.*

I can understand President Bush might have felt queasy about what he was seeing on the tube. But when he turned to General Powell, the Chairman of the Joint Chiefs should have stood tall and reasoned with him. Colin Powell caved in because his political judgment was stronger than his military judgment. I am not saying he should have defied the President. But his duty was to offer the best possible military guidance, not to make political life easier for George Bush. Instead of going along to get along, he should have argued harder. He chose not to. General Powell and General Schwarzkopf both deserve great credit for being the architects of Desert Storm. But the hundred-hours decision was 100 percent dumb.

President Bush, General Powell, and General Schwarzkopf should have delivered a KO punch as we did with Hitler and Tojo. Back then, George Catlett Marshall didn't stop simply because he had succeeded in kicking the Germans out of France. He went in and destroyed Adolf Hitler. We haven't really won a war since World War II.

In the mid-1950s we developed a new kind of soldier-statesman, the warrior-diplomat. The warrior-diplomat is to war what martial music is to classical music: too much oompah instead of mastery over the discipline. What we have now are mainly slick politicians in the Pentagon E-ring and very few warrior generals. That's why United States military strategy has been flawed from Korea to Bosnia. Desert Storm did prove we have a good military, no longer riddled with Vietnam syndrome. But having defeated such a weak opponent, it would be dangerous to draw the wrong conclusions or to preen ourselves for too long. The war was like crushing eggs on a concrete pad with a sledgehammer.

We simply shot ourselves in the foot. In the beginning we followed the so-called Weinberger rules of engagement. They were sensible. You must go in with a force strong enough to do the job, make sure you have the support of the people, make sure the fight concerns the nation's security, and be absolutely certain you have a clear exit plan. The unfortunate reality is that we started well but finished lamely. The President and his political generals didn't have the vision and fortitude to follow through. As an old soldier, I have seen this happen before. We didn't finish the job in Korea; we couldn't finish the job in Vietnam. We went into these two wars with halfway measures. But there's no such thing as a limited war, no such thing as a limited fight. It seems to me if you do all you can sensibly do to avoid going to war, then discover that you must fight anyway, you must fight to win. That means bashing your enemy with every bomb you can drop, every bullet you can shoot, or every rock you can throw until he can't get up again. Or someday, as with Saddam, he *will* get up again. We should have realized this in the Gulf. But we no longer have enough fighting generals. What we have are generals who flip-flop all around the E-ring—political right down to their toes.

The President and the Perfumed Princes worried about the political polls, about international reaction, about whether the allied coalition would fall apart. I think most Americans, informed of the atrocities inflicted on the Kuwaitis by Iraq, would have supported a tougher finale to the war. As for the coalition, the Arabs didn't do much in the war anyway but hang back and watch and sing their favorite marching song: "Onward Christian Soldiers."

The fuzzy thinkers in the policy crowd were convinced it was wiser to leave Saddam Hussein in Baghdad since Iran was just over the horizon and full of Islamic fanatics. There was something to be said for this line of thought, but not much. We always seem to follow the analyses of power brokers who are too subtle by half. Their rationale was if we rode too hard on Saddam Hussein he wouldn't be able to serve as a buffer against Iran: Wouldn't it be better to cut off his legs, but leave him standing on his bloody stumps swinging his sword so he might still scare the hell out of the Iranians? That was like thinking you've defanged a rattlesnake when you've only chopped off a chunk of his tail.

I think this line of reasoning missed the point of modern warfare, especially on the terrain in the Middle East. We could have terminated Saddam Hussein without inviting an attack from Iran. If the rebels we encouraged—then sold out—had succeeded, they might have been more willing to get along with the Saudis and Kuwaitis, even if they didn't particularly like us. If the Iranians had made the mistake of turning aggressive, we could have swatted them down with air and sea power alone in about sixteen minutes. We were fighting in one of the last places where conventional tactics and high-tech weapons actually work, tank country, where you have wide open skies. With stealth bombers and cruise missiles, all of them damned hard to stop, we could have leveled downtown Teheran by long distance. Everything we needed was already there.

The frightening thing is how long it took us to get everything in place.

Desert Storm raises some hair-raising questions about the true state of our military readiness. Before the crisis, the 1st Cav Division and the 24th Mechanized Division were sending the Pentagon C1 and C2 good-to-go readiness reports, the highest. Their fifteen-year-old M-1 tanks deployed only a 105-millimeter cannon with a defective gun computer. The Iraqi T-72 outgunned them. Still, their generals were saying, "We are ready to go." Then in early August when they were ordered to deploy to Saudi Arabia, they started throwing up their hands and saying things like, "Our tank barrels are worn out. The tracks on our tanks are shot. None of our chemical protection gear works. We desperately need spare parts."

Once reality set in, it was "We are *not* good to go."

They were scrounging around, cannibalizing other units, flying in spare parts, even borrowing equipment from the theater out in Korea, which was face-to-face with a genuine, hard enemy. All this even though we had just spent over $3 trillion during the Ronald Reagan-George Bush military buildup.

Those readiness reports were lies from the senior brass. Those lies meant we were taking worn-out tanks, putting them on ships, sending them to Saudi Arabia, and unloading them. Some of the first tanks to reach the battlefield were not our best. From Fort Bragg, North Carolina, Schwarzkopf flew in worn-out Sheridans that had failed in Vietnam and broken down during the invasion of Panama. They were rusted buckets of bolts, and they used up critical airlift capacity. We sent that junk just so we could have machines for CNN to film running around the desert and looking like tanks even though they wouldn't have been worth a fighting damn against the Iraqi armor. But many Iraqi top brass—who were watching CNN—had attended our schools and knew our equipment. No wonder they were all crowing about the imminent mother of all battles.

It wasn't until December, quietly at night, that the troops in the field finally got a stock of the brand-new M1A1 Abrams tanks—in first-rate condition with a 120-millimeter gun—and the new Bradley Fighting Vehicle. The Pentagon had to sneak them out of Germany and slip them to the XVIII Corps units that were equipped with worn-out armor that would have been death wagons. The right weapons didn't start arriving for almost five full months. So much for all those fake gung ho readiness reports, signed by generals who give only lip service to duty, honor, and country.

Senior officers with their ass-covering lies put the lives of our warriors in danger. For months they let brave troops with worn-out and inferior weapons stare down Saddam Hussein. When the 82nd Airborne led the buildup, they arrived in no shape to be anything more than a speed bump. They didn't have decent antitank weapons, and if Saddam Hussein had had the logistical back to attack, he could have roared through Saudi Arabia faster than shit through a goose. The hastily deployed paratroopers, Marines, and USAF flyboys would have been bagged up within a few days and spent the rest of the war behind barbed wire.

It's not that those soldiers wouldn't have fought. They would have fought and died valiantly just like their grandfathers did at the Kasserine Pass and Bataan and just like their fathers did in the beginning rounds of the Korean War. It's always the same old body-bag-filling story: Our grunts never have the right stuff when the guns begin to bark. Initially, in the desert, they would have been slapping at tanks with fly swatters.

The M-60 machine gun was worn out. Soldiers in the desert were using condoms to cover the barrels of rifles too sensitive to dirt, sand, and grit, just as they had done in the Pacific during World War II. Their chemical detection and protection gear was also worn out or nonexistent. Why did our commanders allow all this in peacetime? Real warriors would have cared and raised hell.

We did no better on sea mines than we did on gas masks or chemical detection equipment. We had the Marines tasked for an amphibious feint, a vital ruse meant to draw Saddam Hussein away from our main line of attack. Two ships hit mines, and that put an end to the ruse. Saddam Hussein was able to stop a mighty United States armada simply because we had neglected to pay attention to mines and minesweepers when all was quiet.

Although the SCUD missile proved to be inaccurate in Desert Storm, we can't count on such good battlefield luck for the future. We need a better theater air defense weapon, which can take out the SCUD or other missiles. The campaign to destroy SCUD on their launchers took too many air assets. It was badly conceived and badly executed and it wound up as a total failure.

We didn't have enough roll-on, roll-off transport ships to get our forces in place quickly. We were reduced to shanghaiing foreign ships. To move forces out of Germany, we were using German ships, Belgian ships, Italian ships, Spanish ships. What if they had bucked us and said no? We wouldn't have been able to move our boys and toys to the battlefield.

In addition to our failure at developing a sensible strategic sea lift capability, we also had plenty of other problems in the air. We were moving troops on a fleet of C-5s and C-141s that was twenty to thirty years old. The air fleet just wasn't up to the job. The Secretary of Defense himself said we had flown the wings right off those birds. The real air transport heroes were civilian 747s. They carried the bulk of all supplies delivered by air.

A country that goes to high noon showdowns barefoot and with a water pistol is looking to lose. It is suicide to count on the next bad guy being such a fuckwit as Saddam Hussein.

During Desert Storm, whenever you turned on the tube you saw spectacular footage of smart missiles and other Flash Gordon weapons. The media manipulators made it look as if all the whiz-bang stuff was winning the war, pointing toward a future where we could fight antiseptic wars with almost no casualties. That, of course, was exactly what the briefers wanted you to believe. Whizbang always looks fantastic. The idea was if you were impressed enough, you might go on paying through the nose for it whether or not we really need it and whether or not it's the right kind of weaponry for the wars we will most likely be fighting from now into the next century. The truth is, smart weapons represented only a minuscule fraction of the hardware we used on Saddam Hussein. Some of the most effective and reliable weapons in the desert were among the oldest and dumbest in our arsenal.

Before we let the weapons industry or the U.S. Air Force bedazzle us, let's examine a few facts. Of the ordnance we dumped on Iraq, 97 percent fell in the form of the same dumb, dumber, dumbest iron bombs we dropped on Berlin, Pyongyang, and Hanoi. And I mean

97 percent of all the bombs dropped by the Air Force, the Navy, the Marine Corps. Those smart bombs and missiles you saw on television added up to about 3 percent of the ordnance used.

Smart bombs didn't crack Iraqi morale out in the desert. The real nutcracker was an absolute antique—the B-52 with its incredible psychological value. You looked up and saw this old war horse from even before Vietnam flying through the skies with the grandsons of the original pilots in the cockpits. In desert warfare, the B-52 was terror itself. It could lay down devastating area fire, and if anybody had the bad luck to be underneath those deadly patterns, they were gone. You would hear a loud BOOM, then BOOM-DAH, BOOM-DAH, BOOM, BOOM, BOOM. Huge flashes would sear across the night sky. The survivors could only watch and think, *There but for the grace of Allah go I.* During the day you would look up and see a flight of six B-52s headed into a target area leaving behind a deadly vapor trail, and when they pulled away all that was left on the ground was a giant waffle. It was a tremendous performance by a grand old airplane that had pulled enough hitches to retire with forty years' service and a gold watch.

And while the B-52 was winning the war, where was that turkey the B-1 bomber? Grounded, even though the meter was running at a rate of around $39 billion, grounded because it didn't work, because it wasn't safe to fly, a lemon that some genius in the Pentagon had sworn was a diamond. And good old Ronald "Bomb Them Back to the Stone Age" Reagan brought it back from the dead. What the hell—it was good for his big-spending California supporters. But it wasn't good for America. We had spent ourselves blue on it, but not one of those sci-fi toys went to war in the Gulf.

At the same time, our reliable A-10 Warthogs were superb, particularly at the Battle of Khafji. During the war I spent time with Major Richard Pauli, an airplane driver from Louisiana who told me what the battle looked like from his cockpit.

"The A-10s went into a feeding frenzy," he said.

His outfit's target was a column of Iraqi tanks and armored vehicles 17 kilometers long. Somewhere between 800 to 1,000 vehicles were on the road that night heading south for Khafji. When Major Pauli looked on his infrared screen, he saw Iraqis running from their vehicles and pulling bodies from burning tanks.

On the way in, Major Pauli saw gray and white puffs of antiaircraft flak bursting off his wings. He took eight hits. One piece of shrapnel zipped from one side of his plastic Plexiglas canopy right across his nose and out the other side. When he looked out, he saw a piece of his wing had been blown away. But the plane flew on. Just as in World War II, the mechanics patched it together and two days later Pauli was again bringing death and destruction from the sky. Try doing that with a fragile F-16.

The Warthog, modeled after the World War II German Stuka but bigger, is one of the best ground attack and close air support weapons ever built. After Desert Storm, the Pentagon geniuses announced they were going to put it out to pasture in favor of airplanes that can't hover over a target and fly so fast they can't see what they are supposed to be hitting. God must be on the side of the military industrial complex.

Although it pains me as an old groundpounder to say it, I do believe that airpower did one hell of a job in the Gulf. But I still think that from World War II, through Korea, on to Vietnam and right up to the present, the U.S. Air Force has never lived up to its own inflated claims. Airpower alone doesn't win wars. Period. Thirty-nine hours before the main ground assault, the Marines brilliantly infiltrated large units behind Saddam Hussein's main lines. They found plenty of weapons, ammo, and equipment all intact, untouched by airpower. And no Iraqis to be seen. "Something else must have driven them out," says Colonel Mike Wyly, a retired Marine fighter and thinker. "I think it was the realization that the Marines were behind them." And if airpower had really destroyed the Iraqi command and control, how could the Republican Guard commanders have organized and executed such a successful battlefield retreat?

Stormin' Norman's directive to the Air Force was that once the boys in blue took out 50 percent of the Iraqi armor, artillery, and air defense, the groundpounders would be good to go. In early January, General Buster Glosson, the chief Air Force planner, briefed Schwarzkopf, telling him airpower could achieve 50 percent attrition of the Republican Guard within five days. After the fight, a Marine Corps study conducted on the battlefield, where it was possible to count destroyed tanks, artillery, and other gear, discovered that at tops the destruction rate was only 15 percent. This means that forty-three days after General Glosson's briefings, the Air Force only managed to score 15 points. What really happened was that the Iraqis were such piss-poor soldiers that even a 15 percent rate was enough to make them bug out. This wouldn't have been the case with the Japanese Army or Hitler's Wehrmacht in World War II, or later, the North Koreans and the North Vietnamese. After the war, the Air

Force had the nerve to puff up its claims even more. General Merrill McPeak, then Air Force Chief of Staff, said his boys had done better than 50 percent, pushing the Air Force's favorite theme—who needs the other services? Rising high into the blinding sun of delusion, Air Force generals always say they can do it by themselves. But they never have—and they never will.

Even so, if the Iraqis were already hauling ass for home, the question remains: Why the ground attack? Out of quite a different kind of desperation, I think. The Army and Marines had to justify their presence and role because of the tremendous amount of money they spent and want to spend in the future. No one in the world is going to admit this, of course; but that's how the game is played. Had the ground grunts not attacked, when the Air Force started beating its drum, the other services would have lost face—and, more important, defense dollars.

This is a very cynical business. What the Air Force can do in the Middle East it can't do in Korea, for example, any more than it could win in Vietnam, or a lot of other places where we might have to wage war in the future. But there was no way the Army and Marines were going to stand aside and let the air guys win all the glory. We had sent more than twice as many troops to the Gulf than were needed so we damn sure meant to use them. Just as in the flawed operation in Grenada and the failed Iran hostage mission, everyone had to get in on the act. The idea among the service chiefs is, if you don't use 'em you'll lose 'em—defense dollars that is.

That brings us back to General Colin Powell's famous press conference. He said Iraq was a snake and we were going to chop off its head and kill it. What we really did was chop off a piece of its tail. We completely missed the head. The snake, with its tail bobbed, slithered off to heal and fight another day.

General Colin Powell is a fine political officer with a distinguished career as a military bureaucrat. I respect him. He's a fair man. What he represents in the modern Army is the best of political generals—a decent Perfumed Prince with remarkable leadership ability. When all the Powell for President uproar began, I phoned Ben Willis, one of my best officers in Vietnam—and the man for whom I named a son. He's a real truth teller who just happened to have served in a battalion in Korea under Powell. So I called and asked what he thought of the general.

"Best commander I ever served under," he said.

Pow—right through the heart. *Ben, say it ain't so,* I was thinking. *I named my son after you. Remember me, your skipper? Tuy Hoa, the Plain of Reeds. Three tours in 'Nam together.*

"Uh, really?"

"Yep. The best."

"Well, uh, what kind of war fighter is he?"

"Oh, hell, Hack. He's not *that* kind of soldier."

What a relief.

General Powell's record as a lieutenant and captain shows he started out bright and had a fast learning curve. To become a company commander as a young lieutenant, he obviously had to impress a lot of his seniors, because companies commanded by lieutenants in 1961 were as rare as tits on a bull. His early selection for Vietnam also suggests the Army's career management people had their eye on him as a real fast-burner. After General Powell's memoirs came out, Joseph Persico, the writer who worked with him, said he had to push a little to get him to tell about his combat experience in Vietnam. General Powell is an honest man and I suspect there wasn't a whole lot to tell. His ARVN infantry battalion didn't get into any serious fighting, and after five months in the bush he stepped on a punji stick and was reassigned as adviser to the regimental operations officer. Good duty with clean sheets, but not the kind that produces many war stories.

It was after that first hitch in Vietnam that General Powell's career really took off. He graduated number 3 out of 200 students in his class at Fort Benning's Infantry Career Course. That tells me he was one hell of a student, smart, very ambitious, tactically together. He also went to Command and General Staff College early, another major station of the cross and a sure sign of a comer. If you don't do well, you are usually doomed to low rank and an early pink slip. General Powell graduated number 2. During his 1968 tour in Vietnam he did a knockout job as a staff officer running plans and operations for a division on a tough battlefield. He was looked upon as a boy wonder. And then he turned to politics. He became a White House fellow, met Caspar Weinberger and Frank Carlucci, both of whom served as Secretary of Defense. He was smart, hardworking, talented. A lot of power players stepped forward to become his mentors. For the rest of his career, the time he actually spent with troops fell off to a bare-bones minimum.

The interesting thing is he never commanded a division, which would have been a career killer for anyone else. He commanded a battalion and a brigade, for the minimum hitches. As an assistant division commander he got nailed with a bad efficiency report. His

CG smelled Powell's perfume—and didn't like it. That, too, should have killed him. But he managed to return to the Pentagon, with help from friends in high places, sweat out the damage, and come back strong. He later got a corps in Germany, normally a sure shot to four stars, but he served only five months of what is normally a two-year tour before returning to the Beltway. His mentors had called upon him to become Ronald Reagan's deputy national security adviser. His last command, another major honor, was also cut short for the final speed trip back inside the Beltway.

All along the way General Powell was hard driving, politically shrewd, a good man. But the résumé shows he was also a consummate ticket puncher. From the early 1970s to his appointment as Chairman of the Joint Chiefs of Staff, he was a political general from his short GI Joe haircut right down to his high-glow shoes. He was never a romping, stomping war-fighting general.

This experience clearly shaped his performance during Desert Shield and Desert Storm. Had he been a war fighter, he would not have dispatched the 82nd Airborne Division, light infantry with no decent antitank capability, to face an Iraqi Army that was rated by the CIA at the time as a bad mother.

A fight in the desert in August or September 1990 would have been like Korea, 1950, where another light infantry force, Task Force Smith, with zero antitank capability, was tasked to stop a North Korean Army spearheaded by tanks. Within the first few hours of the fight most of the task force became casualties. The North Koreans brushed them aside like a bulldozer cutting through a sand pile. If the Iraqi tanks had charged, it would have happened again. We can thank Saddam Hussein's bad logistics and his unique stupidity it didn't.

The line President Bush said he was drawing in the sand was a weak one. Our first units were not ready for tanks and they were not ready for chemical attacks. Their gear was worn out or didn't exist, even though Iraq was known to have a large stock of chemical weapons and a history of using them. History is going to show we practically invited Saddam Hussein into Kuwait by underestimating his intentions and doing too little about them early in 1990. Then, when he caught us with our pants down, we had to throw light infantry at him—and pray. This is a decision a real warrior such as Al Gray, the U.S. Marine Corps Commandant, would never have made.

For the first two months, our soldiers had to live hand-to-mouth. The logistical machine had sputtered and then broken down. Units quickly ran out of everything. The MRE rations for some infantry battalions were gone in a couple of weeks. There were stories about grunts living on cold hamburgers and soggy french fries supplied by the Saudis.

You won't find any of this in the books that General Powell and General Schwarzkopf wrote later, nor really a word about the grunts, the sergeants, the captains, and colonels who made it all happen. Their books are both all about *me, me, me.* It burned me up when I read one little passage where General Powell offers a cheeseburger and a little bucking up to the Secretary of Defense, Richard B. Cheney.

Here's what I think General Powell should have said in his book: *We took a hell of a risk when we sent the 82nd and we are damn lucky Saddam didn't take us on. I learned a lesson. As soon as Desert Storm was over, I ordered the services to get their shit together, develop an antitank arsenal and effective chemical protection and detection gear and beef up the 82nd so they will never be left exposed that way again.*

Instead, we got stories about cheeseburgers.

There is a moment in General Schwarzkopf's memoir that says it all. Stormin' Norman wants to delay the ground attack because of bad weather and General Powell flips. He says, "I've already told the president the 24th. How am I supposed to go back and tell him the 26th?"

Perhaps more than any other off guard glimpse, that one captures Powell as the ultimate Beltway warrior, someone who could flat ignore the basics of war fighting: weather, terrain, and the enemy. At that moment, General Schwarzkopf was quite right in asking for the delay. General Powell wanted to attack because George Bush wanted to "get on with this thing." He refused to accept that the guys on the ground needed good weather for G-day. Without it, a lot of people, mainly our young people, could have come home in body bags. But delivering bad news to the President was worse than the thought of those body bags.

Don't get me wrong. I'm not saying generals should give orders to the President of the United States. That is not the way a democracy works, and I believe as strongly as the founding fathers in civilian control of the military. But I do think we won the war only to lose the victory on the final day because political leadership and military leadership failed us. Both men say in their memoirs their orders were only to expel Saddam Hussein from Kuwait. They did do that. But they also congratulate themselves on their performance with sly little comparisons to the generals who were our giants at Normandy. All I can say is Desert Storm was no Normandy, General Powell was no Eisenhower, and General Patton is upstairs in Valhalla somewhere, shaking his head over Stormin' Norman and waiting for his arrival to have a little chat.

CHAPTER 3

JUNKYARD DOGS:

THE BALKANS, 1992

Serb mortars had turned the airport at Sarajevo into a hot landing zone. The pilot couldn't descend the normal way because Serb snipers up on the hills took potshots any time they felt like it. I had bummed a ride on an Italian C-130. The freaked-out pilot came down like a fighter jockey in a gut-churning pedal-to-the-metal combat dive. The plane started falling like a paratrooper with a bad chute. Adrenaline pumped through my body. I took a look at a young *Newsweek* photographer named Antoine Gyori who was sitting across the aisle. The look on his face showed he felt the same electric current sizzling through me.

The plane nosed over even steeper.

"Do you feel it?" I said.

"Oh yeah, man, I feel it. I love it. I'm an adrenaline junkie."

No wonder war won't go away. Young guys get off on it.

The plane pulled up with the engines screaming, skimmed across the runway, and landed. The second we stopped, the cargo door slammed down and the cargo pallets shot out. Antoine and I jumped down and the C-130 was gone, wasting no time hanging around that killing field. As we tore across the landing strip, incoming dropped on all sides of us. CRUMP, CRUMP, CRUMP. Hot steel whizzed past our ears. We piled into a bunker. I hit the floor wondering what the hell I was doing here: *Welcome to the Balkans, motherfucker.*

Tito was dead, the Soviet Union was in ruins, and Yugoslavia was being hacked apart by the inmates running the asylum. In the spring of 1992, the Serbs, Croats, and Muslims began carving up the carcass, and that August I decided to see how far they really meant to go. I flew from New York to Frankfurt and caught a plane to Zagreb, where I set up my base of operations. While I was standing at the car rental agency, a rangy young guy named Jim Bartlett walked up.

"Are you David Hackworth?" he said.

"Yeah,"

"I got your book. Man, I love you."

He told me he was a freelance photographer. He had been on the ground for a couple of years. One look and I could see he was an action freak.

"Can I buy you a beer?" I said. After a couple, I had a new sidekick. Jim knew the roads, the fighting positions, the back ways out in the countryside and he had good contacts among all the combatants. I hired him as my stringer and we were in business.

Civil wars are always rotten, and this one was the craziest I had ever seen. The first morning I took a walk in Zagreb's main park. The city was as clean and pretty as a botanical garden. The people were as fat as the pigeons. It was late August; the sun filtered peacefully through the trees. You couldn't even tell a war was going on, although only a few weeks earlier incoming had been falling like hail on Zagreb. But if you drove out of town, in less than half an hour the country looked like Vietnam in the middle of the war. Villages were blasted, buildings and homes gutted, everything bullet pocked, shrapnel torn, shell blackened. Out there the people were already chopping down trees for firewood.

Jesus, how are they ever going to get through the winter?

Everywhere you looked you saw a fresh graveyard.

It was plain nuts. The rules of war had all been blown away along with everything else in sight. But the people were incredible. They were rebuilding, working their gardens, trying to grow their corn, patching up what had been trampled or destroyed. Even as they went about these chores, they were being mortared. I felt as if I were in a disjointed time warp. In some areas rich fields of corn, pastures, orchards, stretched off as far as you could see. It was like being back in Pratts, Virginia, in 1700. Mules and horses were pulling wagons. Since there was little fuel and the tractors couldn't go, people were back to farming the way they had done it three hundred years earlier.

So far as I could see, there was no military logic to the war, no real front, no artillery exchanges across the lines at opposing breastworks. But the mayhem wasn't really guerrilla fighting, either. What was happening was pure butchery—random, exotic, bloodcurdling. The Serbs were the most efficient killers in the slaughterhouse, but the Croats and Muslims were also standing in blood up to their waists. When the Yugoslavian Army fell apart, the Serbs grabbed the keys to the supply depots and snatched the best stuff. They were able to put together a well-equipped military force with modern tanks and artillery and plenty of ammunition. What was left, the Croats seized. Any remaining scraps went to the Muslims.

To get a fix on the balance of terror, I paid a visit to the headquarters of General Anton Tus, the Croatian Army's chief of staff, a four-star who also served as minister of defense. The interview was at nine o'clock in the morning. As I was setting up my tape recorder, we began to lay out the ground rules. The general looked tough and I needed to break the ice.

"You know," I said. "I've seen your army once before because we fought against each other. I was in Trieste from 1946 to 1950."

General Tus was a big man and he suddenly looked very interested. He got up and grabbed a tactical map and slapped it down on the desk in front of me.

"Where were you?" he said.

I studied the map. I hadn't seen that area for nearly half a century, so it took some time to place everything. Suddenly I could pick out where the old American defensive positions had been near a little village named Banne. The rocky high ground stretching between Yugoslavia and Italy had been my nursery as a soldier. In 1946 I had started out as a private in a recon unit of the 88th Division. We were deployed along the Morgan Line to keep Tito from gobbling up Trieste. The setup was World War II Army surplus: foxholes, bunkers, and big-titted easy girls.

Near the end of the war Tito's partisans had taken Trieste. Earmarking everyone who had been a Fascist or supported the Germans or just got in the way, the partisans lined them up against a wall and wasted them. As a kid soldier, one of the first things I did was go to Garibaldi Square to see the bullet holes in the execution walls. That taught me all I needed to know about Communists. They were bad, worth fighting against. After the war, Trieste went back to Italy. But the Yugoslavs were still right outside town. I also knew from my personal encounters with them along that mean border they were goddamn good soldiers. They were not going to back off from a fight.

My recon platoon regularly patrolled the Morgan Line. I was just a little kid, 135 pounds with my 9.5-pound M-1 rifle locked and loaded, a basic load of ammo and hand grenades hanging down, and my steel helmet on whenever Captain Kenneth Eggleston was around. I had a good friend, who was also fifteen, Jimmy Sparks from Indianapolis. Jimmy had taken reconnaissance basic with me at Fort Knox. Since we were both little kids, we paired off, stuck together, went out on patrols both for pussy and adventure.

The terrain was rocky—high hills with sparse vegetation, some trees and brush, but mostly stony ground and many rock walls. Ducking along behind those walls, Jimmy and I would work our way right up to the border where we could hear the Jug patrols advancing. We used to make a popping sound with our mouths, then lob a rock over the wall. In the old days, a hand grenade popped, then made a fizzling sound. A spark came out when the firing cap popped. The Jugs in diamond formation moving forward in the rain, ponchos and pots on, heard us. "*Granate*," they yelled and hit the deck. THUMP, CLATTER, BANG. They had to smack down on the rocks, skinning their hands and puncturing tender places, and they would get up covered with mud. No explosion. Then there would be nothing. They knew they'd been had. You could hear them swearing all the way to their patrol base. And we lay back behind the rock walls laughing up a storm.

It wasn't all fun and games. While my unit was deployed on the Morgan Line and around Trieste—1945–1954—it took about twenty-five dead and over a hundred wounded or captured. But to two 15-year-olds in steel pots—when Captain E., as we called him, wasn't watching—it was a Huckleberry Finn and Tom Sawyer adventure in olive drab.

The memory flashed through my mind.

"Here, right here west of Banne," I said, stabbing the map with my finger. "I was right here."

General Tus looked at me and a huge grin spread across his face.

"I was there, too," he said. "My regiment was four miles from you, opposing you, right across from you."

Was this good or bad? A little defensively, I said, "I was only fifteen."

"Good," he said, thumping his chest with his big hand, "I was fourteen."

He came over and grabbed me and gave me a bone-crushing bear hug.

Among professional soldiers there is seldom hate down at the dying level. We don't blame the man in the foxhole across the line and he doesn't blame us because we know that something or someone up above us has caused the war. We must kill him, of course, before he kills us. That's only common sense as well as our duty. But when the shooting is over, most grunts always have more in common with other grunts, even old enemies, than they will ever have with the politicians who sent them.

General Tus was of the old school and he wanted to talk. He sat me down and told me everything about his army and how he was fighting his enemies. He was outgunned and he seemed pessimistic until he started talking about armor.

He had a secret weapon.

"We have killed six hundred Serbian tanks and captured two hundred twenty," he said.

He wouldn't tell me how. The numbers were outrageous. It sounded so much like William Westmoreland's old razzle-dazzle with the Vietnam body count that I had to suppress a laugh. But something about the face of the former CG of Tito's air force said he wasn't just bullshitting the troops. I made a mental note to look into the claim.

"Can I do anything for you?" he asked at the end of the interview. I thought for a second and I remembered a trick or two from Desert Storm.

"Yes, General," I said. "Would you give me a letter so I won't have to go through all the hassle of finding spies to take me into your positions."

"Good," he said with a laugh. Then he called in a general and directed him to give me the letter right there. The officer wrote out a document that basically said, "This is a friend from *Newsweek* magazine. He can go wherever he wants. He can see whatever he wants. Help him."

Not even "General" Clifton had been quite so precise—and these orders were real. From then on, whenever I was stopped by the Security Police or a hard-nosed officer as I was whipping around the front, all I had to do was pull out that letter. The cop or soldier would take one look and snap to attention.

Then he would salute me.

I called it "my magic letter."

The countryside outside Zagreb was Ambush Alley—rolling hills with forests, roads that twisted through woods, rivers and creeks snaking under light bridges that couldn't handle armor—ideal turf for guerrilla war. One glance at the landscape and you were transported back to World War II. Everything was shot to shit. Walls were pockmarked by bullets, houses blown away by artillery or air strikes. Debris and mines covered the ground. No logical military purpose explained the havoc. It was chaos. In battle, the normal course of things is to start pounding the enemy's main line or trying to break up his logistical network or to blow apart his artillery. When you apply military force, you are supposed to have a rational purpose and clear objective. But there was nothing normal about this war. This was about firing for its own sake. The targets were villages and cities. The targets were civilians. The point was terror. The Serbs were on one side of the Sava River trying to drive back the Croatians on the other bank. But it didn't really matter in this sector whether you were a Croat or a Serb. The idea was to blister the other guy so badly he would never dare to return.

Every day was an orgy of wild gunfire. Plums were plentiful, and for the locals plum brandy was the drink of choice. In the morning a guy got up, had his breakfast, popped off a few rounds from his mortar or machine gun, then stopped for a little eye-opener. As the morning progressed and his alcohol count increased, the shooting got crazier and crazier. By noon everyone was out of his gourd.

Eilhys had given me a silver cowboy buckle embossed with a horse for my Chinese birth year. The belt had zippers in the back where I could tuck in a thousand dollars' worth of bills. I used the buckle as a character tester on those cutthroats, pirates, and crooks; Muslims, Croats, and Serbs. Every time I stopped at a checkpoint, I would see how long it took them to scope out the belt. The average was about fifteen seconds. After one guy started pointing to it and talking to his mate and monkeying with his AK, I began thinking, *These guys are so bad they're going to knock you off just for your belt buckle, especially after they've had a couple of plum brandies.* So I stuck it in my pack and let my pants ride low.

The drunkenness made me look twice at the ammo supply of these gunslingers. I had never been on a field of battle where the supply of ammunition wasn't rationed. Usually you are told to restrict yourself to just so many rounds per day, be it rifle, mortar, or artillery—unless you are about to be overrun. Here nobody held back. The drunker these guys got, the more stuff they threw. It was mindless shelling and mindless killing with no purpose other than to terrorize people or to take revenge. The erratic shelling was like nothing I had ever seen before. In Korea and Vietnam you got thumped all the time but you knew where it was coming from. When you moved, you moved fast. You always kept your eyes peeled for the next good hole to jump into. In the Balkans the incoming was completely unpredictable. You never knew where or when it would come crumping in. In the Army I came from, a stud would never consider wearing a flak jacket. But in Bosnia I couldn't get the thing on fast enough. Man, I almost *showered* in that flak jacket. I wasn't trying to prove anything to anybody anymore. Luckily, most of the houses in the Sava River sector have deep cellars. Every night before I fell asleep I would think how much I loved my cellar. It was safe and the rent was free.

Now I had another visit from my guardian angel.

I know it sounds nuts, so let me explain. Throughout my life, I should have been killed a bunch of times. But whenever I've come right up to the edge, something has pulled me back. Being face-to-face with death no longer upsets me. Besides, my Gram, who was convinced I'd be hanged like my horse thief great-uncle from Missouri, would always assure me, "If you're born to be hanged, don't worry about drownin'." But sometimes I wonder why I've been snatched back so many times when I should have died.

One morning Jim Bartlett and I set off for the battlefield in our little Hertz rental. Jim had arranged with a friend of his, a Croat sergeant, to meet us, escort us up forward, and infiltrate us through all the checkpoints and into a front line unit. We drove down the highway leading due south of Zagreb until we came to a crossroads. To the left ran the main freeway; to the right was a small one-lane secondary road that meandered cross-country to the village that was our objective. One road was fast; the other would take three or four more hours.

"Your choice, Hack," said Jim.

If we went straight we could move fast but we would have to pass through Serb territory and the Serbs could screw us. Or we could take the meandering little road through Croatian territory and we would be cool. We would get to our rendezvous but we would be cutting it close in terms of the time we were supposed to meet up with our sergeant.

"We'll go the fast way," I said.

The expressway was empty. A lot of heavy fighting had passed this way. Shrapnel was lying all over the road. We had to drive about seventy miles. We were going so fast the poor little Toyota was shaking. I heard a thump and then a clump, clump, clump, and the car went all spongy. A flat. So we took out the spare, put it on, resumed our high speed for another two miles before it was clump, clump, clump again. By now we were only about eight miles out from our goal.

"Let's just ride the rim," I said. I knew the accountants would be upset when they got the bill for new tires, new rims, new wheels, the works. But I wasn't going to stop and get stuck out there in the middle of a killing zone.

It didn't work. Riding the rim never does. By the side of the road I saw a little village, so we stopped.

"I'll hike over there," I said. "Who knows, maybe there'll be a wrecked car we can crib a tire from."

Just as I was ready to start, a truck drove up.

It was a big ice cream truck.

We waved it down. The driver spoke no English, but we got by with my fractured German and hand and arm signals. I went around to the trunk where I had stashed several cartons of Marlboros for just this kind of emergency. Cigs always make the best trading materials on a battlefield. I laid a couple of packs on him. Then I took out a 100-deutsche-mark note. I ripped it in two, gave the ice cream man half, pointed at Jim and the tires, signaled for him to take them over to the village and bring them back, at which point I would give him the other half of the banknote and some more smokes.

He got it. I had found his button.

"You go with him," I told Jim. "Get the two tires fixed and I'll see you as soon as you get back. I'll stay with the car." The driver gave me a couple of ice cream bars.

The two of them set off. I crawled into the backseat, turned on the BBC, and munched ice cream as I started reading a favorite book, Sun Tzu's *The Art of War.* Time passed. No Jim. More time passed. Still no Jim. We were cutting it right down to the microsecond on making the hookup with our sergeant and hauling ass out of that bad place before dark. Finally, Jim came back. He had two tires, used, but plenty good enough for us. We slapped one on the car and headed down the road toward Slavonski Brod, a town across the Sava River from our destination.

The closer we got the hairier things looked. As we were entering the town, I noticed that all the trees were blown away.

Shit, they've received a lot of incoming.

A farmhouse was the meeting point. We drove on slowly. All the houses were shot to hell and several were burning brightly.

Man, some bad stuff has just happened here. We pulled 'round a corner and up to where our farmhouse was supposed to be.

And there it wasn't—totally destroyed, still smoking.

We got out and made our way through the ruins to the cellar. Six people were lying there dead—including our sergeant, who was now reporting in to Saint Peter.

It turned out the Serbs had just shelled the holy shit out of the town. Then an air strike had smacked it. Then a large missile had crashed down.

"Jesus," Jim said. "Thank God we were late."

"We weren't supposed to be here," I said.

"What the hell are you talking about?"

"All my life I've had this guardian angel," I told him. "Every time I get myself in a real corner the angel rolls me over, turns me around, and changes whatever is going to happen. That's why we got the two flat tires. We weren't supposed to reach this place."

Jim looked at me like I had lost it, became very quiet, and walked away, probably his way of dealing with looney tunes. But this sort of thing has happened to me so often I don't know how else to explain it. An hour earlier and we would have been as dead as the six folks in the cellar. In my mind's eye, the guardian angel is a woman. Maybe it's my grandmother, who raised me. Or my mother, who died when I was a few months old. Whoever she is, she always changes the dice and lets me walk away from death with a big pile of chips.

The Croatian field hospital near Bosanski Broad was well marked with red crosses. Serb gunners were using them to draw a better bead. From up on the high ground they were lobbing in the heavy stuff. The pin-striped politicians had declared a cease-fire in London a week earlier, but if anything the incoming had only gotten hotter. I hadn't seen such a relentless attack since the Korean War.

Parked outside in the firestorm were a couple of tanks and a flak wagon with four antiaircraft guns that fired in unison. The field ambulances looked as if they had been working downrange at a turkey shoot. As soon as I pulled up, the hospital started receiving a lot of heavy incoming artillery fire. It had taken at least a dozen hits or near misses. The first floor was all blown to shit—shrapnel had slashed every wall—and the Croats had it barricaded. The operating room was in a deep cellar. You came down through a sandbagged entrance to where the doctors and nurses were working.

While I was standing there watching, the medics brought in two young guys who had been badly shot up earlier in the day. They were both kids, about nineteen. One guy had taken a six-inch sliver of steel in the forehead, the other had a sucking hole in his chest. Dying material. Obviously they were not going to make it through the night. One of them had a nice Austrian pistol tucked down in his belt. A Croat officer reached down and took it. Pulling out a cloth, he wiped the blood off and stuck it under his jacket.

"You got a good pistol there," I said.

"That's eight hundred dollars on the black market in Zagreb," he replied, not missing a stroke.

Jesus, I thought. *What kind of army is this?*

The artillery fire lifted for a few minutes. Then a Serb MiG-21 came in really hot and started strafing us. Around the hospital, there were just enough military vehicles along with my poor little rental to justify the place as a target, I guess. I began keeping score. *Let's see. We've been here thirty minutes and we've been shelled, we've been mortared, and now we're being strafed.*

With that, Major Kincl Miroslav, the surgeon, came running up the stairs. The flak wagon was one of his personal weapons. When he wasn't cutting you open to dig out steel, he was throwing as much of it as he could at the Serbs. As the MiG-21 came in for a second pass, he jumped into the gunner's seat and started blasting away until the pilot thought better of it and turned tail. He also had his personal tank. Between operations and air raids he would get in, cruise the front and zap any Serb who had the bad luck to get in his sights. I was looking at an army of renegades with a surgeon-warrior everyone loved because of his gut and caring.

While all the boom boom was roaring over me, I looked over at a Croat captain who was crouched up against the stone wall next to me. He had been wounded in the face, not badly, and he had come in to have the mad doc pull out the splinters of steel and stitch him up. His name was Derislau Sipicky and he spoke English. In civilian life he was an architect.

As the barrage continued, we started shooting the shit and I pulled out my notebook. To get his attention, I showed him the letter from General Tus.

"That's an important letter," he said.

I told him how pleased the general had been with the score against Serbian tanks but added I doubted the numbers.

"No. They're correct," he said. Reaching into his pocket he pulled out a worn notebook. In it he had written down the performance of his own tank hunter-killer teams. The first note covered the previous forty-eight hours.

"Killed eight tanks," it said. He leaned over and translated the notes for me. During the month that had passed one team alone had "killed thirty-two tanks." Another had blown away forty-five Serbian armored vehicles.

Then came a fascinating scribble. "Eight missiles killed two T-72 tanks, one Praga track vehicle with 30-millimeter antiaircraft guns and five T-55 tanks. No friends hit. Average kill range, 1,500 meters, 95 percent first-round hits."

General Tus did have a secret weapon. But what were those missiles? I pressed the captain.

"Top-secret antitank missile," he said. "It creates a radical pressure change inside the struck vehicle." Sounded better than anything we had. The weapon was "fire and forget," meaning the gunner could duck after he shot. The missile locked onto the target and cut through the protective armor with a double warhead. During Desert Storm our own tank busters had to guide our TOW weapons with a wire to the target. That meant the gunners were fully exposed for at least ten seconds, forever on a hot battlefield, more than enough time to be blown away.

The captain wouldn't tell me what the weapon was or where the Croats were getting it. When the shelling slowed down, I asked him to show me some hard evidence. His driver, no Rambo, wouldn't come back to pick him up because of the shelling.

"I'll take you in my Hertz," I said. "Anywhere you want to go."

He got behind the wheel and whipped down the shell-torn roads toward the front.

We kept driving until we came to a Croat position that had just taken a direct hit from the Serbs, a house fronting the Serb lines. One of the captain's men was dead and two were wounded. But before the Serbs crippled the five-man team, they had taken out a T-55. I could still see the tank smoking and there were three other dead tanks nearby.

When I examined the tanks, I found each had been killed by a missile that had burned a hole through the armor. The hole was about the size a pencil would make if you stabbed it through a sheet of paper.

The front of the building that had been the Croat position was still smoldering. I was standing in the smoke, glass all around, not knowing exactly where the fire was coming from.

"Where's the Serb position?" I yelled.

"On that knob," the captain said, pointing to a ridge about seven hundred meters away. I could just make out the green silhouette of a pair of tanks that from the distance looked like giant turtles.

"Shit," I said, ducking quickly into the back bedroom and behind the fireplace.

When I caught my breath, I joined the captain, who was patching up one of his guys. When finished, he held something up to show to me.

"This is a Serb antipersonnel mine," he said. He was holding a Bouncing Betty.

And the captain dropped it. He accidentally dropped it right at my feet.

All I could think was, *This is it. Holy shit, it's landed between my legs. I'm really gone. There go the family jewels, and there go I.*

It didn't go off. Perhaps my guardian angel saved me. But there was no limit to how reckless these guys were, drunk or sober. Working our way around to the back of the building, we put his wounded in a shot-up car and raced back to the field hospital.

After a few days, I knew I had a good thing going with him. We went back up to the front. I was still skeptical about the kill count and I wanted to find out what the secret weapon was.

"You're knocking out all these tanks," I said. "But I only see a few tanks. How can you be knocking out so many if I'm seeing so few?" What I was really angling to see was the antitank weapon. All the tank talk was smokescreen. Then he said something that knocked me out.

"We are pulling them back and rebuilding them."

"Great. Let's go look."

Jim was with us, so I slipped him one of those tiny pocket cameras.

"See if you can get some pictures of this tank rebuild factory on the sly," I told him, not knowing what we would find.

Four miles behind the line, the captain showed us the workshop—and it was for real. Inside he had about twenty Serb tanks at different stages of repair. The mechanics were cannibalizing one tank to refit the others. They were not soldiers, but they were at the front, working right under the noses of the Serbian gunners.

In any American operation the same sort of factory would have been fifty to one hundred miles behind the lines, with a sophisticated maintenance depot, all the proper spare parts and equipment, probably air-conditioned. Not the Croats. When they finished rebuilding a tank, all they did was stick their own flag on it and send it up to fight with a new crew. These kinds of fighters don't play by our rules. Yet we always assume they do. That's why we so frequently get our asses kicked.

When I examined the tanks, I found more of the little holes. The missile had melted its way through the armor, changing the air pressure. That alone was enough to kill the crew inside. The metal from the hole is called spalling. It's melted, it's hot, but the minute it's forced into the tank it hardens again and starts zinging around at a tremendous velocity, slicing through all those young, tender, fleshy bodies.

This was General Tus's secret weapon.

It had to be a modified French Milan missile. I had seen those same telltale holes on Iraqi tanks in the French sector during Desert Storm. But that still left the question of where General Tus was getting them.

The next day I was sitting with another Croatian officer in a café just behind the lines. I bought a bottle of the local rotgut and we killed it plus a couple more. When we were both hanging onto the edge of the table, I showed him my magic letter from General Tus and told him how much the tank killer program impressed me.

"It's the Milan," he said.

The system cost $80,000; each missile cost $20,000. They were getting them from South Africa. The officer raised his glass.

"The South Africans hate the Serbs," he said. "And they love the profits."

The real winners in this civil war were the gunrunners. The United Nations blockade had more holes in it than the New York subway system. Everywhere I went I saw stacks of military equipment: Russian Sagger missiles, new T-72 tank engines still in the crates, grenades, mortar shells, bullets. The weapons came from every country in the world: from Japan, from Italy, from the United States, from Germany. You could read the origins on the sides of the crates. Only the Muslims were short of ammunition, but the Iranians were helping them play catch-up. While I was there, an Iranian transport plane was nailed in Zagreb loaded with four thousand

AK-47s and millions of rounds of ammunition. The Croatians and the Serbs both had plenty of hardware and were building the support systems a modern army needs. *Where is all of this coming from?* I wondered. *How come that big blockade we are hearing so much about isn't working?*

Finally, I met an officer from NATO who explained everything.

"The blockade leaks like a sieve," he told me. "It's a joke."

We weren't boarding ships. Any captain could say, "Hey, I've got a load of bananas," and he would be waved through. The reply was, "Fine. Pass." And another shipment of rockets or mortar shells was on its way to the killing fields.

No one could keep the peace under these conditions. The French and the Canadians both sent crack regiments under the United Nations flag only to see them become targets or hostages. In this crap game the locals had loaded the dice. The mission of the UN troops should have been peacemaking, not peacekeeping. There was no way to execute a peacekeeping mission because there was no peace. The UN troops should have had the tanks and artillery they needed to separate the warring parties and rules of engagement that had teeth. As one sergeant put it: "If that son of a bitch shoots at you, we want to blow him out of the saddle."

To test my theory I hitched up with the men of November Company, a unit from the elite 22nd Royal Canadian Regiment that had arrived in the spring of 1992. In mid-April, the company had come down into Croatia from Germany, where they were doing NATO duty. The first night they formed up in their assembly area, expecting to be greeted as peacekeepers. Most of the regulars, the NCOs, and the older officers were veterans of other UN missions around the world. They had just gotten off the train and were standing around wondering when they were going to eat. And suddenly, PLOMB, PLOMB, PLOMB, PLOMB, PLOMB, a dozen 120-millimeter mortar rounds landed all around them.

They had no foxholes, no prepared positions. Hey, they didn't need them, weren't they peacekeepers? In some cases it was the first incoming fire the regiment had received since the Korean War. So these guys were a little bit rusty.

"It was one hell of a wake-up call," said Sergeant Major Richard Clark, who herded the 235-man rifle company to safety. Later I met Platoon Warrant Officer Hall, who was stoic about it all. "The shelling was a blessing in disguise," he said. "It told us that this wasn't peacekeeping as we'd known it and that we were going to be up to our eyeballs in a very hot war." From that night on, the company was all heads up and tails down and alert. "In the long haul," he said, "that incoming fire saved a lot of lives."

Sergeant Major Clark was the only man wounded in the shelling. Flying shrapnel nicked his thumb. He had the cut bandaged, and the men renamed their outfit Lucky Company.

The Canadians moved out into Sector West near the city of Pakrac in central Croatia, where they might as well have been in the middle of World War II. Incoming artillery and mortar fire mixed with plenty of small arms fire to scourge the rolling countryside night and day. Snipers, ambushes, and mines added to the danger of patrolling the rugged, heavily forested terrain. Every turn on the road offered a potential ambush. Most of the Canadian armored vehicles I saw wore scars from bullets and shell fragments.

The Canadians were particularly good at peacekeeping. Many of the NCOs had done tours in the Sinai Desert and Cyprus. Peacekeeping is not an infantry role, but only the infantry can do it because the infantry is down on the ground and in position to see and control things. The job is a bitch. These troops were looking at ill-disciplined militia gangs posing as soldiers. The Balkans were a patchwork of deadly ethnic hatreds and no two sectors were the same. One day Lieutenant Kevin Cameron's platoon was manning a roadblock on the Serbian side of the line. To while away the time they were listening to the music and propaganda of Croatia's Zagreb Zelda. A drunken Serb soldier walked up and thrust his weapon in Cameron's face and said, "If you no shut off, I kill you."

I would have whacked the son of a bitch but Cameron, good peacemaker that he was, zapped Zelda, slammed a tape of Guns 'N Roses into his ghetto blaster, and offered the drunk a cigarette. They wound up looking at each other through tobacco haze instead of gunsmoke while Axel Rose kept the peace.

On another patrol, a second soused Croat accidentally fired a burst of submachine gun fire into Lieutenant Cameron's armored vehicle, wounding Corporal Lloyd. The lieutenant disarmed the drunk, patched up the corporal and evacuated him, and went on with the patrol. This kind of work took great discipline. The sad part was good efforts could restore peace here and there but sooner or later the winds of war blew elsewhere and everything withered.

One day I was out in no-man's-land when I ran into a young woman, half-Serb, half-Croat, who was visiting her father near the temporary safety of a UN roadblock. She looked up at Corporal Nick Dorrington, twenty-five, a young guy from Halifax, Nova Scotia, and said, "Thank God for the UN. You are saving us from ourselves."

Next to staying alive the biggest problem for the UN soldiers was staying neutral. "If we took sides we'd be caught in the middle," Sergeant Jim MacDonald told me during one break between patrols. "We'd become a target for every side and we'd lose our effectiveness at getting all parties to the table to talk." He might as well have been a prophet. The first reports on the shelling that greeted November Company blamed the incoming on the Serbs. Then another report said the Croats had done the job. The troopers discovered that they were trapped in a beaten zone between cunning enemies who had spent decades stacking the deck and hoodwinking neutrals. Brave men from Canada were giving candy to kids from all sides. All sides were giving them the finger.

There is something in human nature that seduces us into reducing war and its offspring to clearly divided teams of good and evil. Sometimes, as with Adolf Hitler, the division is accurate. In the Balkans, ethnic cleansing was also evil, no matter who practiced it. But on any given day it could be hard as hell to tell the good guys from the bad guys. Hatred was passed down to the children by their grandmothers. The mom and dad were out in the fields or in the cities working, but granny was at home telling stories to the kids, educating them from two generations away, reminding them of what they might have seen if they had been in Sarajevo when the archduke was killed and the First World War came, or what happened in World War II. To change the mind-set of these people you would have to get rid of all the grandparents. I've never seen hate run so deep.

White-hot emotion made it very hard to be fair to all parties, to report objectively. It surprised me to discover how hostile the Canadians were toward the press, the international press, not just American reporters. They complained about reporters taking sides. Their home base in Germany was only a few hours away by air. One corporal told me when he went "home" he would see TV reports from the Balkans and think he was watching a different war. Another Canadian said, "The press wants a clear-cut good guy and a clear-cut bad guy. They won't dig deep. They just grab what's on the top." Warrant Officer Joe Parsons told me he had been in the middle of one fracas in the morning, flew to Germany that afternoon, then flicked on the tube for the evening news. "I couldn't believe what I saw," he said. "It was mind blowing. I had just lived through that event and the distortion was incredible."

I returned to Sarajevo in mid-August. Landing under fire was the easy part of the trip. It didn't take long to see everything else would be a lot more complicated. The ancient city sat dead center in a valley. Serb artillery and snipers commanded the high ground. It was not like Dien Bien Phu or Khe Sanh, where trained armies fought until one side or the other pulled out. This was a killing zone where mad butchers were blasting away at anything and everything that moved. The Serbs rained death on all living things in the city and villages below. The people were like goldfish in a glass bowl. A drunken lout with an AK-47 was leaning over them, peering into the bowl and shooting them. They couldn't run, they couldn't hide, and they couldn't make the relentless fire go away.

On my first trip I spent equal time with the UN peacekeepers, Muslim troops, and the Serbs. Sarajevo was a zoo. To cover all sides fairly you had to move from the UN peacekeepers—the Canadians, Malaysians, Swedes, and French—to the Muslims and over across the lines to the Serbs. There were no tidy front lines. The battle area was like a checkerboard with the reds and the blacks crazily intermixed. To the eye of a regular soldier, nothing made Army sense. I wasn't looking at World War II-style warfare; nor was it the Desert Storm, Air-Land Battle of fire and maneuver; nor was it guerrilla irregular war. Something deadly new and corrosive was at work. The city was turning into a crater within a valley the way Beirut had turned into a crater by the sea.

To me the oddest thing was that people did so little to dig in or protect themselves. They went about their daily lives as if the war were just a bad dream that would be gone when they awoke. One afternoon I was drinking coffee with a Muslim commander in his rose garden. The day was sunny and warm and I had almost forgotten that I was at ground zero. Suddenly five 82-millimeter mortar rounds crunched in and exploded not more than fifty meters away. The commander and his staff calmly set down their fine porcelain cups and moved with the grace of ballet dancers, behind trees and walls. Saying to hell with pride, I crawled inside a French armored carrier and looked thankfully at the thickness of its steel skin.

I spent a lot of time with the Muslims, a lot of time with the Serbs. When I traveled with the Muslims, I was on my own hook. Then I would join the French. The French commander was an old paratrooper who had trained in the United States. He had also served in Desert Storm and he said this duty was far more dangerous than anything he had seen in the Persian Gulf. As the high sheriff, he could drive through all the lines, and when I was with him so could I. We would stop and talk to a Muslim colonel at one position, then move on and check in with a Serbian colonel a little farther down. One day I ran into a Serbian commander who said, "We pray to God the war will be over by winter. If it were up to us, we'd stop this hour, but the politicians at the top won't let us stop fighting. They have their own secret agenda."

We were standing at a spot overlooking a Serbian village. During the previous three days the village had taken five hundred rounds of artillery and mortar fire.

"Before the war we played football with them," the commander said. "Now we throw mortar bombs."

No one could be pro-Serb given the murderous way the Serbs practiced ethnic cleansing. But I couldn't be pro-Croatian or pro-Muslim, either. The Serbs were so violent they no longer gave a rat's ass if the rest of the world saw them as murderers and rapists. But the Croats and the Muslims were no slouches either. They had worn Nazi black during World War II. They were responsible for a million Serbian, Jewish, and Gypsy deaths in that round of ethnic cleansing. They had gotten along with Adolf Hitler just fine. Even the Nazis were startled by their brutality. The irony now was that Washington and Berlin were both supporting them, that NATO was on their side.

The Bosnians were excellent propagandists. The first people I met in Sarajevo were enormously friendly, well educated, sympathetic. The press briefer spoke English and had a doctorate from an American university. I liked him at first, as I was meant to. Then, when he was taking me around, I noticed we were not going along any direct routes. We would stop by a cemetery where I was told ten people had been killed and buried the previous night. Then there would be a stop where a woman had been raped. Then a school nursery vignette with children and their books and their teachers while their mothers were supposedly out in the fields digging trenches to stop the barbarian Serbs. And so on. He had choreographed the whole tour. If I had not been streetwise, I would have been saying to myself like so many naive visitors did, *Oh my God, look at this. Fresh graves. Mortars coming down. We have to do something.*

Stringing along a visiting reporter was innocent enough but some of the other situations I saw were far more sinister. While I was in Sarajevo, people were saying that the Serbs were behind a mortar attack that had killed several dozen innocent civilians. I went down to examine the physical evidence. When a mortar hits concrete it leaves a very clear signature. But I couldn't find any signs of mortar shells, even though all the reports said the bloodshed had been caused by a couple of 120-millimeter rounds. I found something completely different: the big, roundish, unmistakable scar left by the concussion and backblast of a large claymore mine. A claymore can be fired by command or by trip wire. From twenty or thirty feet away you can press a firing device and BANG. No evidence, unless you know how to look for it. Someone had set the thing off. I suppose a Serb could have sneaked in, planted the mine, and then triggered it. But why go to so much bother and take such a risk when all you had to do was lob in mortar rounds? Some of the Canadian peacekeepers suspected the Muslims had zapped their own people to draw sympathy and blamed the Serbs. No one could prove it, but given the past record of all the warring parties it was enough to put me on my guard. Nothing in this sump of violence could be taken at face value.

On another outing with the Bosnians, I went to a military command post in the sandbagged cellar of a ruined house. I found a group of soldiers watching television. It was noon and I wondered what they were watching, so I slipped up behind them and looked over at the tube. It was a Mozart concert. I thought, *What a war. The Muslim soldiers are so civilized, cultured, incredibly decent. When I was a soldier and back in the rear we would be seeing who could get laid or who could get drunk. And here they are listening to Mozart.*

Then I walked outside. On the roof of the bunker a sniper was lying underneath a camouflage net. Just as I appeared, BAM, he fired. I turned and saw his face and his rifle sticking out. Then I whirled around and saw a man go down in a field about four hundred meters away.

The next day I happened to be out with Serb troops just across the demarcation line from the Bosnian bunker.

"Yesterday I was over there and I saw a guy being shot," I said to the Serbian commander. "What was the deal with him? Was he one of your soldiers?"

"No," said the Serb. "He was an old man, eighty years old, out gathering plums for his lady love. They shot him in the gut, not the head, so he was out there moaning. But we couldn't reach him because of the sniper. He was dead when we finally got to him."

A very old sniper's trick. The Japanese were very good at it; so were the Viet Cong. You cut a guy down so everyone's trying to help him and if you are lucky you can zap eight or nine more. The target had no military value. He was just an old man, a plum-picking lover.

Murdered by the sniper on duty above all those civilized Mozart lovers four hundred meters away.

Not long afterward I went out on another patrol to a little Serb village a few kilometers outside Sarajevo. At one point as we were walking along, we stopped at a cemetery. The patrol leader shouted a few directions to his troops and they started digging up a grave. At first I thought they must be looking for weapons. It is an old trick to cache weapons in a grave. It looks like you are going to dig up Omar but what you are really after is a case of AK-47s. While they were digging, I wandered around the graveyard. Checking the marker they had thrown aside, I saw the guy had a Muslim name and had died in 1944. This puzzled me.

Finally, the shovels hit a coffin. The soldiers scraped away the dirt and climbed out. Then they all lined up around the hole and pulled out their dicks and pissed on the coffin. After that, the patrol leader snapped an order. This time the troops slammed rounds into their chambers and emptied their magazines into the disintegrating casket.

"What goes on here?" I said to the patrol leader.

"That man was very, very bad man," he said. "He killed many, many Serbs during World War Two. He is being punished."

If you go to that kind of trouble to punish the dead, I thought, *what are you doing to the living?*

I left Sarajevo the first week in September, just as the politicians in London announced another cease-fire. The same day thirty-eight people were killed. Any way you looked at it, the only growth industry in the Balkans was laying out new cemeteries. These people were mean as junkyard dogs. I thought we should stop trying to housebreak them.

CHAPTER 4

GOD'S WORK IN HELL:

SOMALIA, 1992–1993

It was Christmas Day in Heat Stroke City.

The carols floating through the hot, humid air over Baledogle came from the psy-war unit's loudspeakers.

The Christmas tree was a thornbush specially trimmed by Sergeant Tyrone Townsend with empty Tabasco bottles, used flares, painted plastic water bottles, and tinsel from a shredded metallic ground sheet.

The Christmas cards came from All Saints Catholic School in Charlotte, North Carolina. Sergeant David Jones, Company A, 2nd Battalion, 87th Infantry, 10th Mountain Division, got one from Bryan Soffera, seven years old, who misspelled the spirit of the season with the same good-heartedness and naïveté that had propelled the United States into Somalia.

"The people down there are dieing," Bryan wrote. "Fix it."

I received orders from New York to move out for Mogadishu in early December of 1992. I have always liked spending Christmas with troops, so when the call came I went out and got an armful of shots, a basic load of malaria pills, and a kit for the desert: mosquito net and bug juice, a pad to sleep on, candles, batteries, hot-weather gear—some of which was still in my closet as war surplus from Desert Storm.

President George Bush thought the time had come "to do God's work" in Somalia. It always puts my teeth on edge when our political leaders begin to talk that way. Before you know it, someone gets killed. The President had won his Pyrrhic victory in the Gulf only to lose the November election of 1992. Famine was ravaging Somalia, the pictures of starving children were heartbreaking, and the stories about local warlords hijacking relief supplies were infuriating. I had no problem with his desire to do good. But in organizing Operation Restore Hope that December, the President forgot the very first principle of the New World Disorder. The road to hell is paved with good intentions.

The operation was a mercy mission. Who could believe how badly we would get wrapped around the axle before it was over? But from the very beginning I felt my sixth sense screaming. Even on the trip in, I saw clues that Operation Restore Hope was turning the heads of a certain kind of senior officer. Odd little things.

I had to stop in Nairobi to catch a flight into Mogadishu. While I was standing there waiting for my plane, a U.S. Navy captain walked up. You could see his eagle, but he had gotten himself fitted out like a warrior. He had a steel pot, jungle fatigues, the right boots. He was even wearing a combat pack. Really pumped. A total phony. He wasn't going to be doing any fighting anywhere soon.

"Hello, sailor," I said as sweetly as any Tu Do Street bar girl.

The captain turned an incredible shade of bright red. This white-haired guy had really punched a hole in his balloon. You could hear the air hiss out of him all the way to Mogadishu.

I flew into Somalia the day after the Marines Corps landed to find the beach at Half Moon Bay already secured by the press corps. The landing was like a comic moment out of the movie *Airplane*. It was a travesty, an amphibious landing on beaches entangled with reporters, camera lights, and microphones, a photo op gyrating like a chopper with its rotor shot off. The grunts were calling it a real Romeo Foxtrot aka Rat Fuck. Under a full moon, a few minutes after midnight, with the world watching by satellite TV, elite U.S. Navy frogmen and Marine recon grunts crept silently out of the Indian Ocean to the shoreline. With camouflaged faces and hands, armed to the teeth, they were prepared to fight for that chunk of beach.

Instead of locking horns with an armed enemy, they found the entire planetary media waiting to get the drop on them.

The warriors whose lives were on the line that night told me afterward they were mad as hell because the lighting had exposed them to possible hostile sniper fire. Photographers were acting like paparazzi chasing Rod Stewart and a sweet young thing down the beach at Malibu. Under the hot white lights, the assault beach looked like a movie set minus John Huston's directorial flair or Clint Eastwood's *Heartbreak Ridge* cool.

This happened because the Pentagon, reversing field completely from Desert Storm, decided to throw the gates wide open this time around. The U.S. military didn't think it could get into trouble in Somalia. It couldn't lose. So there was nothing to hide, no minders, no Thought Police. Never in my life have I seen such total openness. The generals told you exactly where our troops were, what our plans were. When you are doing the Lord's work, I suppose, you want the world to see it. It didn't occur to anyone that the flip side of all this goodness just might be overconfidence, and that overconfidence was just an angel's hair away from arrogance, and that when we wander into the third world, arrogance always spells disaster.

An amphibious operation is dangerous even in a peacetime setting at Camp Pendleton, where the meanest things are blank shots and a starched colonel's barks. Only the training and sharp discipline of the SEALs and Marine grunts prevented members of the press corps from being shot or crushed by the vehicles that stormed onto the beach. At first I thought the reporters were to blame; then I discovered the Pentagon had media-managed the fiasco. Every detail of Operation Restore Hope was given to the media: mission, assault beach, objectives, troop strengths, intentions and assembly plans, and commanders' names. Marine Major Frank Libutti even briefed reporters near the invasion beaches and invited them to cover the assault. All that was missing was an open bar and hors d'oeuvers.

I'm not opposed to close working relations between reporters and the military, but there is a right way and a wrong way to get cozy. When I was a kid in Italy in 1946, Robert Ruark, one of the old shoe leather war correspondents, stayed with my 351st Infantry Regiment recon platoon for the better part of a day. He tagged along, occasionally taking notes and talked to the footsloggers, as grunts were known in those days. Our skipper, Captain Kenneth Eggleston, didn't tell us what to say or not to say, nor did he or any of the regimental brass stand in Ruark's way. Ruark could go and do whatever he wanted to do. He was respectful of our mission and he got down and did grunt things when the situation called for them. He even wore a uniform that was the same as ours except for a war correspondent's insignia. I was always struck with what a good guy he was and we were all proud when he published a story that said we were "the nation's best." He got these stories by being down on the ground with us, by being one of us.

During the Korean War, reporters were treated much the same as they were in World War II. They wore uniforms, and all dispatches and photos were censored to ensure that the troops' security was never put at risk. My 27th Infantry Regiment became the darling of the public because Maggie Higgins of *The New York Times* stayed with us for months and bragged about us in her dispatches. Some said she took a fancy to our fighting spirit, and others said she took a bigger fancy to Iron Mike Michaelis (our John Barrymore handsome commander). Between Maggie, Mike, a great bunch of fighting men, and an effective group of reporters, our story got told to the American people, who had every right to know how we and the war were doing. And believe me, in 1950 we were doing badly.

But in Somalia, the Pentagon lurched from tight to slack reins and the buggy started jolting so crazily it brought out the worst in everyone.

"You want car? You want car?"

The guy hustling me looked like Khadafy's idiot cousin. His eyes were red and darting around all points of the compass. His shirt and pants were worn out, covered with dust. But you had to give him respect. He was carrying that reliable old equalizer, an AK-47. And you weren't going anywhere without him or one of his pals.

Somalia wasn't just another third world country. It was so bad it didn't have a rating. The most primitive spot in Vietnam or Laos seemed like an advanced civilization by comparison.

Mogadishu was set on flat land with no hills. Everything was broken. The port was a bad joke; the cranes didn't work, the airfield was lousy, everything had to be repaired. The capital itself was an outlaw town. The streets were jammed with tens of thousands of people, but there was no sign of government. It was total anarchy.

No one was directing traffic, but then no one had to. The only civilian vehicles were the half-wrecked pickups driven by freelance gunmen. You couldn't go anywhere without someone riding shotgun. The standard operating procedure was to pick up one or two rent-a-guards with their AK-47s. These guys would hang out in front of the little dump where we were staying because they knew reporters had money and needed protection. When you came out the door, they'd nail you. And try as you would to dicker over the price, it was never any less than a hundred bucks. Wherever you went, the car set you back at least another hundred dollars, not counting gas. Generally the guards came from whatever gang controlled that part of town. The gunmen knew the passwords to get you in and out and they spoke a little English, so you got a translation service on the side. You would set off down the street, your shotguns screaming at the crowds, the car nosing slowly through a human mudslide.

Without their guns and contacts other dudes would strip you cleaner than a dog's bone. You wouldn't make a hundred meters before somebody robbed you with a gun or knife. It wasn't safe to carry a laptop, my 1990s basic weapon, or a camera. Thieves would rip them right off the shoulder straps. They would yank out your gold fillings if you were foolish enough to open your mouth. The drug of choice among these violent dudes was called khat. Each morning little planes flew in the day's stash from Ethiopia and Kenya. The drug runners landed at 7:00 A.M., and by 7:30 everyone had scored. The locals who chewed worked up the equivalent of a twenty-cup-of-coffee high. The more stoned the Somalis got, the faster they chewed, and the faster they chewed the redder their eyes got. By three in the afternoon they were stoned out of their minds—Mogadishu's version of the plum brandy heads in the Balkans. Sneeze and they would open fire.

Luckily, my beat was the grunts and I could hotfoot it out of range. There was no way I was going to hang around and listen to the PAOs bullshit at the Daily Follies about how well everything was going. I hired a couple of shotguns, piled in their pickup, and headed for the boondocks. When the gunmen dropped me off, they looked sort of disappointed.

"Don't you want to come back?" one of them asked me.

"No, man," I told him. "I'm joining the army."

A company of the 9th Marines had secured the airfield at Baledogle, where flat windless plains covered with camel-high thornbushes stretched away as far as you could see. The sun-baked earth was as hard as rock, the heat seared you like a blast furnace, body and soul. Everything was bone dry. Walk out anywhere and the wait-a-minute bushes would get you; their sharp thorns snagging your uniform, forcing you to stop and clear the little mothers. I got together with the Marines and waited for the 2nd Battalion of the 87th Infantry, 10th Mountain Division, to fly in from Fort Drum. During the regime of Siad Barre, the airfield had been a Soviet fighter strip.

It had long runways for MiGs. When the Soviets finally said, "Let's get out of here," they cratered the field. Then the Somalis came in and scavenged it, ripping off everything that could be carried away. Now it was crumbling around the edges with mines, booby traps, the works.

It was blazing hot that day. The first plane in was a C-141 transport carrying A Company and part of the battalion's headquarters company. The strip had not been built for heavy aircraft. After five days the Air Force had to shut it down to anything heavier than a C-130. The combination of no maintenance on the strip and the size of our transports, not only too big but loaded to the max, put us over the edge. After twenty C-141s landed, it was as furrowed as a farmer's field.

Military planning at its finest. Air Force engineers had undoubtedly flashed word that Baledogle had a super field, forgetting to say that it was for fighters and perhaps the lighter Soviet IL-76 transport. The assumption was that a good-looking strip must be more or less like our own. When we lay down a runway, it's for everything we've got in our inventory. You can trust it. You couldn't trust anything in Somalia, especially if it was Soviet built. So they had to close down the field for repairs. This meant the first troops down were caught real shorthanded. If there had been a real enemy on the ground who wanted to fight, Company A would have been in a world of hurts.

The grunts came off the transport with their weapons at high port, expecting to shoot their way in. The only enemy that day was the sun. Bending under one-hundred-pound rucksacks, they trooped out of the cool airplane into a blast furnace and began to melt. Straight from Fort Drum, in upstate New York, these guys had gone from snow to sand. Their faces were pale; they were in no condition to absorb such a body blow from the climate. "It was the old zero-casualties mentality," one officer told me. "As a result, we boiled." Luckily, their skipper, Captain Gordy Flowers, had the sharpness and good sense to tell them to throw down their packs. If he hadn't, the outfit would have been in the hospital from heat exhaustion a long time before it got to the assembly area.

The unit had been ordered to be self-sufficient for five days, so the commanders had ordered their men to pack in four of everything. They figured their resupplies should be on the soldiers' backs, but no one could hump all that stuff in such heat. At the Pentagon the brass had decided to deploy the 10th Mountain Division because it was their turn. No one had considered the weather or who might be better trained or acclimated for the mission. It was left to the 10th Mountain grunts to stagger off their transports and do their duty. They were lucky they didn't have to do any fighting that day.

Company A had to move out, set up camp, secure a perimeter, and stack ammo, rations, and water in the 110-degree heat. The first casualties were from the sun. But the troops were very professional and did a good job. The 10th Mountain Division's commander, Steve Arnold, and tactical assistant commander, Greg Giles, now a brigadier general, and once a Ranger lieutenant with me in the Zone in Vietnam, were first-rate warriors, outstanding leaders. You could see it in their troops.

I asked who was in command and a sergeant pointed to a big guy who was looking at maps, talking to people, giving orders. When I walked up to him, I saw that it was Major Marty Stanton, my old friend from Desert Shield and Desert Storm. During the Gulf War he had won the Silver Star, and the grunts now called him Marty of Arabia. He was commanding the advance force, tasked with providing local security at Baledogle, until his battalion commander, a lieutenant colonel, arrived with the rest of the unit. General Arnold had wanted to bring in the whole battalion, but he was told by the planners to limit himself to three planes. This made no sense. At the time the Pentagon was using vital air assets to fly in unnecessary clerks to provide a courtly entourage for the Perfumed Princes. The same mistake was made during Desert Storm when the brass flew in the XVIII Airborne Corps headquarters people before dispatching all of the corps' parachute infantry.

I slowly inched up behind him as he busily snapped out orders. I was less than a foot away from him when he finally looked up, absolutely astonished. "Christ, Colonel Hackworth, what the hell are you doing here?" Right there on the spot I decided to stick around for a week or so with this big lug and his fine outfit.

I arranged to stay with Company A full-time and to go out with all of their platoons. For a while the grunts were worried about having a reporter along, but after a spell, I evolved into kind of an old sarge, the guy who was in all those other wars, who didn't talk much, who slept on the ground with them, ate their chow, asked for no privileges; and after a few days they began to trust me. I would go out with them when they were doing a ground sweep or a motorized patrol.

Then it became, "Well, where would you put the machine gun?" Or, "Do you think this is a likely avenue of approach?" They had the good sense to learn from anyone who had something useful to teach.

Out in the bush the dust turned everyone brown, as if they were covered in Pan-Cake makeup. During the day there was no way to escape the relentless sun. You welcomed the end of the day because at least it was cool. A cup of coffee at night was one of life's few pleasures. The soldiers would blow over their cups, dreaming about a cold, cold shower, clean clothes, and hot food. But darkness brought squadrons of mosquitoes, always on the attack. An Army medical officer came by and said he wanted the men to sleep under mosquito nets to protect them from malaria, but the grunts didn't have nets and were sleeping on the ground as soldiers have for thousands of years. Despite the well-meant medical advice, the grunts fed the mosquitoes, who must have had a feast.

Under these conditions the soldiers conducted patrols, stood guard all night, slept when they could. Down where the rubber meets the road, I could see how well we were doing with the command structure, the operational aspects of the mission, the logistics, and the flow of information. The rifle company is the heartbeat, the very essence of an army. If everything is cool there, then an army is in good shape.

One day I was out for my four-mile walk when I ran across two very unhappy troopers.

"Hey, Colonel," one of them said. "We've been here over two weeks and what a bunch of shit; we're not getting any mail."

By then the airfield had been repaired and planes were arriving from Fort Drum regularly, but no one had had the brains to organize a mailbag for each company. All that had to be done was for someone to arrange for wives and families to drop letters or packages at headquarters and the mail could have been in Somalia the next day. I got word to Marty of Arabia, who had been too busy planning and doing to even think about mail. He told battalion XO Joe Occhuzzio, who fixed it quick smart. The difference in conditions for the fighters and for the support troops was terrible for morale. The fighters were the have-nots and the support troops were the haves. When one platoon replaced another at a roadblock, it passed through an Air Force logistical area. The Wild Blue Yonder fat cats had tents with lights, cots, mosquito screens, and cases of mineral water flown in from Nairobi.

One hot day I was out with a platoon controlling access to the airfield. The heat was so brutal that I was dressed in my shorts and a T-shirt and my jungle hat, but the grunts were in full uniform, sleeves down instead of rolled up, wearing pots and flak jackets. They were carrying their weapons and their ammo and they had their chin straps just right. They looked recruiting poster perfect in case CNN came around to shoot them, which is how the top brass wanted it. But they were frying out there in the heat. It was well over 100 degrees and with all that shit on, their body temperature had to be hitting 110-plus. The rule was, the troopers couldn't even take their ponchos and make them into little lean-tos to get some shade.

"Why don't you organize some shade for these guys?" I asked their lieutenant. "They're gonna fucking melt."

"They gotta look good, look smart," he said, shaking his head. "All the press comes in and out of that gate."

"That's a bunch of shit," I told him.

So he tried. He went to his company commander, who went to the battalion commander, and up the chain of command and back it came—the policy of General Robert Johnston, U.S. Marine Corps, the head man of Operation Restore Hope, was: Look good. Be smart. Stay in uniform.

Later that day I was in the shade behind a machine gun bunker when a chopper dropped down from the sky. Out stepped General Lawson W. Magruder III, the perfect image of the perfect modern general. He had a square jaw and he was wearing shades. His cammies were all starched—and his sleeves were rolled up. He was the 10th Division's assistant commander for maneuver.

He came whipping around the perimeter in his cool threads, checking things out, patting soldiers on the back, working the perimeter as if he were at a cocktail party. I was sitting in the shade thinking this over as the general started walking back to his chopper. Then he saw me, so he came back and said, "Dave, how are you?" I didn't know him from a case of MREs. I guess he must have asked someone who the old guy was. Anyway, I got up.

"Hot, General, just goddamn hot."

"If you find anything wrong here, you let me know," he said. "We're doing an important job. I want to know everything that's going on. You be my eyes here."

Patronizing son of a bitch, but what the hell, since he was right there, might as well get some good out of it.

"Well, General," I said. "If I were you, what I would do, since you're not exactly in the middle of a wild battle and these kids are not receiving incoming fire, is I'd let them take off their pots and body armor. Let them strip down, roll up their sleeves, get rid of all that shit, kind of lighten up."

He turned to the captain who had been trying to accomplish that very objective and snapped, "Why don't you do that? Anyone with any common sense would do that."

"Yes, sir," the captain replied, covering for General Magruder and the rest of the blind senior brass.

The general hadn't seen a thing, hadn't noticed a thing, hadn't heard a thing. He just flew in on a helicopter from about three thousand feet up, very beautiful and cool, hit the ground for a few minutes, did his cheerleader thing, got a bit hot, maybe; then it was back up to a refrigerated three thousand feet and off to the next unit. You saw the same thing in Vietnam. The reason the top brass knew so little was they didn't stay anywhere long enough to find out. If you run a corporation that way, the corporation bellies up. If you do it with an army long enough, sooner or later it will belly up too. But General Magruder was wearing Superman's cape that day and it was up, up, and away.

Marty Stanton had tasked Company A to be the air mobile company, the Division Reaction Force. Their job was to marry up with helicopters and move quickly to put out fires. Since no one in the battalion had any experience using choppers for such an operation, Stanton asked me to give a class to the battalion officers. On Christmas Eve, we all assembled in a bombed-out building the 87th Infantry was using for a command post and I passed on the techniques and tactics we'd used in Vietnam, all learned at an incredible price in lives and ten thousand choppers—and now forgotten. As I was riding back to Gordy Flowers's CP after the class, I dug a bottle of Jack Daniel's out of my pack. I'd been saving it for just such a moment.

"Let's all have a Christmas nip or two," I said to Gordy and his leaders.

"Thanks, Hack, but we can't drink on a mission," Gordy replied.

What a different Army, I thought. *In the old days one guy would have emptied the bottle for all of us.*

East Africa was light-years away from snowbound Fort Drum in New York, but when the calendar said Christmas, the 112 Lightfighters of Company A did what they could to get into the season to be jolly. Battalion commander Jim Sikes had to scrounge something special from a Canadian UN paratrooper unit, trading ammo for chow so the grunts would not have to eat MREs for Christmas dinner. Our grunts wound up eating Canadian field rations—better than ours—while the Canadians got a hot dinner for Christmas. Even so, to our troops it was a night at the Hard Rock Cafe. Johnston had sent out word no one was to make a big deal of Christmas since they were in a Muslim country where people were starving. After that, of course, the brass and the Rear Echelon Motherfuckers sat down to turkey with all the trimmings.

Without getting much help from anyone high in the chain of command, the leaders of Company A busted their butts to make Christmas Day special. Starting Christmas Eve, they organized church services, volleyball games, continuous video movies, and phone calls home.

"Is it a boy or a girl?" Sergeant Willie Black asked his pregnant wife.

"It hasn't come yet but it'll be your Christmas present," she told him. "I love you." And then the satellite line provided by Canadian paratroopers went dead.

But all the work paid off. Before the day was over, I heard a group of soldiers singing, "Rudolph the red-nosed camel."

It was my twentieth Christmas overseas with troops. The spirit of our soldiers in good times and bad never ceases to inspire me. From Depression downers to baby boomers to Generation Xers, when they become grunts, they become America's finest.

Company A was a good unit made up of professional soldiers who said the Army was a 24-hour-a-day, 365-day-a-year job and that being away from home at Christmas came with the territory. Few of the old-timers had clocked much Christmas time in the United States, and the newcomers knew better than to get exercised over Christmas duty.

"In the Army you do what you're told," said Private Jason Flaherty. "They call and you haul."

If these troops were not off on a mission at Christmastime, they were on a training exercise at Fort Irwin or jungle training in Panama or somewhere else. Their biggest bitch besides the absence of the turkey and the mail was not being away from home at Christmas. Rather, with the Cold War over, they no longer felt as if they had an appropriate mission defending America.

"We feel like firemen," Private Flaherty said. "The alarm rang in the night. Now we're sitting in the middle of the desert."

"We're doing George Bush's 'God's work' in hell," one sergeant told me. Another griped that he had joined up to do combat stuff only to wind up on mercy missions to Florida after Hurricane Andrew and now to Somalia.

"We've been baby-sitting and feeding people," he said. "If I wanted to save people, I'd have joined the Salvation Army or the Peace Corps."

Some of the griping was routine. "When soldiers are not complaining, that's when I worry," said Platoon Sergeant Herman Dozier. "Soldiers will bitch if they're sitting on a gold mine." But many of the bitches were dead on target. The troops wanted to know why they didn't get turkey on Christmas Day, why they were not getting their mail, why they had to sleep on the ground, drink water so heavily chlorinated it tasted like it came from a swimming pool, and live the grubby life while REMFs, who lived just down the road at the edge of the airstrip, slept on cots under mosquito nets in insect-proof tents, drank bottled water, and ran around in shorts and floppy hats looking like Santa Monica surfers.

"All my men want are the simple things," said First Sergeant Steven Choinard. "Like mail, newspapers, occasionally a different ration to break the monotony—at least to have the same as the USMC and USAF pukes."

A few days before Christmas, the Marine headquarters had more clerks and perks than any one of four infantry battalions in-country. It was spiffy and clean. They were bringing in plenty of fans and showers, all the modern conveniences, everything to impress an ambassador or the President. The headquarters people were the haves. The guys in the field were the have-nots. They didn't have zilch. Operation Restore Hope, under the command of General Johnston, a Marine Corps three-star, was organized according to Perfumed Princes' rules. It took one rifle company of the twelve we had in Somalia to defend the boys and girls living around General Johnston's flagpole. The 3rd Battalion of the 14th Infantry had three rifle companies, the 2nd of the 87th Infantry had three, the 1st of the 7th Marines had three, and the 3rd of the 9th Marines had three. That made about 1,200 fighters. But Johnston's headquarters strength was 1,141, and the REMFs almost outnumbered the fighters. In all, there were twelve American generals in Somalia, one for every American rifle company.

Never had so few been commanded by quite so many.

Or almost never. The same was true in Vietnam, where we had 60,000 warriors in the bush hunting and being hunted by the enemy, and 500,000 back at the flagpoles. If you consider the ratio of tooth to tail, we were getting down to one little tooth while the tail was getting bigger and bigger and longer and longer every day. The same principle was at work in Somalia.

Our modern generals put first priority on their headquarters. In days of old, General Ulysses Grant would hit the field with six or seven aides and they traveled light and slept on the ground. The rest of his men were fighters. Today, inflation of military brass and headquarters staff is so bad it should embarrass us. At the end of World War II we had a military force of 13 million. Today we have a total of 1.5 million active soldiers and sailors. But we have more generals now than we had during World War II. We also have more bureaucrats so that all those generals won't be lonely. In 1945, with 13 million under arms engaged in a multitheater, multinational alliance, the War Department had about eight undersecretaries, assistant secretaries, and special assistants. Now with about those 1.5 million in uniform, we have somewhere in the neighborhood of fifty undersecretaries, deputy secretaries, assistant secretaries, and special assistants. All draw six-figure paychecks and have aides, offices, and all the other trappings of Pentagon royalty.

This is a bureaucracy, a giant costly bureaucracy, and the governing spirit of any bureaucracy is personal safety, featherbedding, job security, and bloated, ever-growing staffs. All of our military ranks are inflated today. The way we're going now, two-star generals are going to be running what colonels used to run. Their paychecks are well padded. They have mocked up the general officer's good conduct medal (Distinguished Defense Service Medal) to look amazingly like the Silver Star, which is only awarded for "gallantry in action." The high-brass slogan today is, If you can't win one, why not spin one. It's only natural for the Perfumed Princes to surround themselves with help for the castle. This absorbs them so deeply they don't notice how few warriors are up on the ramparts armed with swords.

In Somalia, the grunts in the bush didn't have water for showers. What little they had they used for shaving. And no one on high even visited Company A on Christmas Day. But the grunts were slamming ahead and doing their job superbly. At the beginning of December, Mogadishu had been an armed camp as lawless as Dodge City without Wyatt Earp. Everyone including twelve-year-old punks had AK-47s. Operation Restore Hope quickly stopped the shooting and looting and disappeared the weapons. By the end of December, the mission was weeks ahead of schedule. After Company A secured Baledogle, a combined force of Marines and French troops from the

United Nations took Baidoa. Next came Kismayu, Narbera, and Hoddur. Then Belet Huen and Jalalaxi were opened up as humanitarian centers, completing the seizure of the major relief sectors. Once a new center was opened, it became a kind of humanitarian oil blot. With food distribution and medical assistance and engineer support slowly seeping out, the blot grew larger and larger until it connected with others and spread across Somalia.

Mogadishu served as the launching pad. After Baledogle was taken, the advance turned into an art form. We dropped leaflets telling Somalis the good guys were on the way. The next day Ambassador Robert Oakley would visit and negotiate a nonviolent passport for the troops. Then a battering ram of CIA types with big bellies would Rambo in like B-movie extras, closely followed by a fast-moving and hard-hitting rifle company covered by helicopter gunships and fighter aircraft.

Cheering crowds of Somalis met the advancing Americans. But even by a week or two after New Year's Day, I began to sense the popular mood changing. The guns weren't really gone. The warlords weren't toothless. They had just stashed their weapons and they were biding their time. It seemed to me the better course of valor was to finish distributing the food and get out of there as fast as we could. But President Clinton didn't see it that way.

My sixth sense began screaming again. Something sinister was in the air. *If we stay too long, and keep dicking around, somebody's going to get killed.*

CHAPTER 5

BAD HOMBRES, GOOD JOSS:

VIETNAM, 1993

The tree is still there. Among the green branches that concealed the sniper, the fruit now hangs ripe in the tropical sun. From that tree the sharpshooter got Charlie Reese in the leg, Elijah Frazier in the arm, and Ron Sulcer in the belly. I can hear the crack of the shots. Everything comes back. High noon in the little village of My Hiep. March 25, 1969. Two days earlier a company of my Hardcore Recondo troopers had dropped from the sky on the Viet Cong's Dinh Tuong provisional headquarters. The VC, knowing they couldn't run since U.S. airpower would zap them in the open paddies, hunkered down, took what we dished out, and fought until dark. Now, like ghosts, they were back with reinforcements and all over us. Night sweats. Bad dreams. The 3:00 A.M. knot in the gut. But this time out in the bright sunlight and open air. When I joined the fight, the first thing I saw were the point scouts lying on their backs. My gut told me they were dead. In a hot firefight a warrior keeps his belly as close to the ground as possible, but when the death rattle is in his throat, nine times out of ten he rolls on his back. Now, twenty-four years later, I can see once again the burst of fire from a banana grove cut down Lieutenant William Torpie, who falls trying to rally the company. Dan Moran, right in the center of the VC bunkers, is firing his single-shot M-79 grenade launcher as if it were an automatic. Two AK-47 slugs tear into him—one in the ankle, one in the shoulder—and knock him flat, but he gets up and keeps firing until he runs out of ammo. In the chopper overhead, a glory-hungry major is playing tin soldiers with the lives of real warriors down below. On the ground, red flames are licking at the ammo supply chopper, where the trapped crew chief is screaming as he's burned alive. Rockets, bombs, artillery shells come thundering down, and across the rice paddy floats the stink of cordite and death. And then all of it gone, vanished. It's the summer of 1993 and I'm standing next to Colonel Le Lam, commander of the 261A Main Force Battalion and Colonel Dang Viet Mai, commander of the 502nd Main Force Battalion, looking at the paddy where we tried to kill each other, where more than one hundred warriors—yellow, black, brown, and white—fell. The scorched earth is now rich with rice; the old bunker line that hid the VC fighting positions is a peaceful banana grove. And we the living are looking at each other and wondering what the fuck 25 March 1969 was all about.

In July of 1993, monsoon season, I flew to Ho Chi Minh City from Bangkok aboard a thirty-year-old Russian passenger jet that was so defective even the seat belt didn't work. Everything was rattling, falling apart. The air conditioning was sending out eerie white plumes, the airplane was actually wheezing.

I looked out the window. Down below was Vietnam, the dark hills, the rice paddies, the long stretch of green bordered by the sea. Seeing it again for the first time in so many years, a weird thought suddenly hit me. *If I were the President of Vietnam, I would close my borders and close my ports and not let anyone in. Why get into the rat race?* I was feeling more protective than hostile.

The plane banked, dipped, and nosed down into the descent for Tan Son Nhut airport. My knuckles were white and my heart was pumping so hard I could hear it. I felt like a man about to jump from a slick into a hot LZ. The plane touched down straining at every rivet, shuddered, then started the long taxi toward the same terminal I had left behind twenty-two years earlier. When I pushed my nose up against the grimy window, I could see the bunkers and revetments where we used to park our fighter aircraft, faded, stained, but still there. The first in the set of concrete horseshoes was painted olive drab and it was decorated with a yellow-on-black peace symbol ten feet high, the final statement of some airman who had seen through the light at the end of the tunnel. My brain clicked. *Good morning, Vietnam. I'm back.*

The plane weaved and banged its way down the runway until it reached the front of the spotless, air-conditioned terminal. I stepped off and as I headed for Customs the knots in my stomach got even tighter. Standing there were some familiar-looking little guys wearing North Vietnamese uniforms with green helmets and that red star blinking at the top. They were carrying AK-47s. The last time I had seen those guys with the AK-47s, they were shooting at me. The old fear came rushing back. I felt short of breath; my palms were sweating. My gut clenched like a fist. Had anyone noticed? Nope. They didn't want my life. All they wanted was to stamp my passport.

The woman in line in front of me was Vietnamese-American, and the Customs officer was giving her a hard time. Her papers were out of order; there was no way she could enter. She was rattling away at him in Vietnamese that everything was cool and he was getting more hostile by the moment. For five minutes they stood there toe-to-toe, yelling at each other. Then she stepped back, looked at me, and smiled. Slipping an American twenty-dollar bill in her passport, she handed it to the guard. His face went from cloudy with 100 percent chance of rain to clear and sunny. He smiled back, scarfed up the twenty, stamped her passport, and waved her on. Everything was in order. *Hey, nothing's changed. Everything's the same. Only thing different is the flag.*

The monsoon rains were bucketing down outside the terminal. I walked into a crowd of Vietnamese mustering on the line for transport downtown.

"Are you a Russian?" asked a face in the crowd.

"I'm an American," I said. He seemed almost relieved.

"Oh, we love Americans," he said, and friendly people came pushing up around me, smiling, eager to talk, the little kids brushing the hair on the back of my arm with the same wonder they had shown a million years before. It was hard to believe. There had been a war, but where was the resentment?

Saigon now called itself Ho Chi Minh City, but it looked more like a poor man's Bangkok, the Bangkok of the early 1960s, poised on a launching pad, ready to lift off. All it needed was for the United States to lift its economic embargo. The place was a beehive with all the energy of millions of people crammed into one bustling and hustling town. Tourist hotels and trendy restaurants were going up. Building cranes loomed across the sky. The food stalls were still there on the side of the streets but even the gridlock had changed: bicycles ridden by lovely girls in ao dais and motorcycles weaving tight patterns, but very few cars. Tu Do Street was forlorn, renamed for some Communist war hero. The bars were gone, the Saigon sex and sleeze parlors replaced by neat little tourist shops selling fake GI dog tags and Zippos as souvenirs.

The city's raw, frontier feeling had vanished. But some things never change.

"You want a girl? Fifteen years old? Virgin girl?"

An old voice, a new hustler, sticking there at my elbow.

"No, get out of here."

"Do you want boy? Fifteen years old? Virgin boy? You queer?"

"Piss off."

General Tran Van Tra stuck out his hand. He was wearing a plain khaki uniform with no shoulder boards, no insignia, no medals. I suppose he was in his late seventies, though you couldn't really tell. He stood ramrod straight. His hair had gone white; his voice was mild. He had true command presence, great dignity. If you were to pick him out of a group, you might think he was a banker, not the man who had been the romping, stomping commander of all Viet Cong forces in South Vietnam. That would be a mistake. Unfortunately for General William Westmoreland and all the rest of us, the bright light at the other end of the tunnel was General Tra.

I had met General Tra once before in New York. I know a lovely guy named Bobby Muller, the president of the Vietnam Veterans Association, who put us together. Bobby lost both legs in Vietnam as a Marine platoon leader, but he didn't come out of the war belonging to the hate club. Instead, he helped organize the kiss-and-make-up club. Even though he'd left a big chunk of his body in Vietnam, he put all his time and energy into making peace, an attitude completely opposite to that of many vets. The haters can't let the war go. The result is that they're still letting Vietnam wound them every day of their lives.

At the meeting in New York, General Tra had been sitting on a couch across from me.

"You were the battalion commander who caused us a lot of trouble," he said. "Why were you so successful against my forces in the Delta?"

"General," I said, "I stole a page from your book. I ripped it out and used your exact tactics against you."

He studied me for a second, then laughed, a great belly laugh that nearly doubled him over.

As soon as I got into Ho Chi Minh City, I looked him up. He was running Vietnam's war veterans organization. We met in a large room at his headquarters, where he had assembled seven of his senior commanders. I found myself looking at the commander who led the tanks into Saigon and took the presidential palace.

We talked for several hours about what their victory had brought the country.

"We are not disappointed, but we are not fully satisfied," General Tra said quietly. "We have a long way to go. We were good at fighting and bad at economic planning."

I had to suppress a smile. It was supposed to be the other way around: Karl Marx was supposed to be the great economic genius and the United States was supposed to be the invincible warrior, but that wasn't the way it panned out. All the old soldiers in the room were from the south and they clearly thought the north had screwed them. Yes, the northerners had come down to fight and die with them, but their socialist utopia had emerged stillborn. Vietnam was still one of the poorest, most backward countries in the world. One of the commanders described Hanoi's postwar leadership as "drunk with victory and unfocused." The others nodded.

"Our dream of perfect socialism did not come," one of them said. "Life has forced us to make compromises."

General Tra might have been a lifetime socialist, but he was also a nationalist rebel and his fights with Hanoi have become as legendary as his victories over General Westmoreland.

"Hanoi's centralized planning didn't work," he said. "Our government made many mistakes, not unlike Clinton did in his first six months in office."

He omitted that the bungling went on for eleven years of nonstop Soviet-issue bureaucracy and incompetence fused to Hanoi's version of Mafia corruption. They had all assumed that once the Americans and their allies were whipped a great change, a great new life would come to their country. Instead, they just got more of the same. The bureaucrats came down from Hanoi, set up their machine, took everything worth taking, and life ground on. The freedom and prosperity they had fought for never came.

"They still aren't working in the interests of the people," General Tra said. But he argued that the country's will and energy were as strong as ever and the future still held something better for Vietnam. I left that day thinking, *This might not be a bad guy to help.*

Mark Peters flew in and we hooked up on a new joint expedition for *Newsweek*. It had been two years since Desert Storm and Mark was his old piss-and-vinegar self, pure raw sex and manliness, the last of a breed, hard drinking, hard playing, one great hell-raiser, up all night but always ready for duty in the morning, the ultimate warrior-photographer. Wherever we went he was a magnet for women and a lightning rod for trouble.

One night we were in a pool hall and Mark was playing a Vietnamese guy for money. A lot of Vietnamese were standing around the table betting on the game. The Vietnamese was a very good player and Mark was trying to keep the upper edge with a little psywar. The guy leaned over for a shot.

"Get a chair," Mark said. "You're such a little short shit you'll never be able to make it standing there. Here, I'll get a stool for you to stand on."

I could see that the guy was fuming, getting more and more pissed off. Finally, they got down to the last shot. Whoever made it took the game and the pot. The guy looked down his cue and was just about to shoot.

"Come on, you Communist motherfucker," Mark yelled at him. "You'll never make that shot."

The guy spoke English and he went completely nuts. He jumped into a karate position and got ready to tear Mark's heart out of his chest. Then the other Vietnamese in the room figured out what Mark had said and they all went goofy, too. *Oh man,* I thought. *How are we ever going to get out of this Bruce Lee shit?*

Mark grabbed his pool stick like a bat. He jumped up on the table and he stood there towering over everyone else, getting ready to swing, and the Vietnamese guys sort of melted for an instant. Finally, Mark had enough presence of mind to realize he might be in some kind of danger.

"That was great. Just having good fun here," he shouted. "Buy the house a round."

The Vietnamese all turned and rushed up to the bar. Cost us about a hundred bucks—but we got out alive.

After the Vietnam War I had never had a chance to talk to the enemy. Now I wanted to test my theories against the men I had fought, to talk to the commanders who had been up against my guys. During the war I spent most of my fighting time in the Delta about sixty miles from Saigon. Up north I had fought North Vietnamese regulars, but I didn't see much point in going back there. Those I had fought would have returned home long ago and I would find nothing but graves. I didn't want to visit old battlefields just to relive bad memories and wind up talking to the monkeys. The Delta was the right place to go.

General Tra told me the soldier to talk to was Colonel Bay Cao, a fine old man who had started as a private in the 1940s and had risen through the ranks to become vice commanding officer for Military Region 8, a chunk of the Mekong Delta about the size of Rhode Island. He lived near My Tho. His honorary rank was brigadier general, but they had retired him as a colonel.

My Hardcore Battalion had locked horns with the Viet Cong on Colonel Cao's ground many times. I was eager to meet him. Mark and I rented a car and we drove down to the Delta. We found our man at the party headquarters near My Tho. Colonel Cao was seventy-four years old and his rotting brown teeth had seen better days. But he stood soldier straight and looked you dead in the eye and he was still a no-bullshit gunfighter. I asked him what he thought of General Westmoreland's view that he had prevailed in all the battles only to lose because reporters and peaceniks had sold him out along with their country.

"A bright lie," he said. "A big lie. Westmoreland's kind of tactics won in Iraq but failed in Vietnam. We had determination of steel. The Iraqis had none."

We talked for a while and then I took him to lunch by the river. Mark wanted to photograph him in his uniform, so he said yes, he would take us to his place. We all piled into our car—he didn't have one, even though he was a retired general—and drove deeper and deeper into the boondocks until we reached a place where there were no more electric lines, only a little trail.

We pulled the car off the road and started walking down the muddy path that wound through a series of thatched villages and finally gave out near the little thatched hut that was Colonel Cao's house. He was living absolutely as a peasant. There was no running water, no electricity. He had been a Viet Cong, a patriot, a guy who had fought all his life for his country, and this was his reward. When the war was over, the generals from the North came down and took over the big villas. The Viet Cong warriors didn't get any of that. And this guy didn't want it. I liked him from the moment I met him and we spent a lot of time talking. It turned out that in 1969 one of my ambushes had almost wasted him. He had been floating along in a sampan headed directly for a Hardcore trap. About three hundred yards before the killing zone, the local peasants warned him by beating on the water with paddles and he got away.

With the interpreter translating, Colonel Cao took his time and spelled out exactly how he had chosen to fight us.

"Our first step was to get enough experience fighting you," he told me that day. "Our second step was to develop tactics to counter your mobility and machines. The final step was to wear you down. We were patient, we were prepared to fight a long and protracted war. You were not. We studied your tactics, monitored your radios. Americans talked much on radio. Too much. Give us much intelligence. We even knew when your B-52 bombing attacks would come. Spies told us. We have spies everywhere. Spies are the most important soldier in war."

I thought back on all the times I had grabbed the radio and winced.

"We always knew your plans," he went on. "First come your helicopters circling above like hawks. Then your air strikes and then your troops. Our aim was not to stand and fight, but to run away, unless we could win tactically, or, as in Tet of 1968, we could win a great psychological victory." After the Tet Offensive, he said, he knew beyond any doubt the Viet Cong were going to win.

That made two of us.

In the back of the hut, Colonel Cao kept his uniform. It still fit. He got into it, pulled on his pith helmet, and took us to one of the military cemeteries that seem to lie outside every village in the district. The Vietnamese losses were horrendous, maybe three million dead. When we reached our objective, the old man and I bought some joss sticks. The VC had crammed thirteen thousand bodies into this one little cemetery. We sat there and lit the joss and I felt a wave of emotion swell up inside me. I had felt the same feeling at the Wall; now it came back as I was sitting next to an old enemy. The incense smoldered and in that instant there were no enemies in sight: just the living and the honorable dead.

Colonel Cao organized the whole district for me. He got word out I was there and within a few days I had a rendezvous set for this paddy and that banana grove and up that trail over there: all the old fighters, hardcore mothers every one of them, wanted to talk. One day I went to the house of a guy who'd been a Viet Cong captain and spent three or four hours talking to him and his guys. Then we all piled into sampans and went skimming down the canals. It gave me an eerie feeling to be squatting down in those long boats, slipping down channels where my guys had set their ambushes, where they had waited with their M-16s and their claymores. I had the spooky feeling we would round a bend and be ambushed by my own battalion. It was almost as if I had been taken prisoner and was being led off to a POW camp.

In Vietnam you never knew when or where you would lose a leg, a limb, or your life. If you were in-country for a year it was bad enough; if you did two or three tours, you had a thousand days of that worry imprinted on your brain. I kept catching myself reverting unconsciously to the past, scoping for mines and booby traps every time I took a step. One day we went out in the sampan to see a couple of battlefields where there had been some really stiff fights. "We were here, you were over there," they were saying to me and I would go over to where their positions had been and look back at mine. When they saw me picking my way along with my eyes down like a squirrel looking for nuts, my nuts, they knew exactly why.

"No, no, Colonel," they called out, laughing. "No mines. No booby traps. No worries."

When I arrived in the Delta the last time through—that was March 1, 1969—I discovered I had a fire support base called Daisy. *No fucking way*, I thought at the time. To shake up the grunts, keep them alert, I renamed the place Danger. We had pushed it farther forward than any other into the VC stronghold, the zone that had given birth to the revolution. I deployed my units out and around the base, and my cannons sat inside so I could reach and support my units. The idea was if we could control a big wagon wheel around the base, about a hundred square kilometers, we might make some progress. It had been a hot area, very, very hot.

I wanted to see it again, so one day I got out the map and tracked it down. When I got there, the base was gone. No sign of sandbags. Instead, the Vietnamese had cleared everything and put up a gas station. As I was walking the old perimeter, an old guy came pedaling up on a bicycle and when he saw a Yank standing in the middle of the road, he was so surprised he slammed on the brakes. This wasn't easy, because he had a wooden leg, a peg leg.

He limped over and I told him I had been there during the war.

"There?" he said, motioning toward the old fire base.

"There? You were with the Americans there? They were the ones who shot me and took off my leg. Very good shots."

I didn't quite know what to say. So I pointed to a scar on my own leg—I was wearing shorts in the heat.

"And you guys shot me here," I told him. "You very good shots, too."

"Yes," he replied. "But you have better doctors."

We both had a good laugh, then Private An told me that after he lost his leg and got the wooden peg, he went right back to his outfit and fought for five more years. We shook hands as I got ready to leave.

"Welcome to my country. Welcome anytime," he said. "If you come back with guns, we will kick your ass." Then he got on his bike and wobbled down the road.

I went to the village of My Hiep, Long Hiep during the war, not knowing what to expect. I was the first American to enter the village since 1973, and Vo Van Dut, the village chief, came out to welcome me. He was a wiry little guy who'd spent eight years as a VC fighter, finishing the war as a company commander in the 261A Main Force Battalion. As a present I brought him a copy of Hardcore Battalion's unit journal for March 25, 1969. Someone had marked it "Classified."

"But this is a secret paper," he said, shooting me a look of surprise.

"Hey, the war is over, remember."

"Yes, we now friends, good friends."

Captain Dut got in touch with two of his old superiors, Colonel Le Lam, also of the 261A Main Force Battalion, and Colonel Dang Viet Mai, and we went back to the killing fields near My Hiep in sampans. These guys had stung my ass in a big way, and even though a quarter century had passed since the defeat, I still felt the burn.

I had replayed the battle of My Hiep in my mind a million times, examining it from every angle, picking at it where the scabs were, brooding about it until it made my brain stall out. My brigade commander at the time was Colonel John Hayes. We were great mates even though I was only a lieutenant colonel. He was Special Forces, a great warrior and a good man, and a damn good poker player. He had earned four Purple Hearts and a chest full of medals in Korea and more in Vietnam, where he was then on his fifth tour.

On March 23, 1969, one of my Hardcore Recondo companies had dropped in on Charlie's 261A Main Force Battalion and clobbered them. The company had really distinguished itself. To reward that kind of heroism we had invented something we called an impact award. The goal was to pin a medal on a guy's chest quick smart rather than making him wait for three months while all the paperwork was being stamped. Colonel Hayes rang me and said he wanted me to go to the impact ceremony for the company that had done so well. I had some other plans for Charles and I've really never liked medal ceremonies.

"Nope," I said. "I've got the assets today and I can't go." The assets were the choppers, gunships, and fighter aircraft you needed to kick Charlie's ass. You didn't get them every day, and I didn't intend to waste them.

"Oh hell, Hack, have Major Bumstead take them. Nothing's happening on the battlefield today. Come on with me."

"No, I'm not going to do it."

"You'll only be an hour."

"Well, shit. Okay. I'll go."

Major Lewis Bumstead (a pseudonym) was a brilliant staff man but not my idea of a combat leader. We'd served together in the 101st Airborne and I had his number: writer, not fighter. So I took him aside, showed him the setup that day, and told him to keep his eye peeled for VC remnants from the earlier fighting but under no circumstances to get decisively engaged. Having the helicopters and the backup weren't enough. They had to be correctly deployed. I figured I'd only be gone a couple of hours and nothing could go wrong until I got back.

John came by Danger in his bird, picked me up, and we flew off to Dong Tam, where Company D was waiting and the awards ceremony was going to take place. After the medals were given out, I said, "Let's go."

"I want to go to the PX," he said.

"Fuck the PX. I want to get back. We've got the assets and I want to be back there."

John was a big joker, always playing with me, and since I was hitchhiking in his bird, I had to go along. But I was worried in my gut that day. So finally, I said, "Come on, John. I've gotta get back." He saw he'd fucked with me to the limit and said, "Okay, let's go."

So we hopped back in the bird. When we were about sixty kilometers from my outfit, I switched on my command radio and checked in with my battalion. I couldn't make out much at first, but as we got closer, I could hear a lot of reporting back and forth. So I came on and asked for a SITREP. My tactical operations center came right back.

"Bumstead's in a fight with Bravo Company."

"Okay," I said with a sinking feeling. "Give me the coordinates."

The location came in and we wheeled around and flew over to scope out the action. When we got there, I could see Major Bumstead orbiting in my bird. Bravo was down on the ground. I was maybe five hundred feet up, but I could see right away that a big rat fuck was going on. Bravo Company's point was lying within the enemy positions. I could see two men on their backs, Tran Doi, the Tiger scout, and Earl Marshall Hayes, the point, and I knew they were dead. But there were still three guys on their bellies, Sergeant Donald Wallace, James Fabrizio, and Joe Holleman. When I called Major Bumstead, I got a gung ho situation report. Everything was going swell, we were cleaning their clocks, and a lot of other horseshit.

I can read tea leaves as well as the next Gypsy and this ain't a situation of anybody's clock getting cleaned but ours. So I got on the company frequency to get a SITREP from the company commander on the ground. Bobby Knapp was the regular company commander, but he was away on R and R, so Lieutenant Torpie was acting commander. I could see he had lost control of his company. I told him to get the men squared away, to get back from the enemy fire and dig in, to form a tight perimeter so I could see his guys and I could see the Viet Cong and I could bring death and destruction from above down on Charlie.

The way it was, the men and the VC were commingled, which was exactly the way the Viet Cong liked it. They called this tactic "hugging the belt." The idea was to get in so tight with us that we couldn't call in our big hammer because we'd hit our own troops. The only way out of the mess was to get a lot of fire support going down fast. I got on the radio again and ordered a bunch of gunships and as many fighter aircraft as they could send. Since we had hot contact, we got priority of assets. The message came back that the guns and fighters were on the way.

I knew I couldn't control the battle from John's helicopter. He was the brigade commander and he had a lot of other guys to worry about.

"Why don't you just drop me off on that rice paddy over there?" I said, pointing down.

"Okay," he said, wasting no words.

We flew over the field to make sure it was cool. Then I told Major Bumstead where to pick me up. So John dropped me and we made the switch.

By that time the assets started flying in, the gunships and the fighter aircraft, and I really began pouring it to Charles, putting all kinds of fire on his ass hoping to break contact while the stuff was going in so my guys could pull back. I called Lieutenant Torpie and told him to get his people back. But by that time the company had disintegrated. The commander had totally lost control. Key leaders were dead and eighty guys were down there in that paddy getting blown away. So I was really smoking the lieutenant.

"Get your shit together or I'm gonna come down and boot your ass."

The kid panicked. Instead of being cool, employing his chain of command, he jumped up and tried to do it himself. He forgot that all those little buzzing sounds above the paddy were copper-coated steel bullets going three thousand feet per second. The lieutenant started yelling. He ran a few steps in the mud. He presented a target to the enemy, but by that time so much fire was going down over the whole field he may have just walked into one.

I guess at that moment he was more afraid of me than the enemy fire and he forgot all his training and stood up. He was immediately cut down. The medics got to him and tried to keep him alive. He wasn't hit in the head because he lived for a fair while. He was hit in the chest and the gut and the medics couldn't save him. I wish to hell I could take those words back. I hold myself responsible. If I had been dealing with someone more experienced, it would not have happened. Bobby Knapp would have told me to go fuck myself, but this kid didn't and he got blown away.

There wasn't much time left to pull the company together and get it squared away before dark set in. What I needed to do was destroy the enemy's command and control, put him in the same shape as Bravo Company. The only way to do it was to drop the Lord's own firepower on them but I couldn't do it because three of the men were literally right in the enemy's trenches. Fabrizio told me later a VC soldier was popping up in a bunker only fifteen feet away and giving him the finger, pointing his weapon but not shooting. The VC didn't want to kill those three guys. They wanted to keep them alive, so they were making faces at them, pointing their weapons as if they were about to shoot, then pulling away. Pranky stuff. They were really hugging the belt, squeezing the shit out of the whole company.

I tried every trick in the trade, everything I had learned in twenty-three years of playing this nasty game, to get those three guys out. I used white phosphorus. I used smoke. I used all the ordnance in the world. I was interspersing gunships with airplanes with artillery. An hour passed and the enemy didn't get a minute's breathing spell. I really clobbered them. But there was no way I could do what I really had to do, which was to take out the treeline hiding most of their fighting positions, unless I could extract those three guys.

Finally, out of complete frustration, I said to myself, *We're going to have to go in and get them.*

So I said to the pilot, "Will you go in for them?"

"Roger on that, sir. We'll do it."

"Check it out with the crew."

"Let's do it," came back all around. So I had all the helicopter gunships—there were eight of them—stay in orbit off to one side. I told them as we started down I wanted one bird to break off and follow us in with the others coming right behind him so there would be constant fire slamming in right in front of my bird. The pilots were all pros. They could see the setup and they all said they could do it.

So we came in, flared up and the gunships were going BHAP-BHAP-BHAP, really blistering the treeline, keeping the VC down while we zoomed in to get those boys.

I got out on the skids and jumped to the ground when we were still about six feet up. Major Bumstead followed. We grabbed Holleman and tossed him on the floor of the bird where Doc Ed Schwartz, an Air Force doctor who somehow got detailed to the grunts, started patching him up. I grabbed Sergeant Wallace, who right in the middle of that inferno stood up and saluted and yelled, "Hardcore Recondo, sir," and tossed him in the chopper. Next, I grabbed Fabrizio and threw him in the bird.

With the passengers on, it was time to yell, "All aboard." But my mouth didn't work. I was standing on the skids because by now the back of the bird was crammed with people and I was trying to tell the pilot to take the chopper up. He was just looking at me, waiting. I finally grabbed my flight helmet.

"Let's get the fuck out of here," I said over the mike—or thought I said. Nothing came out. Fear dries you out. My mouth was super-glued shut, frozen. I had so much adrenaline pumping through me my tongue had sealed my mouth. So I gulped some water from my canteen and yelled, "Let's haul ass," and the pilot swung the bird around.

I was still standing on the skid when he swung the boom and a Viet Cong B-40 rocket smacked right where the chopper's tail had been. Another two seconds and we would have been back on the ground, burning.

As we lifted off, Fabrizio coolly asked me for a cigarette as if we were in an amusement park, not downrange in a shooting gallery and in a world of shit. We took a lot of hits. I could hear the slugs ripping through the bird. It took a giant leap up and over and came down like a ten-ton grasshopper; then it made another bound or two getting us clear of the fire zone. The bird was finito but we were out and we had the kids.

I called in another bird and went back to the fight. With the men out of the VC trenches, I started socking the treeline with five-hundred-pounders and that was the end of the game. The enemy resistance lightened up. I put Major Bumstead down on the ground to get the company squared away. I wasn't going to break off this fight. I wanted to kick Charles in the ass big-time. But as the bird rose up, we took a bellyful of hits. I caught one in the leg and it felt like a sledgehammer bashed me. We got up all right but someone from the ground radioed that the bird was on fire, smoke was coming out behind us. The thing to do was to get to my battalion where the doc could patch me up so I could get back. The problem was that the kid next to me, an artillery captain, had taken one in the gut. The pilot craned around and took a look.

"We'd better get this guy straight back to a hospital or he's going to die," he said.

John was still overflying the battle in his chopper. I looked at the kid and he was wearing that death look. He didn't have a lot of time left, shock was setting in hard.

"John, I got a bad one here. We're going to MASH."

"Okay, Hack. I'll take over the battle."

The bird made it back to the hospital just as it ran out of steam. They pulled us out and, since my wound was much worse than I thought, before I knew it I had more needles in me than places to stick.

That night the VC mortared the shit out of Dong Tam. My D Company guys were still there from the awards ceremony. When they heard I was hit, the whole company led by their valiant skipper Ed Clark came over. The doctors tried to stop them, but they pushed the docs aside, coming in with cans of beer and lots of noise. I was lying on the operating table sipping beer with all of them around and they were telling me war stories.

Outside, the world was being blown to shit. In the middle of the inferno they put me in a medevac chopper along with some other wounded and we flew up through the rockets and explosions and zoomed over to Long Binh, where the doctors decided to ship me to Japan. I talked them into medevacing me back to Dong Tam so I wouldn't be dropped from the division rolls and lose the Hardcore Battalion.

We had been defeated and I had not been able to get back in

time to turn defeat into victory. Now I was standing by the same paddy with fifteen Viet Cong soldiers and battalion leaders who were pointing to their old fighting positions. I wasn't trying to play macho man. I was trying to be as humble as possible. But how the hell had they done it? I was standing next to Colonel Mai.

"We were a superpower," I said. "How could you stand up against a force that filled the sky with aircraft and could fire more artillery rounds in one engagement than your side used in one year?"

Colonel Mai thought about it. He glanced at the others, then looked back at me.

"At first your helicopters and aircraft were hard to fight—they go so fast," he said. "Much rocket, bomb, artillery fire. Scared our fighters. But we learned. We set ambushes. We knew you would run out of aircraft and bombs before we ran out of spirit."

And after they pounded my company on March 25, why hadn't we nailed them?

One of them turned to me.

"We waited until it got dark," he said. "There was a big hole in your positions. We went right through it. Then we slipped away."

For a while we stood there swapping war stories and lessons on tactics. I told them I'd done everything I could think of to make sure they had a hard time moving on those battlefields. I didn't march in large battalion formations the way most American units did, I broke my companies down into small teams. Instead of stumbling across the rice paddies like a herd of elephants, there might be six guys on a trail, six on a canal, small teams scattered all over the battlefield. The Viet Cong couldn't communicate the way we could because they didn't have radios. I set up astride their lines of communication and knocked off their messengers, their cadres, their scouts, and supply columns. We didn't react to the enemy, they reacted to us, and we cut them up on our terms. Unlike most other American units, we'd figured out how to make them dance to our tune.

While I was rattling on, they nodded with great interest. No bitterness, just the brotherly love that exists among true warriors. They were all my age or older. One full colonel had almost forty-five years of combat experience. First the Japanese, then the French, then the South Vietnamese and Americans, and then the Cambodians. These guys were pros. They didn't command battalions for six months to get their tickets punched like the Americans. They might command a battalion for four years, then leave to command another for a couple more. It would take them that long to get a regiment. By the time they came to the party, they were bringing twenty to thirty years of experience to every firefight.

On our side, the battalion commanders frequently came from the Pentagon, flashed in, commanded for six months, then hopscotched right back. In those six months they never really learned anything. Then they were snatched away and put in some staff job while another group of greenhorns marched in. They all wound up school wise and street dumb. The only guys who really knew what was going on were grunts who stayed in the killing fields for their complete 365-day tour. For them there was no ticket punching. The only way out was a litter or a body bag.

Very few battalion and brigade commanding officers were even willing to talk to the grunts, to make use of what they knew. They were too busy managing the assets and proving to their generals that they, too, were made of the right stuff for stardom. And pretty soon the grunts began to hate the colonels who were in charge of them. I arrived not long after a new battalion commander had gotten himself killed. He had jumped up on a rice paddy wall even though everyone had warned him not to because that's where the VC liked to plant their mines. He didn't listen. When he ran down the paddy wall and got himself blown away, the grunts all cheered.

That's how bad relations can get between grunts and their senior commanders.

Today, I don't know whether to laugh or cry when I hear our senior commanders claiming that they have mastered the lessons of Vietnam. How could they? They weren't there long enough. They were there six months and gone. The men at the top of the military pyramid today were lieutenants and captains, perhaps a few majors, during the Vietnam War. None of them had significant command experience. Those like myself who were battalion commanders after 1965 and on to 1973 have all retired. We've lost their experience, however limited it was, our institutional memory, precisely at the moment we need it the most.

I couldn't leave Vietnam without visiting Cao Lanh, the place where I fell on my sword in 1971 and drove that sucker right through my belly button up to the hilt. What a change. Talk about boom city! Motels, restaurants, new buildings, and TV antennas poking out of every structure including thatched huts. There was a school where my CP used to be. Beautiful girls dressed in delicious ao dais were learning the wonders of dressmaking courtesy of Singer, the sewing machine company. They'd soon be sewing stuff for The Gap or Banana Republic, getting the scratch together to buy a boom box or one of the ubiquitous motorcycles now replacing all the bikes.

When I left Cao Lanh, it was a war town. Barbed wire, booby traps, and bad guys. My outfit, the 44th Special Zone, ran all the special operations along the border and into Cambodia. We owned the rice paddies and the town during the day, but every night, when the sun went down, the Viet Cong unceremoniously took everything back. Everyone called it "accommodation."

It was at Cao Lanh that I figured out the whole fucking war was an accommodation. There was no way Uncle Sam and his flunkies, the South Vietnamese, would win. I was into denial—it took five years in Vietnam for me to get the word. It was here I ended my Army career.

I drank a beer with a former Viet soldier I ran into who'd been with me in the Zone. He'd returned home after spending a decade in a Communist reeducation camp. He said his time there was "hard duty" and it would have been shorter, but the Cong didn't like former South Vietnamese Green Berets. He remembered my swan song to the press. "Big trouble came our way," he said. "Every inspector in Vietnam come look. They talk to me about what you stole and half the women in village to see how many you have."

Those were tough times. Westy went nuts when I told the world the truth about Vietnam. He did his best to darken me and to bring me down, both to punish me for sounding off and to discredit what I said—that he and his fucking crew were as polished as liars as they were incompetent as war fighters.

I visited the spot where I gave my exit TV interview to ABC's *Issues and Answers.* I could almost see Howard Tuckner and Nick George leaning forward asking their endless questions and Ben Willis sitting there stoically, puffing on an unlit stogey and telling me to "cool down" when I was over the top.

The interview almost didn't begin. As the cameras were setting up, I received a message from my old pal Hank Emerson, then a hotshot brigadier general. Westy had handpicked Hank to head up a leadership study to determine what the hell was wrong with the Army, and Hank wanted me to come home immediately to be his number two man.

I was standing at a fork in the road. Should I take the short track to stars and official glory or should I take the untraveled path, which could be mined.

I took the long, untraveled road.

Looking back, I see the decision was easy. Few of the top bananas gave a shit about the troops. The generals used the war to scurry up the ladder, and the politicians used it to increase their power and help their weapons maker pals get rich while young yellow, black, brown, and white people bled and were forever scarred and died. The night before the interview one of my Special Forces sergeants had died trying to rally a besieged Vietnamese company while its officers had hidden. His death was the final straw.

The cameras rolled . . .

We have not had a proper military anatomy of the war. Too many people are trying to rewrite the history of the war the way the Germans did after World War I: We were stabbed in the back, that is the only reason we lost. According to this school of thought, we won all of the battles but we lost the war. What these people are saying is that at the end of the fight, come the dawn, who held the land? In almost every case the Americans did. So we were king of the mountain. We must have won.

That line of thought might have been okay in World War II. It did make a difference to take Berlin. If you took territory, eventually you did win. But in guerrilla warfare, ground doesn't mean shit. It has no value whatsoever until the end comes. What guerrilla warfare is about is punishing your enemy. Making him bleed. Making him bleed to the point that he quits, as we did. They would fight us all night long. But come dawn, they were not going to stand there while our mobility and firepower came down on their heads. They were not going to fight toe-to-toe. Their whole concept of war was that when the enemy attacks, you withdraw, when the enemy withdraws you attack.

The ignorance in the American military is frightening. I saw it so clearly when Hank Emerson and I gave our lecture on counter-insurgency in 1989. We were two highly experienced commanders briefing the high commanders of the Pacific, along with their staffs and unit commanders, on what we had learned in Vietnam. Hank would make a few remarks, then I would pick up, then he would take over again. We batted everything back and forth for quite a while, giving them odd bits of experience they were gobbling up as if it were magic. Suddenly it dawned on me that all of the tactical riches we had piled up, everything, all of it had been completely lost, completely forgotten as if we'd never been there. We are not talking simple CRS syndrome, we are talking terminal epidemic. All that blood, all the dead, and all the wounded—and the Army *Can't Remember Shit.*

When I got to thinking about it, I realized after Vietnam, the American military was so ashamed and humiliated about what had happened to the Army, they decided to shove it all under the rug. Then they nailed down the rug and swore never to mention Vietnam again. The new emphasis was, "We're going to fight the Russians at the Fulda Gap. We've got to think about the big-battle war." The V word wasn't popular around the schools. But as things turned out, we didn't meet the Russians at the Fulda Gap. We have been thinking for twenty years about big-battle scenarios when our future problems are in places like Somalia, Haiti, and Bosnia.

It was bad enough to lose in Vietnam, but it would be pathologically stupid to keep on behaving the same way. Out in the boondocks with those old Viet Cong warriors this possibility really began to get to me. When I stayed in their villages or rode in their sampans, I found myself looking at the world through their eyes. I remember thinking, *God, we were arrogant to think we could win this war. When it came down to a fight, we were equals, a guy with a rifle against a guy with a rifle. Guys with hand grenades tossing at one another. All of those magic things we had, all those things up in the air, didn't make one damn bit of difference. When it got down to the real nut cutting, we were equal. How arrogant we were to think we could win on their ground when they knew every fold. They could hear a dog barking two miles away and they knew we were coming. They could see movement in a bush a mile away and they knew we were coming. How stupid I was, thinking I could whip these guys. I tried to get my guys to use their tactics. We got pretty good at it. But we weren't the hometown guys.*

Back in Saigon one day, I walked into a place called the American Bar. I looked around and saw it was full of American Viet vets. These were guys on veteran's disability with maybe a few hundred bucks a month coming in, enough to live well in Vietnam. It was like walking into a time warp. Half of them were wearing jungle fatigues and jungle hats. I don't know whether they had been stunted as eighteen- and nineteen-year-olds or if the war had just been too exciting for them, but whatever had happened, they were still trapped in that time. They wore beads and bracelets, all the little emblems of rebellion, and their dialogue came right out of the sixties. It was all, "How you doing, man? Far out. Suppose they gave a war and nobody came." They couldn't let go. They were still grunts with the Army or the Marine Corps, all of them half-cracked.

When I finished my trip to the Delta, I had a few days left so I decided to go to Hanoi and talk to the people responsible for our MIAs. I flew up in a worn-out American plane that hadn't seen a spare part in a long time. We landed in a beautiful city, far more civilized than Saigon and without its frantic buzz. The people of the north clearly lived a harder life than those in the south. It was colder, not a sunny zone where mangoes fall off the trees every three minutes and you don't have to spend half your time worrying where you're going to find something to feed your gut.

In Hanoi, no one hustled me on the streets. I was left alone. But corruption had seeped down from the top into the smallest daily transactions. My guide was a beautiful young woman and dedicated Communist. She spent half her time trying to convert me, while Mark Peters spent half his time trying to convert her. At one point I decided to buy ten pith helmets to send to old Vietnam buddies as souvenirs. They cost about ten dollars. After I paid the peddler, I turned around and saw my little Communist friend getting her little Communist kickback.

As we strolled by the war museum, where pieces of downed B-52s were on display, I looked into a window and saw a beautifully framed photograph of Jane Fonda on the wall of an art studio. She was in an antiaircraft position wearing a pith helmet.

"What's the deal with that?" I said to the interpreter.

"Jane Fonda is a people's hero," she said. "She came here during the war to oppose it."

I thought back to how furious I had been at the time. Now my reaction was different. I wasn't angry so much as I felt pity for Jane Fonda. She had been in her twenties, idealistic. She had lost her balance and jumped into the enemy's trenches when her countrymen and countrywomen were dying just a few hundred miles to the south. I, too, knew the war was bad, not winnable. I, too, wanted to stop the killing. But I didn't go to Hanoi, enter the enemy's ranks, and comfort them. Still, the anger was gone. What I saw was how she had scarred her life forever because that particular sin could never go away. She would always be remembered for doing something truly dumb.

Before I left for home, I was able to arrange a meeting with the American recovery team, mainly Special Forces guys who had been seconded for the MIA search because of their language skills and their jungle expertise. All of them were top quality guys. We spent half a day talking and they laid everything out on the table. They said there was not a POW left in Vietnam. The crazies who come over and chain themselves to the old Hanoi Hilton, where the POWs were locked up, were wasting their time because not one prisoner was left alive in the entire country.

What we sometimes forget is that Vietnam is no longer a war. It's a country, a country trying to recover. It's quite true that hundreds of American POWs are still missing in action. It's also true that over 300,000 Vietnamese are missing in action. Their mothers, fathers, and other loved ones don't know where they are, either. That's terrible, but it's also the nature of war and it's been that way since the beginning of time.

What is keeping the issue alive is the MIA lobby, a business, a multimillion-dollar business. We were spending millions and millions of dollars every year trying to dig up the bones of a few thousand people missing in Vietnam. Yes, it is right to take care of your wounded and not to abandon your dead. But it is insane to make a fetish of the dead. I don't believe anyone who has lived in those jungles, anyone who spent a night in the bush, believes that anybody could survive there as a POW for twenty-five years. Very few people actually know what it's like to live in a little bamboo cage with disease, malnutrition, rats. To think you could go on that way for twenty to thirty years without dying or going stir-crazy is absolutely nuts.

It is entirely possible that at the end of the war some POWs were left behind. In the early 1970s when we got 650 back home, I'm sure Henry Kissinger, Richard Nixon, and Alexander Haig knew we didn't get every last one. But politics had created a burning urgency for a deal. I have no doubt that the Vietnamese kept B-52 pilots who knew how to penetrate China and the Soviet Union with nuclear weapons, sent them on so the Sovs and the Chinese could pick their brains. There were technicians who knew about the latest fighter aircraft and other technical things. My guess is these guys were kept and then dribbled back secretly over the years as bargaining chips in one deal or another. I don't think any more are left alive.

The haters, the truly obsessive haters, still can't let go of the MIA issue because if they do they won't have anyone left to hate. Others remain motivated because the MIAs have become their business. They have their offices, their association. They can fly their black POW flags, sell bracelets at ten bucks a throw, and draw good salaries. When I first came back from Australia and was very naive, one of the kids who had been in my brigade was a leader of one of those groups. I came to Washington to march in a parade of veterans with him. We gathered at the Wall. In the front ranks were a couple of Medal of Honor winners. As we were marching down the street, the kid from my outfit and a lawyer who was president of the group were talking about Senator John Tower, who was in a big push to become Secretary of Defense. They were using vet organizations to lobby for Tower all over the nation.

"I don't understand why you give a shit about John Tower becoming Secretary of Defense," I said, in my hopeless naïveté.

The president of this "Save the POW" black-flag-waving group, a former REMF in Vietnam, was an inside-the-Beltway lawyer. He looked at me as if I was the dumbest shit on the street.

"I'm going to be an assistant secretary of defense," he said. "It's a promise."

It made me sick. I told him he was as bad as the generals in Vietnam who had gotten promoted while our young warriors got body bags. But by that time they had blown the whistle and started the parade. *What am I doing here?* I asked myself and I started dropping back a rank at a time. Before the parade had moved a block, I was in the last rank and then I cut out. I wasn't going to walk with those guys and their dupes.

I guess the truth is we all managed to lose the war together, those of us who fought it, those who ran away, those who protested. Those same people are now becoming our presidents, managers, foremen, leaders. They now have another war on their hands, to put our country straight.

It made me feel good when Robert McNamara owned up to his mistakes, when the Clinton administration moved to recognize Vietnam, and Congress finally saw fit to do business with Hanoi. By nature the Vietnamese are capitalists, a hard-driving, competitive people who will never remain content within a socialist straitjacket. I came back convinced Coca-Cola-ism will one day win over communism.

As things turned out, I wasn't the only guy thinking about the Viet Cong that summer. So was Mohammed Farah Aidid. When our Rangers tried to nail him in Mogadishu one night, he escaped a few minutes ahead of the surprise party. Next to his bed the raiders found his night reading: the works of Chairman Mao and Vo Nguyen Giap.

CHAPTER 6

"UNFORTUNATE CASUALTIES": SOMALIA, 1993

"It was WHAW, WHAW, WHAW, WHAW—coming from everywhere."

Grenades exploding, bullets tearing into flesh and splintering bone, the cries of the wounded and the dying—that's how Staff Sergeant John J. Burns remembers the U.S. Rangers dancing with death along the Mogadishu Mile.

"We stopped in an alley and started putting down suppressing fire. That's when Joyce got shot. He got shot about twenty feet from my vehicle and he went down. I ran over to help load him up on one of the cargo Humvees, and we started moving again. I was hit in the shoulder. The door was open and the bullet went straight through the ballistics window. It hit right on the bone and came straight out. Didn't bother me. I had two guys down in my vehicle plus another from 1st Platoon. It was getting where there was no more room.

"Then they said, 'We're moving.'

"I jumped on the tailgate of the cargo Humvee. I had my leg over across the top of Sergeant Joyce and we made a left-hand turn down

Armed Forces Road. It was so eerie. All quiet. A big gunfight and things just got quiet all of a sudden.

"Then there was all this black smoke—you could see the tires burning—and now it was like WHAW, WHAW, WHAW, WHAW from every direction. We took about four RPGs. I mean we were suppressing the whole time, but they were just like WHAW, WHAW, WHAW, WHAW from inside the buildings.

"We were shooting people that were stupid enough to run out. But all we could do was try to suppress them. I saw them drag their wounded off. I seen people scream. Criminy, those people are savages. They probably left half the wounded laying out there in the street or burning in their own house.

"Out on Armed Forces Road we took fire again. There was about ten guns running out of a house and the SAW gunner opened up but they fired and I got shot in the leg. I pulled myself up on the vehicle and said, 'Pressure dressing.' The guy goes, 'No, I don't have one.' But we kept driving and we was getting shot at so much I just forgot about it. It didn't hurt no more. It hurt and stopped. The cargo Humvee I was on had four flat tires and smoke going out of the engine and it was starting to slow down, and I go 'Whoa, this is getting hairy, hairier than heck.'

"As soon as we hit the K-four Circle it was just like all hell broke loose again. We got WHAW, WHAW, WHAW, WHAW. That's when our driver got hit. And Charlie got shot in the leg and our vehicle stopped dead.

"I just rolled off the tailgate next to the tire and there was a big crowd forming. All of a sudden one person shot from the crowd and I just fired with the SAW. The guy next to me said, 'Are you all right?' I said, 'Yeah, my leg's started killing me, but I'm all right.' And he was like ripping a piece of my pants off trying to tie a tourniquet. It was bleeding pretty bad.

"Then this five-ton pulled up behind us. I didn't see them coming. It pulls up behind us and it's got all the prisoners on it and a bunch of casualties. And it started to push the Humvee. It was like, 'Hell let's get on it. We'll burn the sucker. It's got four flat tires and it's already on fire.'

"The guys jumped off the other Humvees. They took Sergeant Joyce and got the other wounded and helped me up on the truck and we drove right through K-four Circle. When I got back to the airfield I stood up on my right leg and I got real dizzy and I sat down. Then they put me on a stretcher. I wasn't sure where I was shot. There was blood everywhere and they couldn't tell. A couple of nurses came over and they took my shirt off and they started two IVs. They said, 'Okay, you're shot right in the back.' Then they cut my pants leg open and there was an entrance wound and a compound fracture.

"The medic stuck my leg in my boot. That's when the pain started. I was like, 'Cut it off,' and he said, 'Yeah.' They sent me to a hospital and I had surgery. I woke up and it was like two hundred degrees outside and I was freezing cold. My leg wasn't hurting no more. My shoulder didn't hurt. It was like four in the morning. I had no idea what was going on. High as a kite. When I woke up the first thing that went through my head was like, 'Boy, do I got my legs?' "

In the late fall of 1993, I flew down to Fort Benning to reconstruct the way our Rangers in Somalia got cut to pieces when President Clinton and his top military and civilian advisers sent them into combat without the armor they needed to survive. The U.S. Army wasn't eager to talk about what had happened. But one of my guys smuggled me classified after-action reports, and the nightmare came spilling out of the young Rangers who sat down to talk to me. The kid who felt cold at 200 degrees had a Ranger's close-cropped hair and a set of old man's eyes, the eyes of a guy who has seen combat and knows the score. He came back with his legs. But up at Walter Reed Army Hospital I had seen a ward full of our best warriors, missing legs, arms, half a face, or whatever else Mohammed Farah Aidid's gunners had been able to shoot off them—all this because of dumb generals and an even dumber Commander in Chief.

I knew we were heading for trouble the previous June when a company of UN soldiers on station in Somalia was ambushed and twenty-four Pakistani soldiers went home in body bags. The UN Secretary General and the President of the United States started acting more like the Law in Tombstone than statesmen. They thought the thing to do was put a price on Aidid's head. Our C-130 gunships bombarded his headquarters for four days, trashing it. Our mission crept from feeding into fighting as Operation Restore Hope became hopeless and went up in flames.

I suppose the President's advisers believed that high tech and firepower would bring Aidid down. Instead, they fell into LBJ's kind of Viet-think. President Clinton said the United States was "striking a blow against lawlessness and killing." But unlike George Bush during Desert Storm, he couldn't use his military muscle to make good his threat. By then, we'd pulled out most of our combat forces.

The situation began to disintegrate like a brittle Chinese fortune cookie. In August 1993 the Somalis struck back with hit-and-run raids; then four American soldiers were killed by land mines. At that point, I guess, even Aidid knew he was going too far, because he sent a message through Jimmy Carter offering to talk. But once again we saw a serving President who thought he could nail the coonskin to the wall. President Clinton sent a task force of Special Operations troops with clear orders: Get Aidid. Mission creep had changed to galloping Ramboism, and the President forgot that Rambo only wins in the movies.

It seemed to me the Special Ops force had the right warriors to snatch Aidid but the wrong intelligence lash-up to find him. Under Major General William Garrison's inept leadership, the Ranger task force stumbled from one dry hole to the next. They did nail one of Aidid's top lieutenants. But the first six raids were carried out according to carefully designed "templates," and Aidid got the hang of them in time to kick our teeth in on the seventh. The $28 billion the United States spends each year on intelligence wasn't much help against a low-tech opponent who could hide among his people. Aidid just chilled out while General Garrison smoldered. After a Somali mortar shell wounded five of his men, he swore "to kick somebody's butt."

So the President, his political counselors, and his generals all wound up striking heroic poses, but they wouldn't put their weapons systems where their mouths were. On September 14, Major General Tom Montgomery, the U.S. Commander in Somalia requested armor "to get to bases if any were threatened." His bosses, Generals Joe Hoar, Commander in Chief of CENTCOM, and Colin Powell, Chairman of the Joint Chiefs of Staff, relayed the request but did not support it strongly enough to keep Secretary of Defense Les Aspin from spiking it. Aspin's reasoning was that sending the armor would be an unacceptable military escalation. In other words, after telling our warriors to get Aidid, we were saying, "Oh, by the way, we want you to go in there naked."

In mid-September of 1993, I decided to go back to Somalia. I called one of my contacts in the Pentagon and asked for permission to go out with the Rangers. The guy said great, good story, but we've got to clear it with the force commander. About a week later I got a call back.

"Nope. You've been turned down," the public affairs officer said.

"Why?" I asked him.

"General Montgomery doesn't want you with *his* Rangers."

In early October, I was in Fort Carson, Colorado, to give a talk to the NCOs and officers of the 4th Infantry Division. Colonel David Hunt, a true stud and one of our finest warriors, had invited me to speak to the division's officers, so I gave my dog and pony show, fired the Bradley, and took a look at the troops. I was in Dave's office when he got the flash report on the Mogadishu raid. It had been a disaster, a lot of people had gone down, we had lost choppers, the whole thing was a bloody mess. I immediately plugged into the military circuit, and a group of Special Forces warriors began feeding me intel. My field position in Colorado was almost as good as downtown Mogadishu and a hell of a lot safer. For about an hour I was up to my neck in commo. Then the phone rang again.

It was Tim Grattan calling from Whitefish.

"Look, Hack, I don't know if you've heard yet, but if you haven't, Casey Joyce, Larry's boy, was killed."

My eyes began to sting.

Larry Joyce, Tim, and I had all served together in Germany in the 1960s. Tim and Larry were young lieutenants with families and I was a captain who had married late, so we all had babies at the same time. I had known Larry Joyce as a first-rate officer, airborne infantry, a fine young stud with a master's degree from Texas Tech and a beautiful wife named Gail. I had bounced Steve Joyce, their first child, on my knee. Casey was born in 1969. He had grown up wanting to be a warrior stud like his dad. He was a bright, handsome kid, extraordinarily gifted. He went to college for three years, then decided to join the Rangers. He wanted to prove something to himself. His dad's combat experience in Vietnam—two tours at the peak of the war—fascinated him. He wanted to know what it was like to be under fire.

Now he was dead. I sat there thinking, *I lost my boy in that operation.*

When the wounded Rangers were brought to Walter Reed Hospital for treatment that fall, I went to visit them. I didn't go to interview them. I just wanted to pay my respects and condolences as a soldier and see if I could find out more about Casey. The wounded were in a big ward lined on both sides with beds. Some were hobbling on crutches, and there were a few wheelchairs, but a lot of them couldn't move around. You could sense their pain and their anger. They were terribly shot up. Many were arm and leg amputation cases. Many had been lying on the ground for hours.

"Why so many amputations?" I asked one of the medics.

"Infections," he said.

Somalia is one of the filthiest places in the world and those guys had been lying out there with bad wounds for hour after hour. One of the infections was gangrene. It was as if conditions had regressed to the Civil War. After I left that day, I requested an interview with men who had fought in the raid. The request had to be approved by the Special Operations command in Florida. Then I flew down to Fort Benning where the Ranger battalion was based and spent a whole week there while the Army did everything to stall.

"Look, I have nothing more important to do," I said. "I'll just mosey over to the Infantry School. It just happens to be commanded by Major General Jerry White, one of my Vietnam company commanders. I'll be happy to go hang out, look at new weapons, check out the instructors, and shoot the shit with Jerry."

I knew I could sweat them out.

Finally, they agreed to the interview, though they brought in a lieutenant colonel to act as head Thought Cop. I met the Rangers in a conference room. That day about ten of them entered in groups of two or three and sat down behind a long table. Several had been wounded, but they were not the most serious cases, who were still in the hospital. Obviously they had been prepped. Lieutenant Colonel Danny McKnight, the commander of the 3rd Battalion, 75th Ranger Regiment, came in with the first group. You don't just get command of a Ranger battalion; first you have to have done a great job commanding an infantry battalion; so McKnight was the cream of the cream. In a word, a stud. Aidid's people had shot out the windshield of the colonel's Humvee. He kept it in his office as a trophy of war, which I found a bit much. No way would I keep displays of the bullets or shrapnel I took or the shot-up shirts or helmets. But to each his own.

I'm not going to criticize Colonel McKnight for what happened in Mogadishu—I wasn't in his boots—but I do think it's fair to say that when he came in to talk, he was in maximum damage control mode. He opened up by saying reporters were always asking him if his troops had lived up to his expectations in combat.

"I very dryly answer, 'No they did not,' " he said.

"People always look at me real funny," he went on, "because I'm sure they think they've got a good story. But the next thing that comes out is, 'These Rangers, nineteen-year-olds, twenty-year-olds, exceeded any expectation that I as a battalion commander or *you* as a (former) battle commander' "—here he shot me a hard look—" 'could ever expect young men to do.' "

I hadn't gone there to give Colonel McKnight a hard time. The shake of the dice is never fair in combat. But to me, he sounded almost aggressively defensive, like a guy expecting his own side to hang him out to dry.

"Was the mission worth it?" I asked him.

"I don't even want to relate to it in terms of risk and gain," he said. "One of my soldiers is worth more than anything else in the whole world. Anybody who thinks anything but that these Rangers and all the Special Operations people and aviators were successful is dead wrong. They are absolutely wrong. It was successful from the time it started till the finish."

That was the official spin—clearly it had been drilled into all of them—but Colonel McKnight had to leave for a minute to take a phone call and I was able to make my own pitch to the younger men.

"Well, we've had the warm-up stuff that I gather Big Daddy throws around a lot," I said. "And I've heard it all before."

I could feel them ease up. They shifted around in their chairs, began to unwind, and pretty soon we were walking down the Mogadishu Mile.

The mission began on the first Sunday in October, a hot day like all the others that fall. A slight breeze was blowing in off the sea. The 3rd Battalion, 75th Ranger Regiment, had made the Mogadishu air base its headquarters. Around noon the Rangers were playing volleyball, hanging out. At about the same moment the spooks were reporting that some of Aidid's top lieutenants were holding a meeting that day over near the Bakar Market area of Mogadishu. An hour or so later, rumors started going around that a hard scrimmage was in the works, sometime after lunch, a new game. No nets.

At 1400 hours the Rangers were told the light was green. "Everything started real slow," one of the young guys told me. "It was 'Get everything. Get dressed.' "

Sixty Rangers and forty-six Special Ops troops started collecting their gear. For the raid each rifleman took his M-16 with seven 30-round mags, and two frag grenades. The M-60 machine gunners loaded up with six hundred rounds. Each man carried two one-quart canteens of water. All wore the new, experimental Ranger body armor, black, not green like the standard issue. The vests had new steel plates, lighter than anything they had worn before, about twenty-five pounds, good gear if you've got the full kit.

The Rangers didn't.

The leaders were worried about all the weight they would be carrying when they fast-roped down to the objective. The decision was made to go in with only half the body armor—the front half.

That left the Rangers with nothing more than their fatigue jackets to protect their backs.

I wanted to know what else they didn't take.

"Packs?" I asked.

"No packs."

"Night vision gear?"

"No."

No rucksacks, no smaller packs. They didn't have enough ammo for a long hot firefight, let alone a night-long siege. And they didn't have night vision devices. The thinking was that this would be a daylight raid, that none of the six earlier missions had taken more than an hour or an hour and a half. No one expected the seventh to run beyond sundown. It would be WHAM, BAM, thank you, ma'am. Quick in. Quick out. The main assault was to come from the air with a ground support convoy to extract the raiders and their prisoners. The support element had eight Humvees, plus Colonel McKnight's vehicle, plus three 5-ton trucks. They took along the trucks because they figured they might find a lot of documents or a safe that wouldn't open.

Already I could see trouble on the way. *They were not half as prepared as they needed to be*, I thought. *They'd forgotten Murphy's Law: "Anything that can go wrong will go wrong."*

The five-tons came from the cooks and bakers and whoever else had one to spare. They had sandbags on the floor and in the back, but no special armor. Bullets and shrapnel would rip right through them. The Humvees were a little better, but not much. The grunts called them Grungies. They were mounted with .50 cals and Mark 19 40-millimeter guns, they had armor plating over the wheel wells, bomblast sheets under the seats, and bullet-resistant windows. But only the doors and wells were armored. One burst from an AK-47 or an incoming RPG could turn the Grungies into colanders and anyone in them to salad dressing. Out on the airfield the crews were organizing a flock of choppers: six MH-60 Black Hawks, four MH-6 and four AH-6 "Little Birds," a search and rescue Black Hawk, and a bird for overhead command and control.

The briefing was quick but comprehensive. They had good aerial photos. The objective was a building up the street from the Olympic Hotel, bad guy country, Aidid's own playpen.

Shortly after 1500 hours the order came to launch.

"Okay. Get it on," said one of the team leaders.

And everybody went out to the birds.

The target was about three miles away as the crow flies. The men filed into the choppers and took their places; the birds rose and flared toward bad guy land. Adrenaline started pouring in, hearts pumped, guts clenched, everyone hoping it didn't show.

The rules of engagement were clear as vodka.

"It was a combat mission every time we flew," one of the raiders told me. "That's what we were there for."

The ride took only a few minutes. Chief Warrant Officer Clifton Wolcott, piloting one of the Black Hawks, gave everyone a one-minute warning, enough time to pull on their gloves. Going in, he limited himself to five words: "Where do you want it?"

The lead chopper hunted for targets, saw none, and began the assault.

When the first Black Hawk reached the objective, the raiders threw out eighty-foot ropes and started fast-roping down. The choppers created the day's first tactical problem. The wash from the rotor blades churned dust and trash into a brownout so dense the raiders who followed could barely see their hands in front of their faces. The other birds arrived in the middle of a dust storm created by high technology. They looked down and saw they'd have to rope in almost blind.

"That's when I started to feel really weird," one of them told me.

Another chopper did a tight orbit to let the dust settle; the last chopper descended using a building and power lines as reference. Then everything vanished. The raiders thought they were still high over the LZ. They started grabbing the fast ropes and sliding down only to jam their ankles. They had been hovering three feet above the ground. Special Ops forces led the raid to snatch Aidid's men. The Rangers pulled security around the target area. They all came in fast and hard.

"You're sliding down pretty fast?" I asked.

"Oooooh yes, sir," said one guy who will never forget that drop. "You're supposed to use your feet and your hands to clamp on quickly. But it doesn't happen. If you are lucky you land on your feet and can walk out. If you're like me, M-60-gunner having all the extra ammo and the gun on you, I just sort of fell on the ground and someone else landed on my chest. Then he got out of the way and they cut the rope and it fell on me."

The brownout produced the first casualty. Ranger Blackburn lost his grip and fell from the fast rope—seventy-five feet, three stories, to the ground. He landed on his back.

"He was hurt real bad," remembered one of the Rangers who saw him. "Internal bleeding, head trauma, busted his right leg and hip."

Sergeant Joyce and another Ranger carried the injured warrior one hundred yards through intense enemy fire to a Humvee. After putting Blackburn safely in the vehicle, Joyce turned and gave a thumbs-up. Blackburn later said, "I owe my life to Sergeant Joyce."

While the Special Ops raiders hauled ass into the target building and started rounding up Aidid's surprised operatives, the Rangers took up blocking positions: nobody in, nobody out. The strike was so quick that two dim Somalis wandered into the drop zone without weapons. A sergeant grabbed them, trussed them up, and dumped them in the street like garbage bags. Soon after that sporadic fire started coming in. It drew first blood.

"Ranger Berendsen got some shrapnel in his left arm, Sergeant Gallantine, the team leader got hit right in the thumb," one of the Rangers recalled. "Sergeant Telscher was jumping around helping everyone. Then I got some shrapnel in my leg."

At that moment, the ground convoy led by Colonel McKnight pulled up. The trip from the air base had taken only about twelve minutes. The convoy arrived just as the last of the raiders were fast-roping down.

"The last vehicle was still on National Street," one of the grunts told me. "About eight to ten Somalis came out from the alley behind the target building. One had an RPG, the rest had AK-47s. They were ten feet in front of our vehicle. We fired the guns up—and from that time on we duked it out."

Colonel McKnight and the vehicle element stopped about two hundred meters away from the target, which was two buildings up from the Olympic Hotel. He took a quick scan and liked what he saw:

"We achieved surprise, speed, and great ferocity in the way we attacked. We did it exactly the way it should have been done. The timing was perfect, right on the money. They hit the target. Within fifteen minutes it was 'Objective accomplished.' We were ready for exit."

Then came Act II, as though Murphy had written the script.

"The medic tells me, 'I've got a serious, urgent casualty. I've got to get him out of here.' "

Colonel McKnight went over and took a look at Ranger Blackburn.

The medic was right: He was in bad shape. So the colonel decided to put him in a vehicle and send him back to the air base. He also ordered two more vehicles to escort the mercy wagon.

"Ranger Blackburn is alive today. He's still got a bruised memory loop in the brain. The negative part was that Sergeant Pilla was killed during the evacuation."

Up to this point everything Colonel McKnight said had tracked smoothly. But here was a note that sounded a little off-key to me.

Yeah, I guess you could call that negative, I thought.

The basic mission had gone well, but it had stirred up the Islamic Green Hornet, and all his little brothers and sisters were gathering to sting back. The medevac vehicles ran into them and Sergeant Dominic M. Pilla, a twenty-one-year-old Ranger from Vineland, New Jersey, was killed fighting them off. So the brownout produced the Blackburn accident and the accident produced the medevac and the medevac got Pilla killed and Colonel McKnight was standing there not yet knowing the score. In the fog of the gathering battle, no man could.

"Everything was just going fine," he told me. "Then we started to load the detainees. That's when everything changed."

All of a sudden Aidid's irregulars loosed a fusillade of five RPGs. One of the rockets killed a five-ton, and Staff Sergeant David Wilson picked up some shrapnel in his leg. They destroyed the truck in place. The ground platoon was now down to four vehicles and a long way from home.

The gunner on the .50 cal clutched his arm and said he was hit. The medics pulled him down and Sergeant Lorenzo Ruiz, twenty-seven, from El Paso, got up on the gun, hosed down the alleys, and bought a short breather.

Then it got worse.

"I heard this WHOP, WHOP, WHOP, WHOP," one young Ranger told me. "I'm running back. I'm looking down National Street over the top of the building and I saw the first bird get shot. We were all saying, 'The bird, the bird is hit.' "

The Black Hawk flown by Chief Warrant Officer Wolcott had hovered over the target, circling, providing fire support. Wolcott was directing his gunners and snipers. At one point the chopper wheeled broadside, and an RPG gunner cut loose and hit the rotor. Wolcott lost control. The chopper began to spin.

"I called and said, 'Hey, a helicopter just got shot down.' I was trying to tell where it was. I was trying to ask, 'Do you want me to go to the bird?' "

The Rangers pulling security had buttoned up their zones. The young Ranger ran up to talk to Colonel McKnight, who was talking to the ground force, telling them to hurry up and load the prisoners.

"I said, 'Hey, sir, I just saw the bird go down. Do you want me to move to it?' And there was a big 'No. No, stay where you are.' So I said, 'Okay,' and ran back to my vehicle."

Colonel McKnight was in a hell of a corner. So far as he knew, his casualties up to that point had been light. He had in hand the prisoners he had been ordered to snatch. He could return to base with the prisoners, escorted by the undivided assault force (minus the medevac vehicles and the dead five-ton). That would give him about three times the strength he had on his successful ride in. Or he could go to the rescue of the downed chopper. The evidence from the grunts is that his first impulse was cautious and sound: not to divide his forces further or to haul ass directly to the downed bird.

Then events began to crackle like bolts of lightning in a summer storm.

The pilots of the other choppers saw Chief Warrant Officer Wolcott's bird get hit. Over the radio they heard him calmly telling his crew and the two snipers aboard they were going down. Then the radio went dead. The bird crashed into a building and courtyard below, killing Wolcott and his copilot. But from the wreck two snipers were firing at a growing mob of Somalis. The Rangers lived by a code: They would never leave one of their own behind, not dead and sure as hell not alive. I feel the same way sometimes, and other times I wonder. When you're hauling ass and you've got four dead and if you are going to take twenty men to carry them, it's better to stash them under a rock. I have had to make this bad choice twice in my life, once in Korea and once in Vietnam. Both times I chose the rock. It made me feel lousy, but I think the priority has to be for the living.

What happened next is blurred by the fog of battle. Some say General Garrison gave an order for McKnight to move the assault team and rally, round the chopper. When I heard that, I found myself thinking, *Where was General Garrison sitting when he made that decision? Could he see the rat's maze of streets and alleys around the Bakar Market? Could he see the Somalis swarming everywhere? Could he see he was ordering the raiders right into a buzz saw?* Or was he just drilling holes in the sky over the fight, puffing on a dead cigar, blind as Ray Charles and nowhere near as talented at his work?

The Resister, the voice of the Special Forces underground, later reported that Garrison was overhead, micromanaging platoon leaders on the ground in the worst tradition of Vietnam. The report said that he ordered a flight of Cobra gunships five minutes away, "sitting on the tarmac, fully armed with engines running," to stand down. It quoted a Ranger platoon sergeant as saying, "I wish I'd had a goddamn Stinger."

By another account, the plan called all along for Colonel McKnight to provide the backup for extracting any search and rescue team that had to go in for a downed bird. Still others say that Lieutenant Thomas R. Ditomasso, who won the Silver Star for his gallantry, took off on automatic for the crushed bird. We may never know the truth because the "After-Action Review of Task Force Rangers Action in Mogadishu" is classified Top Secret/Special Category. The report is locked up at the 75th Ranger Regimental S-2 office at Fort Benning, snug and safe from prying eyes and Freedom of Information Act requests. The Pentagon has also classified eight videotapes of the fighting taped during the battle by P-3 Orion reconnaissance aircraft. The party line is to protect secret technology. The real reason is that President Clinton and the Pentagon want the whole disaster to go away. The shame of it is that these films show the valor and skill and sacrifice of wonderful warriors fighting one of the most valiant, small-unit actions in U.S. Army history. Outnumbered 100 to 1, trapped dead center on the enemy's home ground, they stood tall, didn't flinch, and wasted 1,000 Somalis. Because of cheap political games, their heroism may never be fully known.

Colonel McKnight leaned over and looked right at me.

"I'll say it right now. We did not secure a helicopter with dead people. Too many people ask, 'Why did you go there to secure a dead pilot?' We didn't. There were four survivors out of that crash. We had other helicopters still flying. You could see down. There were survivors. Period. And three out of those four survivors are alive today."

The Americans began to pick their way toward the crash site. By now the element of surprise was gone and the Somalis were fighting on their own turf, an advantage they knew how to use. The Rangers were exposed on every flank.

"Somalis were running in the street, people were pointing to us, there was AK-47 fire everywhere," said one good kid. "Rounds everywhere. Nonstop. The shooting was coming from a tall building. Someone was firing RPGs. A .50 caliber opened up and tore into him—I mean it just ruined his whole day—the guy fell off the roof in three pieces."

The Bakar Market is the bad part of town in a city where everything starts at lousy. The distance between the original objective and the crash site was about five hundred meters—if you were a bird—but on the ground the Rangers had to make their way through a labyrinth of streets and alleys offering zero cover. Every intersection was set perfectly for flanking ambushes and cross fire.

"Where was the fire coming from?" I asked.

The answer came as a chorus: "Everywhere."

One Ranger's memory stuck in my brain:

"They would hang their weapon around a corner so you couldn't see anything but their hands and just fire off a twenty-round burst. Then they would back away and run to another corner and do the same thing. They were not command controlled. But there was a maneuver. Some were in groups of eight or ten, but most were smaller. The ones I think were trained, who did understand a little bit about fighting, were the ones that were using the windows and the doors and the tops of buildings. But most of them, and I'm talking from the ages of thirteen years old up, were running around with weapons. You hear that there were women and children? Well, the children of thirteen and fourteen were shooting at you and me with RPGs and AK-47s."

Suddenly this sounded very familiar. *Hugging the belt*, I thought. *In a city, not a rice paddy. With women and children as shields. Otherwise it works the same way.*

And if you are the guy getting hugged, it is pure fucking hell.

One Ranger who knew exactly how it felt said, "We made a right-hand turn down the alley and another right-hand turn and then the lead gunner on the first vehicle got shot. The next thing I knew Kohler was up there standing on the brakes. I told everybody, 'Get off' and they threw down the M-60 machine gun and we covered back where we had just come from. Sure enough the Somalis were all running from that direction, coming at us from the rear. That's when Ruiz got hit and I got shot in the shoulder. When I got hit, it knocked me out of the vehicle. I dropped the radio and I was trying to get up on the gun and a first teamer said, 'No, I'll get the gun' and he jumped up on top.

"So I found the entrance wound on Ruiz but never found the exit wound because he had a vest on. As soon as I got his vest undone, I felt blood. Okay, this is the entrance wound. I started feeling around and I ripped his shirt open and found the entrance wound on his right side. I called for a medic because Doc Jenrud was working on two casualties in the street from the special teams. A medic from the med vehicle came back and started helping with Ruiz. I don't know if he ever got the IV started because the minute that happened an RPG went over us and my ears haven't stopped ringing since.

"The gunners stayed up on the .50 cals even though they knew the other gunners had been killed and wounded. When Sergeant Ruiz got shot, they laid him where the gunner would normally stand between the seats underneath the gunner's well. So the next guy to take the gun sat on top of the vehicle. Specialist Richie, David Richie. He was steady. He was great. He was running alongside the vehicle on my right flank. His gun got shot. He had an M-60 and a round went through the butt plate and he's like, 'Goddamn it, I can't get the bolt to the rear, it's jammed.' So I went over and I give him the 16 and I pull the bolt to the rear, but when I lift the feed tray everything shoots out. So we just threw the gun in the back and said, 'This ain't firing anymore today.'

"When Ruiz went down, Richie goes, 'I'll take the gun' and he gets up there and by now he's the third or fourth gunner. Then he got hit in the ankle."

Colonel McKnight was rightly proud of the guts his boys showed getting back up on the .50 cal.

"Was there a shield on it?" I asked him.

"No," he said. I could feel my jaw clench. "That is something we've already talked to people about. It needs to be looked into."

It sure does, I thought. This reminded me of the Humvee-mounted TOW missiles in the desert.

By now the incoming had disabled almost all of the vehicles in the convoy. Flat tires, blown gas tanks, smoking engines. About then, Colonel McKnight, wounded himself, decided to move with the prisoners and an increasing load of casualties back to the airfield.

"At the moment he called for the exfiltration, Ruiz was sitting up and he was like, 'I'm all right. I'm all right.' He goes, 'It hurts a little bit, but I'm all right.' You couldn't really tell how bad he actually was. I still couldn't find the exit wound."

Casey Joyce was with the Rangers fighting their way toward the downed chopper. The Humvees drew such intense fire that the rescuers had to unass the vehicles and fight door to door. At one point, a Somali technical vehicle with a heavy machine gun was blistering them. Casey grabbed an antitank weapon and put one right in the gunner's lap. The vehicle disintegrated. Casey turned and shouted:

"All right."

Then he took a hit in the back, where his body armor was AWOL.

The round ricocheted off his front armor and reentered, leaving him with a massive chest wound.

When I heard the story, it hit me like a slug through my own body. Casey wound up on the back of a cargo Humvee with a heap of wounded Rangers.

"I was on top of Ranger Berendsen," one of them told me. "Sergeant Gallantine was over on the right side. I had Specialist Diemer on top of me with an SAW shooting. Sergeant Joyce was a little in between us. He just didn't look right. His eyes were open but they had turned a different color."

"When you got back to the air base, was Casey still alive?" I asked him.

"No. He went into shock. Because it took us so long to get back. Because we kept getting hit and hit and hit, he went into shock and died out there."

My boy. All my boys. I felt tears sting my eyes. Casey wanted to be a stud like his old man. He was. Both of them were cut from the same sturdy cloth.

Most of the casualties took place between 1530 and 1800 hours when the Rangers and Special Ops team were trying to consolidate at the crash site.

Wolcott's chopper was Number 61 in the flock of birds flown by the 160th Special Operations Aviation Regiment. When he went down, Number 62 flown by Chief Warrant Officer Michael Durant came in to replace him. Within fifteen minutes another volley of RPGs sizzled aloft and once again Aidid's gunners scored. Durant's chopper took one in the tail rotor. For a few moments he was able to keep flying, but he didn't have enough power to make it back to the airfield. He too lost control and the bird started to spin. It crashed a little over a mile away from Number 61. Two Special Ops warriors, Master Sergeant Gary I. Gordon and Sergeant First-class Randall D. Shughart, volunteered to fast-rope down to the second wreck. They pulled the crew's bodies out of the chopper and placed Durant, the only survivor, in a sheltered position. Then, single-handedly, they fought off hundreds of Somali attackers, slugging it out in high noon hell until they were gunned down.

Colonel McKnight put the number of RPGs fired at the choppers at around two hundred. "That is not an air defense weapon," he said. "It wasn't aimed. It was just, 'If we can fire enough of these things up here we are bound to hit something sooner or later.' "

Exactly. The Viet Cong used machine guns and RPGs the same way. Anyone who had studied their tactics knew this and Aidid and at least one of his top lieutenants had boned up under the Soviets. A few days before the raid, Aidid's irregulars had downed a chopper with an RPG, so no one should have been surprised when those mean little suckers started whumping up into the sky.

"Maybe that's the standard issue when they turn thirteen," Colonel McKnight said, allowing himself a second of gallows humor. "They all get their own RPG for their birthday. I don't know."

There wasn't any time to figure it out. The convoy had to fight every inch of the way back to the K-4 traffic circle. One of the Rangers aboard the retreating Humvees was a sniper by specialty, but he put aside his rifle and started using the driver's M-16.

"At the ranges we were fighting, a bolt-action rifle wouldn't have done any good," he told me. "We were trying to get to Crash Site One, not making it, and the firing was getting heavy, when the decision came to turn around. We went down one street and turned on another and one of my friends said, 'Oh God, we're lost.' "

A five-ton truck doesn't turn on a dime.

"We just barely made it. There were steps going up to a building and we just rode up the steps and come back down, taking fire the whole time. The prisoners are in the back, so are some wounded Rangers. We just lined the windshield with them and turned around. I saw a guy underneath a pile of junk shooting at ground level. So I was kind of leaning up over the window, trying to shoot down, but you couldn't get at him. You seen the muzzle flash coming out from underneath the pile, but when you got up you couldn't see nothing. So I was just firing at the pile and one come through the door and hit me in the leg, another came through the same door and hit me in the knee. Turned out it was the jacket from the bullet that skipped off the bone.

"At that point, my driver started screaming, 'I can't see!'

"I got one leg propped up on the window, the one that's shot, and I got the rifle propped up, shooting down alleys when we go by. And I had to look up and over my shoulder and his glasses are all bowed and turned down. I'm thinking he's just knocked his glasses off and he's blind as a bat. Two minutes later he stopped for one of the alleys and he yells in back, 'Does anybody else want to drive? I've been shot in the head and the leg.' The bullet went right through his K-pot at the base of his head."

"Everyone was yelling, 'Let's get outta here,' " another of the Rangers reported. "The Somalis had set up roadblocks. I felt a huge jolt. My head was spinning and the cab was filled with smoke. I saw the driver and I knew he must be dead. I reached in and pulled him out. His arm had been severed by the blast. I put him in the Humvee and went back and grabbed a radio and the driver's M-16. The hand guards had disintegrated, the barrel was bent, and the gas tube looked like a tree branch. It turned out my helmet and weapon had been blown out of the cab. The RPG struck just under the driver's side window. The cone of the round had flipped up and lodged in the driver's midsection—the propellant had filled the cab and choked us."

Where was the armor? I wondered as the horror stories kept smacking me like the concussion from incoming rounds. *Where was the goddamn armor?*

"We led the convoy of wounded back to the airfield," another Ranger reported. "We received fire from every corner we passed. We drove between parked and oncoming vehicles scraping the mirrors off those that didn't move out of our way. I looked back to see if my buddies were okay. I saw one slumped over with his brains dumped out in his ammo can and I knew he was dead."

The men and women who had stayed behind at the air base came running to unload the wounded and dead. Earlier that day, a sergeant who worked at the base fire station had waved at the departing choppers. That was the custom at the "Dish" whenever the Hawks took off on a raid. "I guess it was about thirty minutes later I came in and the guys were all jumping on the fire truck," he reported. "So I got the Humvee and followed them out to the runway. There was a UH-60 that had been shot up. We took one guy off it with half his head blown off. We hosed the aircraft down and returned to the station. We had no sooner got back when somebody ran in yelling that a Humvee had been hit and it was full of wounded. I ran over and helped unload the body of one guy who looked to be about seventeen. I still had no idea what the hell was going on. A five-ton pulled up and I ran over and helped them open the back of it. What I saw then was something I was never prepared for. Wounded Rangers were piled on top of dead ones, who were on top of Somali POWs, who were on top of Somali dead. I reached up and helped pull a guy down. His leg and arm were blown off. He was dead. I ran and got another one. He had been hit real bad. He died on the stretcher. A sergeant and I picked up another stretcher and carried it to the morgue before we realized the guy had an RPG stuck in his chest. We had to build a bunker for his body because of fear of the RPG exploding."

By now the original task force was split in two between those who had fought their way back to the air base and those who were still out in bandit country. Rangers from the security detail fought their way to the site where Chief Warrant Officer Wolcott's chopper had gone down. The first Rangers on the scene saw an MH-6 Little Bird land in a narrow alley with only three feet clearance for the rotors. The copilot jumped out and ran to the downed Hawk, firing his 9-millimeter while the pilot covered him with a submachine gun. He was able to haul two of the badly wounded crew back to the Little Bird, which lifted off just before the C-SAR search and rescue chopper arrived, ferrying in fifteen Rangers, all with combat and medical training. The Black Hawk hovered while the Rangers fast-roped down. Just as the last two men grabbed the rope, the chopper took a hit from an RPG. The pilot held steady until the last man was on the ground. Then he and his crew nursed the crippled Hawk back to the air base, where they ran to a spare chopper and returned to the battle.

"The second rescue chopper team started assessing wounded and tried to move them out of the fire away from the Hawk," one Ranger officer recalled. "Every time someone tried to move a litter, he was shot. All we could do was return fire and keep the crowds and gunmen away. The pilot who was killed was trapped inside the helicopter, so we couldn't move too far away from the craft for fear of losing the body to the gunmen or crowds. This man was one of us and we were not leaving without him."

The Americans were able to lay down suppressing fire and move the casualties into a building just south of the crash site. An NCO threw his body against a locked metal gate and broke into a courtyard where the Rangers subdued and trussed up five Somalis. The building became the strongpoint for the night.

"The eleven wounded and one dead were moved into the center of the complex," reported one of the Rangers. "I put two Rangers in the hole in the wall just south of the aircraft to secure the bird. The remaining Rangers went to the south courtyard and set up there. At this point, there are thirty-three soldiers, not counting the dead, inside our perimeter. It is now roughly three hours into the fight around the aircraft and darkness is approaching. We were running out of ammunition, water, and medical supplies. One person went back into the crash and scavenged much of what we needed. We passed half of the supplies to a team who had occupied the building to the immediate west of us.

We then recovered two sets of night vision gear from the bird and passed one set to each team."

Darkness fell and everyone hunkered down to sweat out the night and wait for reinforcements.

As I sat in that safe, peaceful room looking at the young Rangers on the other side of the table, I thought, *We've got the best young warriors in the world. Nothing wrong with them. But who the hell got them into this mess?*

A young medic started talking, returning us all to the battle.

"I was checking out Specialist Errico," he said. "He'd lost maybe ten ccs of blood, about as much as that cup of coffee you're drinking. But he was still alert. No shock. I had I.V.s pumping fluids. Sergeant Goodale was in the back, shot in the butt. He was on the radio."

The worst wounded was a specialist named James Smith, a warrior from New Jersey, whose dad had been a Ranger captain in Vietnam and lost a leg. He had a belly wound, a round had severed his femoral artery, an extremely deadly wound. The medics ran out of IV fluids and morphine. They were giving Nubain and Tylenol No. 3. They held back one I.V. for a final emergency. "Somewhere in the night they told us Smith had died," the medic said quietly. "The medics were worried about Lieutenant Lechner. I was working on Sergeant Goodale. We were just sitting there waiting out the night and listening to the birds overhead."

All night the choppers flew over the Ranger perimeter. On the ground the defenders had SINCGARS FM commo gear and Sabre MX radios. "The RTOs never took off their radios," one lieutenant reported. "One RTO was continuously calling fire missions with lethal accuracy. If it weren't for the aircraft suppressing the enemy and keeping the crowds off us, we would have been in trouble. This RTO was shot in the head, the bullet grazed his helmet and snapped his head viciously, yet he continued to perform his duty." Up above, the Black Hawks had door gunners and the Little Birds had 2.75 rockets and min-guns. They also had night vision gear. All through the night, from the ground to the choppers, it was, "Request fire mission, over."

"How many times did I hear this," one of the pilots recalled. "I will always remember the calm demeanor and professionalism the RTO showed over the radio even as I heard bullets hitting very near his position each time he keyed his radio microphone. He expertly coordinated and called in helicopter fire to within fifteen meters of friendly positions." And from the all-seeing Black Hawks in the night sky to the RTO on the ground, it was, "We've got somebody coming from your left. There's six people. We can't get them from where we're at." Roger on that, the Rangers answered—and spent the rest of the night wasting the Somalis.

As the night wore on, rockets, RPGs, and tracers stitched crazy patterns in the blackness over the battle zone. The fireman back at the base watched transfixed, worrying about three buddies who were fighting underneath the fireworks. "I heard someone yell, 'Wounded coming in,' " he recalled. "The U.S. Army Quick Reaction Force was leaving on a convoy to go help the Rangers. I looked over and saw them and wondered how many of them we would be loading in body bags. I remembered the shocked look on one dead Ranger's face: He had his eyes open and a look of fear even though he was dead. A loud explosion broke my train of thought. I said another prayer for my friends."

Where's the relief column? I thought as the Rangers spun out the story. *Why is it taking so damn long to saddle up?*

To a degree what happened next depends on your angle of vision. Senior officers emphasize the positive: the nerve, training, skill, and endurance of the pinned-down Rangers.

"Some people have said, 'Why did it take so long?' " Colonel McKnight observed, reading my mind. "There was no need to rush it because we were calm. We knew they were okay. They were really in control. We had nobody die of wounds out there that night because they were under good care."

What about Specialist Smith? I wondered if the colonel was just careless or was he bullshitting me? One of his men later stood up for him, saying that he was a caring officer, devastated by the losses. No one can doubt the passion that went into extracting the trapped Rangers. But the fact remains that it took five hours to organize a relief column with enough muscle to punch through to them.

Another young stud frustrated by the delay told me, "They were calling from the objective and they had some critical casualties who were bleeding out there because it took so long. And I was, 'Hey, we are ready to go.' We were just sitting there when it seems like something could have been done. My buddies were the ones that were up there on the target and I wanted to get there."

To get there they needed armor. Since their own Commander in Chief and Secretary of Defense had refused to give it to them, they had to borrow some from the Pakistanis and Malaysians, who were in-country under the auspices of the United Nations. A separate command. Three languages. It all took time.

At about 1830 word went out the relief column was forming. But no one crossed the line of departure until 2330, a five-hour delay. The column consisted of four Pakistani M-48 tanks, twenty-four Malaysian APCs—rubber wheeled—two light infantry companies from the 10th Mountain Division and around fifty Rangers. The Pakistani tanks and Malaysian APCs took the point.

The convoy had to advance street by street, alley by alley, building by building, laying down suppressing fire the whole way. The Somalis were not rabble. They kept pressing the attack when they could have gone home to chew khat and fight another day. "When you shot a guy, the next joker would just come over and try to pick up the weapon," one of the rescuers told me. "The same thing with the RPGs. They would pick up the RPGs, reload and fire."

It took two and a half hours of stiff fighting to even get near the objective. The pinned-down Rangers knew the convoy was on the way because the fire kept getting closer and closer, but in the darkness and confusion, the relief column didn't know where the Rangers were holed up. "We had to have people on the target fire pen gun flares up," said one Ranger who went in on the relief column. "We saw the flare come up and we were able to home in all our weapons systems."

The column broke into two sections. The first went to the site where Chief Warrant Officer Wolcott's chopper had gone down; the second checked out the site where Chief Warrant Officer Durant had crashed. They reached their objective at about 0200.

At crash site one, it took until sunup to load the casualties. "The whole time we were taking fire," said one guy from the column. "The .50 cals and the Mark 19s were great suppressers. But in fifteen or twenty minutes the Somalis would pick a different route to come at you. They figured out that they didn't really need to hit our vehicle. We had some .50 caliber and Mark 19 gunners hit with shrapnel because they were on the top of the Humvees when RPGs were slapping the buildings."

The second column found nothing but a trail of spent brass casings at crash site two. Chief Warrant Officer Durant, his crew, Sergeant Gordon and Sergeant Shughart, were all missing. The column followed the trail for a while, but it led nowhere and they had to turn back. A rear guard was posted on National Street to cover the withdrawal. The plan called for the columns to link and return in reverse order, a tank at the point, then the APCs, then the Humvees, with a tank at the tail. As dawn approached the Somalis started creeping along National Street and the side alleys, pinpointing the rear guard. Just before sunup the column from crash site two turned up at the rendezvous point, but instead of waiting it barreled forward, headed for the stadium where the Pakistanis had their base. Then the second column appeared. Communications broke down, and it too started hauling ass for home.

"Myself and Sergeant Struecker ended up being the last ones," one young stud from the rear guard told me. "We're still on the road and daylight is breaking. I could see Rangers still running down the road. They were receiving fire and some of the APCs were just driving by them. I told my men to hold fire. I knew we couldn't take off. A lot of people were in a big hurry to get out of there for obvious reasons. But I could still see people coming. Some were jumping into the M-113 personnel carriers, some were jumping into the backs of Humvees, some were climbing up on tanks. The tanks had started to roll and we looked back and we saw about six more Rangers coming. I said to Sergeant Struecker, 'We need to go back and get them.' So we drove back in. Actually, we backed down National Street and loaded those personnel. People were climbing up on us. Somebody was sitting on my hood—and we were started back." What they did next was to haul ass up in the column, until they could pull in front of the tanks, putting some heavy armor at the rear. Then they fought their way back to the Pakistani stadium.

"That's the scene that stays with me more than anything," the rearguard trooper told me. "They had bodies of the KIA on top of the APCs and all the wounded were being offloaded. All our buddies had been up there, everybody from our company, and I was going from stretcher to stretcher to stretcher, looking to see who was who. Looking for my fellow platoon leaders and then people from our element. Just not knowing. That was the hardest thing. You didn't know."

Back at the air base, the fireman who had said prayers for the relief column was also loading and unloading the dead and the wounded. "We would take them off and place them in the morgue or the hospital tent," he wrote later. "I helped carry in three more dead. Each litter was a body covered by a plastic bag. The rotor wash blew the bag off one of the guys. He was very young looking. He had a peaceful look on his face. Then I noticed the whole top of his head was gone."

"Unfortunate casualties."

To me, the words the Commander in Chief used afterward are so rotten I can hardly believe he said them. But he did. He sent our best young warriors to their deaths or the gimp ward as if they were his toys, not the nation's finest treasure. He did this in pursuit of an idiotic decision to shift the mission in Somalia from feeding to fighting. Then he and his Secretary of Defense refused to provide the Rangers with the armor they needed to carry out his orders without getting themselves blown away. The generals who should have thrown their stars on the table and resigned rather than carry out this mission went along for the ride. One of them even got another star for being such a good sport.

Bill Clinton has surrounded himself with Rhodes Scholars, but very few of them seem to have a lick of common sense. Most of them are totally naive about military operations. A President needs more experienced advisers, people who can say to the novices, "Hey guys, you may have been great in the pristine surroundings of Oxford, but let me tell you about Mean Street. What you want to do won't work."

Here's what I would like to see. When one of those policymakers gets a soldier killed, that policy geek should have to go with the chaplain and member of the unit who bring the bad news to the widow or parents. We need to do something to make the policy crowd a hell of a lot more careful. The dead are not just numbers. These are young men and women with names that are real, with dreams, with a future. Their lives are being snuffed out because the Olympians in Washington often don't understand the consequences of their acts. We saw the tremendous shock and grief that struck the State Department when three of our diplomats were killed in the Balkans. Later the whole nation mourned when Secretary of Commerce Ron Brown and a number of business executives died there in a plane crash. Sure, that was very sad. But how long did we grieve for our soldiers killed in Somalia?

Why was no one held responsible for this disaster? Was no one to blame? What about the generals in the field? The missing tanks were just four days' sailing time away, anchored at Diego Garcia. They were Marine tanks, and they were ready. They could have come in if the commanding generals in the field had showed enough nerve to keep hammering at the Chairman of the Joint Chiefs of Staff. *Why didn't they say, All right, it's too expensive to fly Abrams tanks from Savannah, so why don't we just send in the Marine tanks? What the hell are they there for anyway*? The reason the Army didn't call in the Marines is that interservice rivalry kept them from asking for Marine help. The tanks should have been brought in before the Rangers went into search-and-destroy mode.

If we'd equipped the Rangers with Marine tanks, they could have gone in, kicked ass, and knocked over the mud huts, put a steel cable around the tail of the chopper, and pulled that son of a bitch, dead pilot and all, out of there. After they towed the wreck away, they could have taken a steel saw and cut the pilot out. The same kind of wrongheadedness kept General Garrison from using the 10th Mountain Division's Cobra gunships. We don't just have interservice rivalry, we get people killed because of intraservice disdain. Special Forces troops look upon regular Army troops the same way Mike Tyson looks at Mickey Mouse. Even in real trouble, the Cobras "couldn't be trusted."

So we wound up trying to play Robo Cop before we figured out the drill. Then, with the entire world watching—and when it was far too late to do any good—those in charge wanted only to cut and run from Somalia, to avoid any accountability. So that's how Operation Restore Hope finally came to its sorry end. One more demonstration of how our modus operandi has become one of bulling in and bugging out. Do we really think these gyrations make us look strong?

On the way back from Georgia I bumped into three Rangers who had been in Mogadishu. When I got on the plane I took a right aisle seat as I customarily try to do. I saw the guys in the next two seats were in civilian clothes, but they had Ranger haircuts and that rare Ranger spirit. The Rangers have a really high white wall with a little strip on top. They're as easy to ID as an Irish cop in civvies. You can't miss them. We got to talking and one of them had read a story I wrote about the raid in *Soldier of Fortune* magazine.

"You know," he said, "that was the first story I read that really hit the nail on the head of what happened."

"Were you there?" I asked him.

"No, we weren't on the operation but we were in Somalia," he said. "If you want to talk to somebody who was involved, got wounded, there's a sergeant behind you who was there."

So I swapped seats and introduced myself and the sergeant and I started talking.

"I understand you were there," I said.

"Yeah." He shrugged, shooting me a glance that said, "What more is there to say?"

But it turned out that he had read *About Face* and we shot the breeze for the rest of the flight. He was a good man. He had enlisted for a four-year hitch after he finished college. He was going hunting with his dad for five days, but we arranged to meet after that and I was able to spend an evening with him. A warrior's intuition usually tells me what has happened in a fight, but I never go by intuition alone. I check everything, double-check it. After we polished off a bottle or two of good wine, he confirmed my worst suspicions. The sergeant had been inside the perimeter the night of the operation. In his opinion the mission had been a rat fuck. He didn't agree the planning had been flawless.

"We telegraphed our punch," he said. "We gave ourselves away."

The President's assigned mission was to get Aidid. The Rangers had to snatch him quickly if they were to get him at all. Choppers provided the speed at the price of an Achilles' heel. As in Vietnam, a thin-skinned chopper frequently gets you into more trouble than it's worth.

Why else did the raid fail?

It was a very bad plan conducted by generals too eager to prove themselves, to win their spurs on a new battlefield. There was bad unity of command, no AC-130 gunship support, no armor, and no go-to-hell plan in case everything fell apart. When Senator Sam Nunn announced that his Armed Services Committee would conduct an investigation, I wrote him a letter, because I wanted to help him ask the right questions. Where was the armor? The brass were sure to tell him tanks wouldn't have reduced the casualties. My reporting showed tanks would have prevented at least six KIA and thirty WIA. What about those U.S. Marine Corps tanks aboard the floating reserve? Why a daylight raid when even novices know the best way to out-G a guerrilla is to hit him at night? Why run seven raids using the exact same tactics? Even a recruit knows you never do the same thing twice in a row on a counterinsurgency battlefield—you never set a pattern against a guerrilla army.

Where was the Spectre C-130 gunship, which could have made it a lot hotter for all those RPG gunners on the ground? Rangers report that the Spectre crew was R and R-ing in Italy. Why then didn't Garrison use the Cobras back at the airfield? He explained later he preferred to use the Little Birds. But these light choppers just didn't have the punch. Garrison also told the Senate inquiry that he had all the airpower he needed and that if he put down another round of steel it would have sunk the city.

Casey's father, a combat-experienced aviator, doesn't buy Garrison's alibi. Larry Joyce points out that the Black Hawk was the wrong aircraft for the mission. It didn't have the right firepower and its unwieldy turning radius left it a clay pigeon for Aidid's RPGs. A

Black Hawk doing tight circles at seventy-five feet might as well be a stationary target. Special Operations people were also upset because they knew the six previous raids had left what they called "a footprint" that Aidid could follow to blow them out of the sky.

Aidid was too good a guerrilla to let himself be zapped by choppers and not figure out how to strike back. In 1992, I watched USMC Cobra gunship pilots hose down Aidid's Technicals and I thought, *Hey, we're hitting these guys with total impunity, but one day Aidid will wake up and blow those slow ducks right out of the sky just like the Viet Cong did.*

There were more tough questions that needed answering. Why were none of the services doing more about mines? In Vietnam 15 percent of our casualties came from those nasty devices. The impulse that led directly to the raid, that triggered President Clinton's temper tantrum, was a mine that blew up a Humvee and ultimately killed six Americans in Somalia. From up in Martha's Vineyard, the President ordered the raid on his own without checking first with Colin Powell, who had been advising him to proceed cautiously.

On the ground, I think General Montgomery was the commander responsible. When the mission in Somalia changed from feeding to fighting, he failed to realize his organization was largely a logistics operation, all tail and almost no teeth. He did know he needed more combat power, I have to give him credit for that. But instead of standing tall and demanding the tanks before allowing the Rangers to jump off, he went ahead without them after Aspin said no.

Aspin's view that sending the armor would be seen as unacceptably escalating the war put diplomatic double-think ahead of military common sense, a deadly mind game that got our Rangers cut to ribbons. Montgomery's written request for tanks covered his flanks professionally, but George Patton would have said give me tanks or send in a new general. Any real war fighter worth his dog tags would have refused to put his troops in harm's way unless they could be protected. General Montgomery's letter offered onion skin, not armor, to Casey Joyce, to Sergeant Lorenzo Ruiz, who died not knowing how badly he was hurt, to Specialist Smith, who bled to death waiting for the relief column, to Master Sergeant Gordon and Sergeant First-class Shughart, and all the other brave warriors who were killed or wounded.

According to a basic principle of war—unity of command—General Montgomery was the boss and he could have vetoed the raid. Why didn't he? Only a few weeks before the mission I had asked to accompany the Rangers. At the time, I was told General Montgomery didn't want reporters with "his" Rangers. That is a direct quote. But after the disaster, they were no longer his. They suddenly belonged to General Garrison and General Hoar.

A few days after the disaster I spoke to Major David Stockwell, a public relations officer for General Montgomery.

"You know, Dave," I said. "It was General Montgomery's Ranger battalion when I couldn't go, but now they've got chewed up, it's no longer his Ranger battalion."

"Hack," he said. "You've got to ease up on General Montgomery. It's a very funny chain of command here. General Montgomery doesn't command Special Operations."

Meaning, I suppose, the mess hadn't happened on his watch.

"Bullshit," I said. "He was theater commander and you know as well as I do that he could have vetoed that raid. He could have told Garrison to cool it because the troops were not protected—because they would be too vulnerable."

"Yes, we could have vetoed it," Major Stockwell said. "But he wasn't in his headquarters when it was planned. He got a phone call just thirty minutes before the raid was going to be launched by General Garrison."

Garrison, meanwhile, was running around playing cops and robbers. The Pentagon couldn't really justify two major generals on the scene with such a small force on the ground. Garrison's job was to run the Ranger task force. Imagine a cigar-chewing two-star ramrodding only 400 to 500 guys. It was as if he didn't trust his very seasoned commander, Colonel McKnight. Talk about rank inflation and micromanagement. Montgomery didn't have enough time to really look into it. According to Major Stockwell, he just said on the phone, "Yeah, go ahead and do it, fine."

I didn't buy the argument that the fault lay with gung ho Special Operations people. Special Operations soldiers are like good attack dogs. If the master holds the leash, there's no problem. We need fighters with someone like Matt Ridgway to hold them back; not dancers and prancers. The problem is the vast majority of top brass are not war fighters. They are salesmen—charming, intelligent, aggressive salesmen, selected because they can sell Congress on tanks, airplanes, missiles, and ships. A war fighter simply would not have allowed such a high-risk operation to go beyond the planning stage. But the salesmen didn't have a clue.

General Garrison's real mistake was the same as General Montgomery's. Earlier he'd asked for tanks and for C-130 gunships to give his men close air support in case they got into trouble. That was the right thing to do. But when the request was denied, he should have refused to conduct the operation. Instead, he tried to do it on the cheap. In this case, cheap cost America 18 dead and 100 wounded.

After I wrote a piece in *Newsweek* criticizing General Montgomery, Major Stockwell wrote a letter to the magazine trying, on his general's behalf, to serve up another scapegoat. This time he argued Somalia was the first U.S. peacemaking mission for the United Nations, and the principles of war didn't apply, especially unity of command. He went on to say General Montgomery had requested the tanks but never received them. He added that while there were 4,000 U.S. troops in Somalia, we had only 600 riflemen. And he concluded by saying General Montgomery couldn't use armor that was UN armor because he didn't command it all, conveniently failing to mention that Montgomery also happened to be the Deputy Commanding General of all UN forces in Somalia. He added that General Montgomery didn't command the Ranger battalion in Somalia, General Joe Hoar did.

That was too much for me. I wrote the major a note thanking him for the letter.

"Did Aidid write it?" I asked him.

General Montgomery and General Garrison were to blame yet they were not held to account. General Montgomery was promoted to three stars. During the Senate inquiry, General Garrison did accept responsibility. Some of his men defended him, calling him a soldier's soldier. As one Ranger sergeant told me, "For the rest of his life he will bleed for what happened."

But Larry Joyce doesn't buy that. He says, "Initially, I gave Garrision the benefit of the doubt; but the more Rangers I talked to, the clearer it became that he had no good reason to launch the raid the way he did. The tactics were completely flawed. Garrison was a cowboy going for his third star at the expense of his guys." So the leader who made every dumb mistake in the book was put in charge of the Army's Special Warfare School at Fort Bragg. The shift was a lateral move, not a promotion. But it was still like taking the guy who flunks bonehead English and making him head of the English Department. Can you believe it? After this disaster, General Garrison wound up *teaching* special warfare, when he should have been demoted to PFC and put in the 3rd Infantry Regiment's stable shoveling out the horseshit.

Somalia offered us Vietnam in microform: a wrongheaded President, generals not standing tall, the wrong strategy, the wrong tactics, the wrong weapons—and the wrong body count. Why in the name of God do we keep doing this, plunking our best young people out in places where they can be turned to mincemeat? Only incredible valor saved the day for the survivors. The Rangers put up a heroic fight, some of those guys taking two or three hits, patching themselves up, advancing, still fighting, never surrendering.

Again, brave young men paid the price for dumb commanders who didn't know what the hell they were doing. They just rolled the dice and threw our Rangers on the battlefield and got them killed.

The mission was one more costly example of the way Aidid absolutely outguerrillaed us. We had arrogantly written him off as a camel driver, a shallow thug who didn't know what he was doing, a third worlder who could be swatted aside. In reality, he was not a fool but a well-trained professional soldier and a student of war. He was a guy who had done his homework. He had gone to Soviet military schools, studied Mao and Giap, and he knew exactly what he was doing. As smart soldiers fared, he was sharp enough to see the pattern in the earlier raids. He was completely capable of thinking, *There's where they are weak, this is when they change the guard, this is their security situation, this is the perfect time to attack.*

He had the Bakar Market area divided into sectors with early warnings when the Americans were on the way. He had broken his territory into a checkerboard with centers of gravity and reinforcement and ambush lines. He had mastered the same kind of bush communications the Viet Cong had used against us. We had our high-tech intelligence system—all the satellites, all those listening devices—and he didn't have high-frequency anything. He used the cheapest stuff going, but with it he could communicate with his commanders. There was no way we could use all our shiny snooping gear to break his electronic communications system because he didn't have an elaborate one. We had no idea what was going on, and we couldn't break him. And like everywhere else in the New World Disorder, we had few spies on the ground. In warfare, the most important player of all is the spy who breaks bread with the chieftain.

But there we went again, going into the third world thinking our technology and firepower guaranteed our invincibility. In the beginning we were able to blow away any Somali gunmen crazy enough to point a weapon at us. But that couldn't last forever. Aidid figured out how to set up helicopter ambushes just as the Viet Cong had figured it out. Back then, they would stake three guys out in a field with a machine gun, tell them to fire on anyone who passed by, promising them the People's Hero Award 1st Class if they didn't make it through the night. The "volunteers" would open up on anything that flew over, drawing in our gunships, while out on the flanks the real Viet Cong gunners were waiting. When our choppers came in, they got zapped—thousands of them. The guerrillas in Afghanistan used the same technique against the Soviets. The Russians lost so many choppers and fighter planes and men they had to abandon the war. Aidid knew all this. He knew it better than we did.

If you want to out-G a G you have got to become a G. The first requirement of becoming a G yourself is knowing when to shut the lights off so the other side can't see what you're doing. To steal the night from him you've got to use his techniques. One great technological advantage we do have is our night vision gear. It may cost a lot of bread, but it lets us rule the night.

"The daylight raid was too dangerous," I said to Major Stockwell at one point.

Stockwell's response was surprising. He was a Vietnam vet, a former grunt, too. Besides being a master spinner of Army-speak, he must be a charter member of the CRS club.

"We ran a daylight raid before and got away with it," he said.

"With an Aidid you can get away with it one or two times if you're lucky," I said. "But don't ever try it three times."

President Clinton, that master of communication, has done everything he can to erase his fingerprints from this disaster. But our new Information Age just won't let it happen. When he sent "personal letters" to the loved ones of the Rangers who were killed, the families—they had formed a network—compared them and found that they had gotten a virtual form letter. Many of the parents hoped to meet with him, but he gave them the brush-off, even though he was meeting with the parents of a little girl kidnapped and murdered in California and a Japanese student who was killed in the South. "He wouldn't meet with the families of the soldiers who had died because of him," Larry Joyce told me. After the President awarded the Medal of Honor posthumously to the widows and families of Master Sergeant Gordon and Sergeant First-class Shughart, Shughart's widow said to Larry, her father-in-law had told Clinton that "he wasn't worthy to be Commander in Chief"—and refused to shake his hand.

The President hunkered down, avoiding the parents. But Larry Joyce stuck to him like a pit bull, writing, phoning, networking, going public on television, working *Larry King Live* and *Dateline*. He demanded answers and Senate hearings. At first all he got was the old bureaucratic runaround: Don't call us, we'll call you. Then, on the very day Aspin took the fall for Clinton, Larry got a call saying the President hadn't forgotten him. His own view is that the call was simply damage control: The aim was to preempt him from going on the tube again and pointing a finger at the White House.

Then, for nearly six months, the line once again went dead. In May 1994, the President saw a copy of a prepared statement Larry was going to make before Nunn's Senate Armed Services Committee. The testimony was devastating. To head off the attack, the President invited the Joyces and the Smiths to meet him in the Oval Office. They insisted that he also invite the Pillas.

The meeting lasted for forty-five minutes. The President, Tony Lake, the National Security Adviser, and George Stephanopoulos, Clinton's major fixer—obviously bored, he sat to the side and kept looking at his watch—were all there in the Oval Office. They sat down around a table. "What sticks in my mind about Clinton was his insincerity," Larry told me later. "He never once expressed his condolences." The President's endgame was to distance himself from the action. "He wouldn't let us say anything," Larry recalled.

The President opened with a ten-minute monologue saying that he was glad the Senate was holding hearings and that he was closely following them. He told them that Colin Powell had advised him on taking office to leave military decisions to military commanders. He said he didn't want to repeat the mistakes of LBJ in micromanaging military decisions better left to commanders in the field.

"When he finally took a breath, that's when I jumped in," Larry told me. Pushed beyond endurance, the grieving father demanded to know whether it was true Jimmy Carter had contacted Aidid well before the raid and reported to the President that Aidid wanted to negotiate. The President nodded and said that diplomatic moves had been undertaken after Carter's intervention.

"If that was the case, Mr. President, why the raid on October third?" Larry asked. "Why continue to hunt Aidid if you were about to negotiate with him? Why would you put American lives at risk if you were working on a diplomatic track?"

Larry said the President looked stunned and added that when he learned of the operation he had turned to Anthony Lake, the National Security Adviser, and asked, "Why did they launch that raid?"

Jim Smith, like Larry a Vietnam war hero—whose son had bled to death—also went on the attack.

"From then on, Clinton was on the ropes," Larry remembered. "Jim Smith ate him alive. He virtually shouted, 'If it's worth doing, it's worth doing right.'

"The whole misadventure in Somalia was not in the national interest," Larry told me. "Its objective was never clear. To lose eighteen soldiers in one day, and say the next that we are pulling out, meant those lives were lost in vain."

The President's main concern was his own image, a preoccupation that led him into a campaign of massive spin control. In the months that followed, whenever Mogadishu came up, both the President and Vice President Gore offered a carefully scripted and rehearsed response. They were into total damage control, chanting their commercials. Dan Rather challenged the President on *48 Hours,* asking him what he had to say to a father who believed his son had died in vain. Larry pressed the same issue home against Gore on *Nightline.* In both cases, the rote reply was that the Rangers had not died in vain, that their deaths had saved thousands of Somali lives. The President, forgetting that *he* was the one who changed the mission in Somalia from feeding to fighting, is now trying to hide behind George Bush's earlier humanitarian operation.

"That is horse pucky," Larry told me. "The Rangers didn't go over there on August twenty-second to save lives. Their mission was to get Aidid. They went at the directive of the President to capture a warlord in the back alleys of a city where he had once been mayor, for Pete's sake. It was an impossible mission. They didn't have the resources to do the job."

That's exactly what President Clinton doesn't want the American people to know. And we are sure to see a lot more of this cheap manipulation as the 1996 political campaign heats up. The cover-up began with Nunn's Senate hearings. The Senate investigators faulted Aspin on the decision not to send tanks. They found Garrison and Montgomery wanting. And they were also critical of Colin Powell. But the results of the investigation were bottled up for over a year. Senator Carl Levin of Michigan, a Democrat fronting for the President, did all he could to bury it. I had a Deep Throat in a senator's office giving me the inside dope. The logjam broke when Senator Strom Thurmond, the South Carolina Republican, went public and attacked the Democrats for blocking the report. In the end, the report was released in October 1995, just a few days before the second anniversary of the raid. Even the White House knew it couldn't let two years go by without running into trouble. The report came out at nine-thirty on a Friday night, buried at the very bottom of the weekly news cycle in a press conference attended by three reporters. If you are wondering why you never heard the details, you can be sure it was because no one around the White House wanted you to.

I don't know which is worse: the attempt to duck responsibility or the image of a Commander in Chief so out of touch that he allowed his foreign policy and military aims to take entirely different paths—both fatally mined. In a conversation with Larry Joyce, Colin Powell blew Bill Clinton's principal alibi. Micromanagement wasn't the issue. For the President, the real issue was whether his left foot knew what direction his right foot was taking. It didn't. He abdicated his main responsibility—troops in combat have got to be sure their actions are consistent with foreign policy. But Powell told Larry Joyce that the President didn't even inform him about the Aidid peace feeler. Because the President never told the military, they could not alter their mission. If the President had not defaulted on his responsibility as Commander in Chief, Colin Powell could have told him to bring the Rangers home, or at the very minimum, the military could have canceled the raid.

Joyce cut the military people in charge no slack. He nailed Admiral Jonathan Howe for pressuring Colin Powell to use military force to back up the United Nations after it put a $25,000 price on Aidid's head. Initially, Powell resisted an armed Rambo posse, but in the end he bowed to political pressure and caved in, the same way he did at the end of Desert Storm. "He is truly a political general," Larry said. To give Powell credit, the retiring Chairman of the Joint Chiefs gave Bill Clinton full warning. Just before the battle, the last time he appeared in uniform at the Oval Office, he said to the President, "Get those Rangers out of Somalia."

The President ignored the warning. At the White House meeting with the families, Gail Joyce broke down in tears and told the President her greatest fear was that her son and the other Rangers would be forgotten. Casey's widow, DeAnna Joyce, showed the President a picture of Casey in uniform. Then she showed him a picture of his headstone at Arlington National Cemetery. At the funeral, Aspin, the guy whose decision to deny the tanks had gotten the Rangers killed, had wanted to present her with Casey's casket colors. DeAnna refused. She told the Secretary of Defense she wanted a Ranger to do these last honors.

"I hope Bill Clinton knows that when he puts American troops in a combat environment he has a solemn obligation to put them at the top of his agenda and not relegate them behind his political agenda," Larry told me. "He simply didn't do that for our brave young people in Somalia."

And it cost them their lives.

CHAPTER 7

FLIMFLAM WAR:

KOREA, 1994

Saturday had rolled around and I was up in the DMZ with a Republic of Korea tank brigade. General Chun Chi Ha was showing me his units, the barracks, the training fields. His Korean-made tanks were not built for guys my size, but I could poke my head in, and I saw they were first-rate. The general was sharp, a stud. His English was better than mine. He had graduated from our armored school at Fort Knox and the prestigious Command and General Staff College at Fort Leavenworth. After showing me around, he asked me to brief his officers.

"What we want," he said, "is for you to tell us everything about the Korean War. You see, we have no one left who fought in it."

That knocked me out. *My God, it has been forty years!*

So I spent the next hour in his office briefing tough young officers who were eager to bone up for the next round.

Suddenly the phone rang.

He picked up the receiver and I saw his jaw tighten.

The balloon just went up, I thought.

He set the phone back down and said something in Korean to an aide. Then he looked at me.

"Kim Il Sung is dead."

I leaned forward.

"What did you say?"

"Kim Il Sung is dead. What do you think?"

"I don't know whether to dance or cry," I replied. Jimmy Carter had defanged old hard line Kim; now we had the son, Kim Jong Il, and the CIA had always painted him as a true loony tune.

The general calmly ordered his units to a high state of readiness. Then, after ten minutes, we reassembled.

"We'll continue now with the Korean War," the general said. "Your lesson for us was . . ."

As if nothing had happened. What cool. We were a couple of miles from the front and right astride the North Korean main avenue of approach.

I played it cool, too, but it was tough—I realized I was probably sitting on the scoop of the year.

Dead or alive, Kim Il Sung and I go back a long way. The first time I fought him and his troops he had only been pumping up the North Koreans for about five years after World War II. But by the summer of 1994, when clouds of war came rolling down the Korean peninsula, he had been in the saddle for nearly half a century. Everyone was in his pocket, everyone brainwashed—from great-grandfathers to nursing kids, and these people were born fanatics. Maybe it's something in the water. The North Koreans have always been superstitious warriors. They like ceremonies, rituals, auspicious dates. June 25 was the anniversary of their hell-for-leather attack on the South in 1950, and as the summer of 1994 approached, a lot of people were afraid they might try it again. The director of the CIA and the Secretary of Defense were saying the North Koreans had the bomb and were in a Pearl Harbor state of mind. The pulse of Washington was fluctuating wildly. But I checked with one former U.S. commander in Korea and he told me he saw a lot more smoke than fire. Then I called a SEAL officer in Korea whose job was to keep an eye out for early signs of war and he told me he was going to Hawaii on R and R to meet his wife.

It got harder and harder to read the true temperature of the crisis. Finally, at the beginning of June, I asked "General" Parker to send me to Seoul as *Newsweek*'s forward observer. A week later, Maynard called. "All right, Hack," he said. "Go over and take a look. Find out what's really going on."

I knew I had to get over there right away. The weather late in June would offer the North Koreans perfect conditions for an attack. The monsoon overcast would still be in their favor. They needed to strike before blue sky summer. "Once the skies have cleared," my best Air Force source said, "we could establish air superiority within the third to seventh day of any attack." So I grabbed my basic weapon, my trusty laptop, and headed for Seoul.

Everything I had read in the papers, seen on Larry King, heard on radio and television, was so hot I thought I might have to fight my way off the airplane and run to a bunker, like landing at Sarajevo. *Wasn't this a bubbling pot?*

So where were the blackout curtains? Everything seemed calm.

On June 21, 1994, I walked out of Kimpo airport into a stinking hot day. I could not recognize Seoul. In 1950, I had been with a reconnaissance company that fought the rearguard action out of the city. We were the last Americans to leave. Our job was to delay the advancing Communists, to deceive them and disorganize their attack. The idea was to make them think they were attacking a regiment when we really had no more than three lightweight platoons. Our orders were to hold out until the main forces withdrew, then scoot across the Han River and blow the bridge. Five minutes later, the Han bridge was concrete in the water and splinters in the air. An entire platoon of my company didn't make it. Only four guys survived, one being Dick Adams, currently president of the Korean War Association, who knew every Korean general—retired and active—and was generous enough to outfit me with letters of introduction that were pure gold.

The last time I had seen Seoul, the only thing left alive were the rats feasting off the corpses buried in the ruins. It was desolate, empty. Now it had become a city of fifteen million people with skyscrapers, wall to wall people, gridlock, a glittering sphere already spinning into the twenty-first century.

I dumped my bags at the Capital Hotel. The place was about twenty minutes from military headquarters and they were nice to old colonels. They gave me a good rate and a chance to bump into military folks—to get lucky. There were no other reporters. The Capital was not a whiz-bang hostelry. We are not talking the Hilton. This was a little, older hotel where a lot of warriors hung out.

A retired general and an old buddy from my Vietnam days who knew Korea like the front of his West Point football jersey had lined up a lot of the top South Korean brass for me to see, along with some aides to General Gary E. Luck, the superstud who wore three hats:

Commander in Chief of the United Nations Command; Commander, ROK-U.S. Combined Forces Command; and Commander, U.S. Forces in Korea. General Luck was a true, no-bullshit gunfighter. Beyond that I had many contacts with retired high brass who had served in Korea and who kept an eye on the peninsula. So before I even made my first recon, I started with a good intel network. I was able to wander around all our military installations. The average reporter would have been intercepted at the gate and turned over to the tender mercies of the Thought Control PAOs. But I could smile and show the guard my military ID. He would glance at it and write me off as a military retiree—there were a lot of old warriors around because you could rent a mamasan, buy cheap booze, and live off the fat of the plushest PX in the world. Any time headquarters offered to give me a briefing, I would say, "No, thanks. I'll get my own briefings." I knew exactly what a canned propaganda briefing was worth. I'd rather eat an MRE in the dark.

I came back to my old battlefield prepped by the Clinton administration and a lot of tag-along analysts who were forecasting real trouble. About the same time the head of the CIA was wailing that the North Koreans had the Bomb, spooks from the Company were leaking word that the North had tested and deployed a new missile called the Rodent. The implication was that the North Koreans only had to strap their nuclear warhead to that missile and good-bye Seoul, good-bye Tokyo, good-bye San Francisco.

My gut was sure they didn't have the Bomb, but I had hard data showing the North had a mean inventory of chemical and biological weapons and the means to deliver them: missiles, bombs, mortars, and artillery shells. And even though the United States had pulled its nukes out of South Korea, they were there just offshore on carriers and submarines. If we got into a war, the minute the North Koreans pushed their NBC (nuclear, biological, chemical) button we would counterpunch with nukes in thirty seconds. The prevailing winds in Korea go from north to south. The fallout from a nuclear-bio-chem exchange would drift downwind, poisoning South Korea, parts of Japan, and the whole fucking worldwide environment Clinton and Gore supposedly spend so much time worrying about.

The bean counters in the State Department and White House didn't understand the first thing about Korea. I knew if we tangled with the North Koreans, we weren't going to see another hundred-hour war. The North Koreans aren't your ordinary enemy. They are crazy. You never want to bait these guys. There's no one meaner than a North Korean with a gun. No one in the world can really read their hand, they are that nuts. I have seen them attack positions when there was absolutely no way they could win. I have seen them barefoot, barely clad, with only a few rounds of ammo, with nothing to eat, sure to lose—and they attacked like Marines hitting a beach. They do not understand fear. This is not a tiger whose tail you want to twist.

To make things even worse, the terrain on the Korean peninsula is a bitch: mountains threaded with rivers, rice paddies that sink armor, roads and ground that are easy to mine and so exposed you are dead the minute the sun comes out. When it comes out. A lot of the time the weather is lousy. The ground fog in the valleys socks you in and the constant overcast keeps you from using your air, especially fast burners, to stick it to the enemy. You could not pick a worse place in the world to fight or a tougher opponent. In summer you melt; in winter it drops down to 30 to 60 below zero. I have been there when all our weapons froze. You had to kick your M-1 bolt back with your foot to chamber a round. At night we set out our grenades, ammo, and trench knives in the berm in front of our foxholes, so they'd be handy when the little bastards attacked. In the morning, they would be frozen, superglued, to the dirt. Over eight thousand American soldiers fell in the grim winter of 1950 to frostbite. In my company, we lost a minimum of one soldier a night to frostbite.

No one could win a new Korean war. It was obviously a war none of the belligerents should fight. But the minute I got on the ground, I could see signs we were involved in a serious military buildup. I saw augmentation teams everywhere. You could tell them from their shoulder patches. I would go into an office or club and suddenly would see 4th Division patches here and I Corps patches there and 101st patches over yonder. Out on the airfields I saw Air Force squadrons that weren't supposed to be in Korea, fighter aircraft from Japan and the United States. We were deploying Apache gunships and other new weapons systems. Munitions were pouring into the theater. The war drums were thumping, and the braves were starting to put on war paint and dance.

Early on I thought, *If these guys are for real, we're going to be bringing in a lot of stuff by ship.* So I took the train down to Pusan, where I found the harbor filled with vessels unloading new artillery pieces, tons of ammo, and other kinds of gear. After checking out the port, I made the rounds of the bars where the merchant sailors hang out. They told me their ships were filled with munitions and war goodies. They were saying among themselves if war broke out, they would never sail these waters, there weren't dollars enough in the world to entice them should the guns begin to toot. No question. The Pentagon wasn't just huffing and puffing, it was bringing in whatever it needed to blow Kim's house down.

Once again we were deploying Patriot missiles. I had real doubts about how well they worked. I kept finding more evidence of the sort I gathered in Desert Storm that we'd been completely snowed about these particular toys. This one was called the PAC-3. A buddy who had been at Raytheon with the Army's Patriot project team wrote it off as a "bad joke that couldn't hit an elephant in the ass at ten paces."

I thought, *Okay, here goes another fortune into the toilet.* But the Patriots at least looked like a big chip on the table. My spies told me after the Patriots came in, the North Koreans started developing an even worse case of bunker nerves.

The North Korean order of battle showed their command and control were still highly centralized along traditional Communist lines. They had 26 infantry divisions, 23 motorized or mechanized brigades, 14 tank brigades, and 3,700 main battle tanks and 500 light tanks. Their air force had 14 MiG-29s, a first-class fighter, and 36 SU-25s, good enough for close air support. The rest consisted of 750 planes, mostly 1950s and 1960s vintage stuff. It was junk. I talked to one general who said we would have the greatest junkyard of Red aircraft in the world after all of these things crashed at the end of their airstrips.

Every time I got on the phone to the United States, I could tell nerves were getting tighter than the slack in a hair trigger. When the Korean crisis lit up, the television people started dragging out the same experts and retired commanders they used for Desert Storm. All of them were back, pontificating about what was going to happen. None of them knew what the hell was up any more than they'd known during the Gulf War. Not that any of the retired brass would tell the truth if they did. They didn't want to lose their places at the bar at the Army and Navy Club in Washington.

When I was in Korea the last time, forty years ago, the worst thing a GI wanted to hear was "Fix bayonets." Now the worst thing would be "This is Tom Brokaw reporting from the DMZ."

Looking back, I was damn lucky to get out of Korea alive the first time through. It has always been an easy place to cash out.

About three weeks into my trip the phone rang at four o'clock one very dark morning. Groggily, I picked up the receiver.

"Hack, Hack, are you alive?"

The voice was a woman's and it was angelic. I thought I was dreaming. I slapped myself everywhere the way you do after an artillery round comes slamming in. Arms okay, legs okay, balls still there. Brain still working.

"Yeah, yeah, uh, I think so."

It was Diane Pearson, *Newsweek*'s charming publicity director. A reader had called to report my death. I felt the way Mark Twain did when the newspapers printed his obituary and he said, "Reports of my death are greatly exaggerated." Apparently the reader had confused me with another infamous Army colonel, Charlie Beckwith, the guy who was the Delta Force commander, an old friend. Charlie had just checked out of the net, not me. I rolled over and went back to sleep.

I knew General Luck was too smart a warrior to slug it out with the North Koreans toe-to-toe. I thought he would try to hit 'em where they ain't, the smart way. During Desert Storm he had commanded the XVIII Airborne Corps. His troops were by far the stars of that show. They ripped through Saddam like a chain saw through cottage cheese. My guess was he would adapt the Air-Land Battle concepts of Desert Storm to achieve high mobility and to develop killing zones where he could shoot Kim Il Sung between the horns.

That was not as easy as it sounds. The first problem was there were too damn many places where the North Koreans could pop up and fool you. The second was the terrain, which was not desert flat, and the North Koreans were well dug in. For sixty miles, they had wall-to-wall Siegfried lines. The third was the South Koreans, supported by our side, had developed an attitude toward never giving up an inch of ground. All that sacred Korean soil. With this in mind they were deployed nose to nose on the border. This meant the North Koreans could pull their lanyards and pin us down with the first shots of the fight.

According to the best estimates, the North Koreans were capable of putting down the greatest barrage in human history—a million rounds of artillery. I worried that during the first twenty-four hours of any fight, the sharp edge of the army, the finest part of the regulars, would be destroyed. In my own view, it made more sense to man the front lightly with second-rate reserves, keeping the best south of the Han River, where the North's artillery could not clobber them, where the likelihood of getting hit by air was minimum. That way you could keep your strong right arm cocked and ready for a counteroffensive. Others had doped out the same approach, but no one had been able to change the mind-set of South Korea's politicians and generals. So our combined forces were deployed in a way that meant we would lose the best in the initial barrage and assault and counterattack with our second string.

Some people were expecting big tank battles. I didn't see it that way. Korea isn't tank country; it's infantry country, crosshatched with forests, rugged mountains, and deep valleys. There's no room for tanks to stretch out. The roads are easy to block with barriers. Refugees will clog the line of advance. During the Korean War one of our biggest problems came from the hundreds of thousands of civilians fleeing south, a sea of people frantically surging forward, running for their lives. At night panic would drive these people into our battle positions. They would run through a wall of fire and mines, getting killed as they rushed to escape from dying.

I knew in any new war, the North Koreans would have the advantage of surprise. They could concentrate their forces and punch wherever they pleased. The defenders were stretched thin because they could not be sure where the North would strike. The attackers were sure to experiment, to try the unexpected, and they could easily punch holes through our forward wall. I thought what we were most likely to see was a fight in the mountains against the best, toughest light infantry in the world. Those guys have little legs like hydraulic pumps. They can run up a hill with a seventy-pound pack on their back and not miss a stroke. And they would be supported by 100,000 special forces troops—the meanest of the mean who would have infiltrated over the days, weeks, or even months leading up to an assault.

The air war possibilities were as murky as the sky in monsoon season. Our side could outfly and outgun the North. I had no doubt about that. In good weather, the North Koreans would be lucky to get their Air Force off the ground. One American general told me our intelligence showed Kim Il Sung could only afford to give his pilots seven hours of flight time a year. Ours got two or three times that every month.

But good weather is not Korea's long suit and no one would be dumb enough to attack while the sun was out. During the monsoon, our air superiority might not offer the immediate and decisive advantage it gave us during Desert Storm. The North Korean air defense system was undoubtedly stronger than anything Baghdad had been able to throw at us, and the air defense over Baghdad had been one of Saddam Hussein's few real bragging points. The intelligence picture showed the North Koreans had almost

9,000 air defense guns and perhaps 10,000 surface-to-air missiles: SAMs from the Soviets, shoulder-fired weapons, 90- and 120-millimeter AA guns, .50-caliber machine guns, ack-acks, the works. During the Korean War the flak was worse than anything our pilots had seen over many cities in Europe in World War II, and the North Koreans had been working for nearly fifty years to make it even more deadly. They had built what the spooks call a "robust" air defense. Very robust. Meaning they could put a wall of steel over their positions.

At one point I drove out to Osan air base about forty miles south of the Han River. I knew the landscape well because I had spent a lot of time there freezing my ass off when the North Koreans were not trying to put it in a sling. And I had picked up my first Purple Heart at Osan on February 6, 1951, a day so socked in even low-flying Corsairs and P-51s could not save my Wolfhound unit from the meat grinder.

As soon as I arrived, a motor-mouthed PAO intercepted me and tried to brainwash me with all the usual stuff about how airpower would prevail just as it had "in the desert." No way to avoid him. He stuck to me like a leech. That was his job: to try to distract me from seeing a fucking thing. But as we were rambling around the base, I noticed a lot of F-16s that weren't supposed to be there. Down the way a colonel was standing next to an A-10 Warthog. So I walked over to find out why all the fighter jocks were in town.

"You've got my favorite airplane," I said, hoping to crank up a conversation.

"It's my plane," he replied, giving away nothing. "I'm an A-10 driver."

That didn't get me very far, but it was a start. It turned out he was the deputy commander of the base. And I instantly liked him and marked him down as a truth teller. A little ladder ran up to the cockpit, so I climbed up and looked inside. The first thing I noticed was the airplane did not have night flying gear. The colonel had told me his A-10s were forward air controllers. They were the eyes and ears of the F-16s that provided close air support to the troops on the ground. But with no see-in-the-dark capability, the colonel was like a blind man telling the fruit pickers where the peaches were.

So I looked back over my shoulder at this very shiny, bright boy colonel.

"Does this aircraft have the LANTIRN [low-altitude night targeting infrared navigation]?" I asked.

He stared at me, then looked down at his shoes for a few seconds like an ashamed little boy caught playing with himself.

"No, sir, no sir. I'm afraid it doesn't," he said. "We don't have night vision capability."

"Then how are you going to fight the night battle?" I asked him. "Because, Colonel, let me tell you something about the guys you're going to fight. They ain't gonna fight at high noon."

The discovery about our night vision capabilities, or vulnerabilities, surprised me. We had better planes and more of them. If the North Koreans attacked, it was a royal flush cinch it would be at night. We had on station F-16s and A-10s that would be blind. After spending billions for satellites and Star Wars and the F-17 and the F-22 stealth fighter, were we really going to wander around Korea blind? This flat didn't make sense. The thinking was our air was sure to stop them. All I can say is, airpower didn't stop them from 1950 to 1953 even though we controlled the skies. Back then nothing flew over our lines except Bedcheck Charlie, a single-engine mosquito that dropped mortar bombs. After the first few days of the war, that was their airpower. South of the peninsula, we swatted them out of the sky all the way up to the Yalu River. In the North, we obliterated their capital, destroyed their railroads, their crops, their road system. And they still kept coming.

Psychologically and tactically, it seemed to me, airpower was making us too confident. Yes, we were sure to have air superiority, but the prevailing overcast that time of year would offset it. We also appeared to be forgetting how good the enemy was with a shovel. Over the years the North Koreans had dug themselves so deeply beneath those rocky hills and mountains we no longer had a clue where they had hidden some of their divisions. They had caves big enough to stash entire battalions and regiments. They also had batteries of rockets and artillery emplaced in tunnels bored into granite mountains and protected by steel blast doors. This is how it worked: The door comes up, the weapon blasts off, and the door slams shut. We return fire, even with deadly accuracy, but it only bounces off those granite mountains and steel doors.

We also had to figure out how to neutralize Kim Il Sung's submarines. "It's the holy shit factor of the war," one commander told me on the sly. The subs lived in deep pens built into the mountains just two days' sail from the major ports. They would come sliding out to sneak down the coast and blow up ships in the ports on D-day. The first sign the North Koreans were on the way would come the moment those steel doors swung open to let the submarines out.

The smart thing to do would be to blow the doors and trap the subs in their pens. Failing that, we could send SEALS to attach early-warning devices that would pick up the sound of the screws turning as the subs glided out. But the war plan didn't allow for either course of action. As things stood, we wouldn't know the subs were heading toward us until it was too late to stop them. A top SEAL planner said, "We're fucked."

The North Korean subs could lay mines, block harbors, spread chemicals, deliver special forces. This threatened our plans for reinforcing the two very lonely American brigades on the ground with the six U.S. divisions scheduled to gallop to the rescue when things got tough. It reminded me of the NATO plan to reinforce Europe in case the bear came charging over the Iron Curtain. Just as mines had kept the Marines from landing during Desert Storm, mines would hold up our sea lift to Korea. Even after the mine fiasco in the Gulf, the Pentagon had chosen not to give high priority to new minesweepers, so we did not have enough offshore in Korea. We tried to sneak some over from the United States, but the North Koreans screamed hysterically that minesweepers meant war, and the four-ship flotilla never got near Korean waters.

Just before I went to Korea I heard the Chief of Naval Operations admit that mines could be a show stopper. He was absolutely right. We don't have a competent minesweeper fleet. We still haven't appropriated money to solve this problem. But even though the admiral running the Navy had said mines could stop us, we didn't do anything about it. Instead, we bought more Sea Wolf subs and carriers. As I have said before, minesweepers are not sexy. Bio-chem gear is not sexy. Spare parts are not sexy. What the hell, the Pentagon must think, when we need them they'll magically appear. *Sure they will. After we lose the opening rounds of the fight and fill a container ship with body bags.*

North Korea's special forces presented a real nightmare. They were well trained and hard charging, the elite of the North Korean military. Kim Il Sung had over 100,000 of these guys, totally dedicated, totally fanatic. And they could arrive by boat, plane, or tunnel. The spooks were saying the North Koreans had predrilled more than two dozen tunnels under the DMZ. We had found only a few of these. I drove up to the DMZ one day and took a look at one so well constructed the North Koreans could run a regiment through them—3,000 armed men, three or four abreast, all pounding forward hell-for-leather, ready to pop out and yell, "Surprise." Behind the immediate front lines, they would create instant chaos, shutting off our artillery, blowing up command posts, cutting support lines, blowing up airfields and radars, all the tricks they had mastered in their postmortem from Korean War I. If they were able to attack us from the rear, we would have to waste a lot of combat power securing 360 perimeters and ferreting out the murderous bastards.

These same special forces could infiltrate by slipping down the coasts in boats. They also had a little wooden aircraft called the Colt capable of carrying nine passengers that could not be picked up by radar. We had always held Colts in contempt, but the damn things could well bring in troops whose mission would be to blow up roads, ports, airfields, command and control posts. They could assassinate leaders, tear up communications systems, and turn the rear into a living hell. During the Korean War, I had seen a few well-placed snipers hold up a battalion of men, killing three guys here, four guys there, stopping everyone else. It did not take much of an imagination to see what the rear would be like with thousands of these guys, all well-trained fanatics, blistering us from behind every rock and tree. Look at the chaos only a few hundred Nazi infiltrators were able to create during the Battle of the Bulge in 1944.

How well would the ROK Army stand up under fire? Our two brigades were not going to stop much. The real question was the fire the South Koreans had in their belly. But getting a good look at the ROK warriors was about as easy as sneaking into the War Room at the Pentagon. No one wanted to talk. As Thought Cops they had more practice than our guys and for a while I was so frustrated I thought they might skunk me. For more than three weeks I fought a running battle while they stonewalled me. Their message was no American reporters got to hobnob with the ROK.

"I have to do this," I said at one point. "I can't go back after looking at two brigades of the U.S. Army, and tell the American public we can defend Korea, when the South Korean Army has four to five million men under arms and I haven't checked on one unit. It doesn't make sense."

It did not get me anywhere. Finally, in desperation, I called a lieutenant general named Won Tae Hwong. An Army friend had given me an introduction to him. It was my ace in the hole, and I was so desperate I played it.

"Please come over, my friend," he said.

He did not need to invite me twice. I got over there as fast as rubber could carry me. After some small talk I hit him with my written request to visit his army. I told him the American public affairs officers had all supported it. I asked for clearance to visit the

ROK 1st Division. I wanted a background interview with the division commander and a chance to meet with officers, NCOs, and soldiers and visit his soldiers both on the DMZ and in training. I told him my primary interests were readiness, training, unclassified information showing how the ROK Army would defend South Korea. Finally, I wanted to speak to the Minister of National Defense, the ROK Chief of Staff, and the Chairman of the Joint Chiefs of Staff.

He raised his eyes from the paper and said nothing. I admit I had given him quite a list, but there was no other way to find out how close we were to war and how well we would fight. This time, the PAO guys had supported me, especially Luck's deputy PAO Jim Cole, a straight shooter who had not hit me with the routine smoke and mirrors jazz. On top of that, I was able to say I had served alongside the ROK 1st Division during the Korean War; it was South Korea's premier fighting division and I reassured him there was no reason to think I wanted to hurt them.

"You guys are not being fair to the American people," I said. "I want to see your units. I fought here. The United States has spent a great fortune here. Its youth have died on your fields protecting your land and now stand ready at your flank to again fight and die for your country."

I was damn mad.

"What am I gonna tell the American people?" I shot out at him. "Here's what I'm gonna say, 'Why are we here? This is a dictatorship.' You're afraid to let me see. You don't want anyone around to tell the truth. There must be something wrong, something you are hiding." I was speaking loudly in short machine gun bursts, spraying every button in sight.

Finally, he said quietly, "I will try my best. I have spoken to the Minister of Defense. Someone is blocking your request. They simply don't want American reporters visiting our units."

I told him the defense of Korea would be determined by the quality of the ROK units and the American people should know what they were paying for and what they were about to get into. Afterward, I thought I had overplayed my hand. But then, bingo, three hours later he rang up to say permission had been granted for my visit. The orders were to let me see whether or not the ROK Army could defend the country.

It did not take long in the field to see the South Koreans had assembled an enormously powerful force. Their officers were gung ho, dedicated, well trained, highly disciplined. The Old Army spirit of the Americans was still alive in their protégé, the ROK Army. The units were made up mainly of conscripts. They were not very happy about being there, but they were disciplined, well trained, and knew how to fight. The leaders, mostly academy guys, knew how to move and groove.

During the Korean War, I had fought alongside ROK units and relieved some of them in the trenches. What I was seeing now represented a 10,000 percent improvement in their army. Forty years ago, most of their units were raggedy and badly disciplined. Blow a bugle in front of them and they would cut and run. The army I was seeing had completely transformed itself into a combat-ready, gung ho force.

They were building their own tank, the K-1, a beautiful machine designed to fight in mountainous terrain. They were hoping to go into the arms export business using it as their crown jewel. But too much of their energy and money was going into high-tech military stuff and not enough to the forward combat troops, just like the U.S.A. Their weapons upfront were older than ours and their warriors just as vulnerable. Most of their artillery was obsolete towed stuff made in the 1950s and '60s. They would be no match in an artillery dual with their brothers from the North. They were spending billions on things they did not really need, like submarines that couldn't cut off North Korea's supplies—the main supply lines all come overland from China—and gold-plated, superpower wonder weapons.

General Won had told me he thought it would be very hard to blow the North Koreans out of their bunkers. They were really dug in. You could pound them with air and everything else you had but as soon as the plane flew away or the artillery stopped, they would be back fighting. His biggest worry was their artillery, which was first-rate. He thought their greatest vulnerability would come from internal economic pressures. The South was betting that when the North Koreans became hungry enough they would revolt. Not me, at least not in the short haul. At present, Kim Jong Il and his Stalinist sidekicks are too well entrenched to fall to an internal revolution fomented by the widespread starvation racking the country. But there's always a chance he could use the U.S.-promoted war talk to attack in a reckless attempt to keep his twenty-three million enslaved people toeing the line.

After a few weeks on the ground, I started to think, *Wait a minute.*

Something was wrong, very wrong. It didn't grab me at first. I had hit the deck running, checking out the worst dangers first. Now my peripheral vision began sending me signals, little blips through the blur.

Seoul was within easy missile and artillery range of the DMZ, right in the beaten zone for any war, only twenty-five to thirty miles from the North Korean heavy stuff. The minute war broke out, it was certain to be hit. But in the streets no one looked worried. In the towns where I was poking around, everywhere I went—looking in the shops, the factories, the garages—everybody was watching the World Cup soccer matches. They were not glued to the tube expecting a war. They did not have any war jitters. They thought the crisis was one big yawn.

Then there was something funny about all the signs of augmentation I was seeing every time I turned a corner. Ordinarily the first thing you do if you are on an augmentation mission is rip off your patches. In Vietnam while on recon, I used to cut off my jump wings so no one could identify if I was Airborne or Special Forces. It's basic security, like taping up the bumper number on your vehicle. Seoul is full of spies and clearly we meant for them to see all the new faces and those patches. The signal was blunt. A warning that we meant business. But there was something about it that smacked of bluffing. It was all a little too much.

I rewound my memory and started rethinking some of the other things I had seen. It occurred to me I had not noticed any significant countermeasures to protect the ports or airfields other than a flimflam Patriot show. We really had no means to keep the vital channels open, the pipelines that would carry in reinforcements by sea and air. We would be just as vulnerable as we had been in the Gulf, more vulnerable because the North Koreans were not toothless and gutless like the Iraqis.

What the hell was going on?

I did not know how to read it. *Were we really expecting a fight or were we shadowboxing in the dark?*

I began to do some recalculating. The South Koreans had used forty years well to harden their defenses. They had moved entire mountains and dug huge ditches for tank traps. Every road leading to Seoul was rigged with two-ton blocks of concrete set with explosive charge. In front of the barbed-wire-stitched lines they had set out millions of mines, enough to blow Kim Il Sung halfway to Beijing if he came skipping across. That is, if the annual brush fires have not melted the fuses. To get at the entrenched South Koreans, the North would have to leave their bunkers and come out in the open. They would be exposed to a terrible potential killing machine. Assuming they penetrated the forward positions and hauled ass, they would still be faced with an obstacle that has stopped dozens of invaders since even before the invention of gunpowder—the mighty Han River. After the Han was crossed they would have to punch down a wide flat valley and fight at the end of a badly stretched supply line. And air could pulverize them all along the way just as during the Iraqi attack into Khafji.

It was still true the enemy could move out of its forward fighting positions with almost no notice. But they would not be able to punch very far or sustain a long drive. Years ago, the North Koreans had organized a campaign to make themselves self-sufficient in ammo and just about everything else, but they were really hurting on fuel. They would probably run out of juice about fifty kilometers south of the Han River. My insiders told me they only had enough stocks to wage war for ninety days. Since the fall of the Soviet Union, the North Korean Army had been coming up short on spare parts for its all-Soviet-equipped Army. Any army runs on spare parts. In our own, an Abrams tank gobbles up one spare part every fifty kilometers. The shortages had forced the North Koreans to cut down their training. They're in almost the same predicament we were in before World War II when on maneuvers we had to write "Tank" on the sides of trucks, send our infantry on attack with broomsticks for machine guns, and drop flour sack "bombs" from rickety biplanes.

Just before I shoved off for Korea I had lunch with my old friend and Whitefish neighbor, Bill Carpenter. Bill, recently retired, had just refused several well-paying jobs with the military industrial congressional complex. One of the rare high-rankers who has not cashed out to cash in, he has held to his view that the purpose of the U.S. military is not to support corporate America or to provide former high-ranking military guys cushy jobs within the MICC, trading on their contacts. Bill Carpenter is not for sale.

During a recent Korean tour, where he commanded all American and South Korean ground forces, Bill had been headquartered in Uijongbu. He told me it was now a booming metropolis. Because I had spent a three-Purple-Heart year within fifty miles of Uijongbu at bloody battlefields such as Kumhwa and Chorwon, I marked it down on my list of places to see.

A few days after I hit Seoul, I drove about twenty-five miles north of the capital to the little town I remembered so well. The last time I had passed that way, the entire place was flatter than a paratrooper's belly. Back in 1951, the only man-made structure left was the bank's big safe. It stood right smack in the middle of that demolished town, a solitary symbol of protest against the insanity of war. When I drove up, I was staggered. Ground zero was gone. What I saw instead was a modern skyscraper city with tens of thousands of thriving people, high-rises, and modern buildings from horizon to horizon. *What a difference peace makes,* I thought.

Beyond Uijongbu, the road divides. The left fork goes to Panmunjom, but ten miles down the right fork is a spot I will always call Objective Logan, the scene of one of the bloodiest fights I have ever seen in a lifetime of exposure to the battlefield. As I passed through Uijongbu, I realized I didn't care about that boomtown. My real purpose had been Logan all the time. My trusty little rental car, as if on automatic pilot, took me to the base of the rugged mountain where everything came rushing back.

March 31, 1951. Fog and mist covered the steep slope that led to the top of Objective Logan. Lieutenant Gilchrist was now the platoon skipper, replacing the lieutenant who got killed after the lieutenant who got killed after Lieutenant John Land was killed on February 6. In Korea, as in every war, rifle platoon leaders were expendable; if they lasted for more than a month, they had rare luck. Gilchrist was a lucky old pro who had somehow survived for five months in this war and several years in World War II. We trusted and loved him. He always got the objective, never talked down to us grunts, and most important, never wasted us.

Taking that steep, boulder-covered fortress would challenge his every skill. As we neared the top, our artillery lifted and from then on it was rifles, frag hand grenades, bayonets, and pure grit. The Chinese were dug in and planned to stay. The hill was so damn steep and narrow only a single platoon could be used in the assault. The other platoons had ringside seats on higher ground about five hundred yards to our rear from where they whacked in recoilless rifle, mortar, and machine gun fire when they saw an opening. The rest was up to us.

The Chinese hit us with a wall of mortar, howitzer, and heavy machine-gun fire, some of it raining down on their own positions. Somehow we breached their line and got a toehold on the crest of the hill, and then our guys rolled hand grenades into their bunkers from behind, giving us a big bite of their line. By now we were bleeding badly and running low, real low, on ammo.

I was a weapons squad leader and managed to get our .30-cal Browning machine gun into a shell crater. A two-war veteran Kentucky boy named Thatcher was at the trigger when the Chinese launched the first of many counterattacks, and he mowed the attackers down like a farmer cutting summer wheat. Short, well-aimed bursts and lots of courage. Ray Wells, the assistant gunner, kept the machine gun well fed while I pointed out targets, and plunked away at the attackers with my M-1, praying reinforcements would hurry up to fill in the firing line and pump us up on ammo.

As with most infantry battles, the fight centered around the machine gun. It was our big stick and Gilchrist wanted to protect it: It was the Chinese's worst enemy and they wanted to destroy it. So it became the focus of the battle. The enemy threw everything it could at us including the kitchen wok, but somehow Thatcher kept that red-hot Jesse going. Shells thumped down all around us, and overhead the incoming bullets buzzed like a swarm of yellow hornets gone mad. A slug or piece of shrapnel smacked into the side of the gun, knocking it over. Thatcher coolly set it back up and continued to spit death and destruction into the enemy ranks.

Help arrived and the first guy I saw was a 2nd Platoon warrior, my dear friend, Jimmy Aguda. He yelled as he ran by the back of our hole, "Hey, Hack, big fight here, huh?" Jimmy was a Browning Automatic Rifle man from Oahu with balls the size of coconuts and he could make that twenty-two-pound killing machine sing. Now the 2nd Platoon guys were flopping on the ground all around us, putting heavy fire down on the enemy and throwing us ammo. It looked like we might stick around for a while. Then another attack came. You could hear the Chinese leaders yelling and whistles blowing. Mortars thumped in, one taking out a complete squad in the 1st Platoon. Artillery blistered the hill. It was bedlam. I peeked over the hole's edge and saw a mob, a steamroller of Chinese troops surging up toward us. Then brave, reckless, lovable, and Hawaiian-crazy Aguda stood up and walked to the crest of the hill as casually as if he were strolling down Waikiki for the surf. He crouched low like a boxer, turned the BAR on its side, jammed the butt in his hip and made that bad mother sing. He fired magazine after magazine, his rounds mowing down the attackers. All the time he was shouting at the Chinese, "Come on, you motherfuckers, come and get me."

I yelled, "Get down, for Christ's sake, Aguda, get down!" It seemed all the enemy fire was now directed at him, but he just did not give a shit. He just stood there facing that hail of fire like some hero in a Rambo movie where only blanks are used and the actors get up after the scene is shot. The enemy attack faltered and then ran out of steam. But suddenly there was a final burst, all aimed at Jimmy. I could see Jimmy get it—a slug in the leg, another in the arm, and all the while he kept pouring slugs into the withdrawing enemy. Then he took a burst in the chest, dropped his BAR, and cashed in his chips.

As the fight cooled down, Gilchrist organized the defense, although there weren't that many defenders left for him to set into holes. The 3rd Platoon of Company G was now cut down to a small squad and we were all exhausted. We evacuated our wounded and dead, redistributed ammo and started enthusiastically employing a grunt's favorite weapon, the spade, going for cover, deep cover, digging down well below ground.

Darkness brought more enemy attacks, but by this time our heavy artillery was pumping out in front of us and we were in good shape. Then came the command "Hold at all costs," the worst thing a grunt can hear, especially if the opposing commander has ordered, "Take that hill at all costs." Next, as is so often the case in the confusion of battle, we were suddenly told to beat feet in retreat. Word was the companies on our left and right hadn't taken their objectives, and the battalion next door on our east had gotten pushed off position. Mighty Company G—G for gallant, that is—had its head stuck way out and, fortunately for us, some smart staff officer figured we might just get it permanently lopped off.

It was after midnight by the time we pulled back. By then the Chinese were pressing our positions. Artillery and mortars crashed down right in front of our foxholes, disorganizing their attack. We walked backward, blasting the bastards with our rifles as they crested the hilltop. It was kind of like shooting lobsters in a fancy seafood restaurant showcase.

I was still about a hundred yards from our new front line when I ran into two men carrying a wounded sergeant in a poncho as if they were about to dump him. I leveled my rifle, gave one of them a jab of cold steel with my bayonet, and said, "We'll take him back, won't we guys?" They both became pretty enthusiastic litter bearers after that, but I stuck with them all the way to the line just in case they had a change of heart. Gilchrist was the last guy off that mother of a hill. He was covered with blood both from his own wounds and those of a soldier who got it by mortar fire and died in his arms. Again he had defied the odds. Several hours later as we were digging into our new positions, his only bitch was he'd still have to wait another hour before he could light his beloved pipe. We had been standing up to our ears in blood for almost twenty-two hours when I sat down. A few minutes later the word was passed. "All platoon leaders to the CP to get the attack order."

On the drive back from Logan I could see Uijongbu skyscrapers in the distance. I had a sudden flash. In 1951 you needed a bulldozer to wade through this town—and then it had been only a farming village. Its road network made the place a strategic chokepoint, the way Bastogne had been so strategic to Ike and the Nazi generals in World War II. Now, more than forty years later, I looked at Uijongbu and thought, *There's no way the North Koreans could ever get through all these built-up areas if they were zapped. Lay down a few artillery barrages, whack in a few air strikes, and this place would resemble Uijongbu in 1951. But the pile of rubble would be more like Mexico City after an earthquake. Nothing on wheels or tracks could push through it. The attackers would be stuck dead in the center of thousands of tons of rubble, and the critical roads would make Kuwait's Highway of Death look like the Indianapolis Speedway. Any ground attack would turn into sheer gridlock. Nothing would get through.*

That night in Seoul I had dinner with a retired Korean general, who told me the real estate boom was an essential part of the South's battle plan. The military was calculating the craze—and pressing need—for real estate would drive the high-rises right up to the edge of the DMZ. If the North Koreans wanted to fight, the government's idea was to blow the buildings and block the way south. In 1950 the North Koreans reached Seoul in a day and a half. The hope was next time gridlock would keep them back at least past rush hour.

I could see some real problems with this line of thinking. Gridlock would also keep the defending army from moving forward to reinforce or counterattack. Mechanized forces wouldn't be worth much. The conditions would be perfect for light infantry. Remember, the North had those twenty-six light divisions. That is why I found it a little unnerving when so many ROK officers, from generals on down, would keep asking me, "How did the Korean War go?" They had forgotten their own history and wanted to pump it out of me. This was not a good sign.

By now I was feeling confused. So I decided to return to Go. I began by taking stock of our armor and what I found surprised me. One day when I was out in the 2nd Infantry Division's motor pool, I saw a trooper washing an M-1 tank. After he finished, I opened the commander's hatch, and there was a puddle of water. *What does that tell you?* I asked myself. We were up against an enemy with a powerful capability to go NBC—nuclear, biological, chemical. If water can get inside a tank, you do not have to be Bill Gates to figure out that all those little germs and nerve gas could wiggle right in there, too. Once again the Pentagon and CIA were talking war and we had an Army equipped with tanks so old they were not NBC-proof to protect the warriors. Bad enough the gas masks in Desert Storm, now in Korea it was the tanks and the gas masks.

I spent about a week with the 2nd Infantry Division checking out their tanks and artillery units. I discovered they all had fifteen-year-old M-1 tanks—the same ones that flunked the course in Desert Shield—and even older artillery pieces.

Why do we have old stuff here if the threat is so hot? I began to ask myself. *Why does the hottest theater get the coldest priority?*

The tube artillery was all self-propelled and at least thirty years old. I spent a day with one battery and found 10 percent of the guns were laid up because they didn't have spare parts. So I went down to the supply depot for a walk around. Things were tidy and military efficient. I asked the captain to trace how long it took to get a part from the States. Replacement time was about three days, which was incredible. But when I asked why the 155-millimeter guns were inoperative for three weeks because of no spare parts, I was told, "Sometimes there's a glitch in the system." But this "glitch" was over essential parts for broken war-fighting gear that got the same or less priority than spark plugs and computer discs. What was needed was a "Bluestreak" system for the important stuff—and, most critical, clearly prioritizing what was the most important.

But there was a glaring underlying weakness as well. The supply system was dependent on FedEx.

I don't want to slam FedEx. They're so efficient maybe we should get them to run the whole Pentagon. But I couldn't help remembering all those smart merchant sailors who said they weren't going to sail into a Korean harbor if the shit hit the fan. The airfields would be taking it too. Would FedEx, so gung ho and cheerful in peacetime, still deliver the goods when the slugs were flying? This looked to me like a super glitch in our logistics system. The only way you could discover it was to go down to the warehouse, but I didn't see either reporters or generals anywhere near a supply depot.

I spent a lot of time in the field with the infantry watching them attack simulated enemy positions with their vintage M-60 machine guns constantly jamming. The troops had the SAW, the squad automatic weapon. But the ones I saw were also jamming. Here again we were spending billions of dollars on Patriot missiles and other cute stuff that might not be worth beans, but the guys who would do the real fighting still had hand-me-down gear from the 1960s, stuff that was flat worn out. The defense contractors and lobbyists for the big-ticket weapons systems were getting rich, the generals who collaborated with them were getting promoted, and our warriors were about to enter the arena with broken swords. This did not make sense to me.

After several weeks in Korea the more I began to wonder whether the pot was really boiling or whether someone was just leading the American people on. The reality was the North Koreans will always present a threat. They've been growling, snapping, and biting for forty years. But was I really seeing something new and dangerous here? What was really going on?

At one point during my meeting with General Won, he had told me that while political tension was very high he thought the military tension was fairly low. I also spent a lot of time with retired Korean officers. They were well connected to division and corps commanders and others in high places. I said, "Look, there's got to be something going on here. I'm looking for it and I'm beginning to think there's more smoke than fire." They nodded and said the war talk was all hype.

All right, I said to myself. *Let's take it one at a time.*

The Bomb scare was hairy; so did the North Koreans have the Bomb or didn't they? Why would the head of the CIA get in such a sweat that his talk show performances had scared the American people half to death? I checked back with one of my Army sources who was tracking the story for me through his own CIA connections. He told me CIA Director James Woolsey's intel was coming from the KGB. *From those lying bastards? Moscow Center is now a good guy? Do we really buy that? Have they forgotten Aldrich Ames already?*

Then I ran into a spook whose job in-country was to track the North Koreans.

"They don't have the Bomb," he said. "It's bullshit."

After that I took a much closer look at Kim Il Sung's Rodent missile. The CIA had made it sound like the rat that would eat Asia. I discovered the Rodent was the mouse that roared. It was actually not much more than an improved version of the SCUD, a little better than Saddam Hussein's boomerang but nothing like an effective intermediate-range pinpoint missile, let alone an ICBM. The CIA was hyping the Rodent just as they were the Bomb.

Who could you believe?

Before going to Korea I had lined up a truth squad. On it were a U.S. Navy commander who knew everything, top guys in the Marines, the Army's 2nd Infantry Division, and Special Forces. All good men. I asked them to sort it all out. Our own battle plan was supposed to be twenty-five inches thick. I did not have it in front of me, but I could reconstruct the essentials. What the Navy guy didn't know, the Marine guy did, and what he couldn't tell me, the Army guy could.

I got together with my logistics guy.

"Our depots are in no shape to fight a major war," he said. "We can't move a lot of the munitions because we don't have enough forklifts. The transportation system is lousy. We have millions of rounds for the 106mm recoilless rifle, but the rifle itself is no longer in the inventory."

I met my truth teller from the Army.

"The supply system here is broken," he said. "We ain't got the trucks, cargo handling gear, and the rail and air facilities to move the stuff that's needed when it's needed. General Luck knows this and is fixing it."

Wars are built on logistics. I had no doubt the two U.S. brigades we had in Korea were lean and mean and could put up a good fight. The ROK Army forces looked good. But if the logistics systems were this sick, there was no way we could fight the kind of war that was raising everyone's temperature back home. It was one more sign that something was terribly wrong. Our tanks were leaking; the artillery was not good, FedEx was bringing in the parts and the war plan was screwy because it let the North Korean subs close down our ports. But then we did have a million and a half rounds of vintage 106-millimeter shells in the supply dumps. No rifles to fire them, of course. But they would make a great explosion if the North Koreans ever hit our ammo dumps.

The American Army in South Korea was there to provide a trip wire, not a defense. Normally, the 2nd Infantry Division had three brigades, but only two were stationed in-country. The third was at Fort Lewis, Washington. That gave us 6,000 fighting men. The rest of our 34,000 soldiers were ash and trash support troops. One day I was sitting in the office of a Marine Corps colonel in Seoul getting a rundown on what our semper fi guys were doing when Jim Cole, General Luck's flack, came bolting in.

"Hack, the general wants to see you," he said. "And the general wants to see you *now.*"

I hadn't asked for an interview because I never hang out around a general's flagpole, but by then General Luck knew I was snooping around and he had sent Jim to find me. General Luck is a real war-fighter, down-to-earth, a grunt's kind of soldier. The troops love him and I thought we were damn fortunate to have him there right then.

I was curious to hear what he had on his mind, and when a general sends you that kind of invitation it's not smart to keep him waiting. So in the car I went and off we scooted to the head shed.

A full colonel, an aide to the general, was posted in the outer office and General Luck was waiting at the door to the inner sanctum.

"Hi, Hack," he said. "How are you doing? Come on in."

He was a short beefy guy with a salty voice and arms like a weightlifter's. He had come up through Special Forces and he had been a Green Beret, but he also had a Ph.D. The complete warrior, not another Pentagon General Play-It-Right. I liked him straight off. He sat me down on a couch and took a chair, a very informal guy, completely confident, no stuffed shirt, he talked like a GI.

"How's the trip? Where you been? What are you seeing?"

We shot the breeze for a while.

"My army's bigger than my army," he said.

What he meant was that the ROK Army can muster five million soldiers. South Korea has 1.2 million men in its regular armed forces and 4 million in the reserves, double what the Pentagon can turn out on a good day.

We talked about the American component of his massive army, the 2nd Infantry Division. When all three brigades were together, the division was the most powerful in the U.S. Army. Major General John Abrams, the division's commander, was leading a division that was not just there to train. The grunts could see a real enemy, just across the DMZ, who might at any moment come pounding down on them. Many were veterans of Grenada, Panama, Desert Storm, and Somalia. But at the same time, 30 percent of the troops turned over every three months. No unit could be a cohesive hard-hitting team with that kind of turnover. So in the overall scheme of a real shooting war, there was no way those two brigades could be anything but a pimple on the ass of Korea, 6,000 fighters among 10 million regulars and reservists mustered by the South and the North.

General Luck saw the division's mission as a deterrent. The hope was these troops would deter the North the same way the 360,000 troops we deployed in Europe deterred the Soviets during the Cold War—with just a little help from thousands of nuclear weapons. This was all standard stuff. I have always thought the deterrent argument was just good propaganda to justify keeping forces on the peninsula; it is good for the Army to have a hot mission, something that justifies another division and a couple of billion bucks a year.

Then we got down to business. I told him I thought the 2nd

Infantry Division was deployed way too far forward. He dodged easily, letting it glance off, changing the subject. Korea's sacred soil was something he was stuck with the same way all earlier American commanders had been stuck. What was the point of getting into it? I had gone into the interview with a full house because I already knew most of what General Luck now told me. He said the North Korean gear was old, vintage 1950s and '60s stuff. Spare parts were a serious problem; so was fuel, which limited how far their armor could roll. They did not have a strong logistics tail. They were short of food.

What General Luck really wanted to do that day was look back over the Korean War for any lessons that might be useful to him.

"How did you close down their artillery?" he asked me. "What about their mortars?"

I told him the artillery back then was not that bad but we had never been able to get to their mortars. They had them dug into little fucking cat holes on their back slope where only the tube stuck out and you had to drop one right down the pipe to shut it off. We talked tactics for a while, and then I decided to drop one on him.

"I'm getting a lot of reports that your logistical system is fucked."

He shot me a hold-on-there-soldier look.

"Yeah," he said, "we've had some problems, but we're shaping it up. It's under control."

He firmly denied any serious difficulties, but something in a soldier's eyes, his and mine, told me he was bullshitting the troops. Maybe I had gone too far. But there are ways of saying things that don't take words and I had been in the Army long enough to translate. Neither side was in great shape to take on the other. North Korea was unraveling just like Cuba. They weren't getting the support, the oil, and other crucial goods from their old sugar daddy, the Soviet Union. They were still crazy, yes, but not that crazy. They were not coming.

The truth, I could see now, was that our political leaders had gotten us into a very dangerous game of chicken. No one really knew what was going on. Not General Luck. Not the Communist leadership. Not the Pentagon and not President Clinton. No one knew. The only certainty was any fight we stumbled into would be the worst since Stalingrad. The ferocity and tenacity of the enemy, the possibilities of nuclear, biological, and chemical weapons, all made it dangerous in the extreme to flirt the way they had been flirting with war. With so much ranting going on, deploying the Patriot missiles was a particularly reckless thing to do. The North Koreans were watching the deployment. They did not give a damn that the missile was supposed to be for air defense. All they knew was a Patriot was a missile. Their spies had to be saying, "Look, they're unloading more artillery, more tanks, all of the ammo in the world. They're coming forward. They're conducting maneuvers. And now they're deploying missiles."

You can bet your boots the North Koreans had spies in our military headquarters, and they were in the bars scoping out all the new guys snooping around. They were not big on high-tech stuff, but were more like Aidid in Mogadishu and the Viet Cong, both of whom used little people to get hands-on information from human sources. The South and the North had become like two huge munitions trains headed toward one another on the same track. All the war talk from Washington was provoking a collision engineered by both the politicians and the media.

It does not take much sense to know that even in the Hermit Kingdom the insiders were watching the tube and hearing all the huffing and puffing and saying to themselves, *We better increase our state of alertness. The only superpower left is about to blow us away. We better go with the first strike.*

The United States was spending $300 billion a year on toys and boys and spies and lies; the North Koreans were spending $2.1 billion. What did that tell you? The North Koreans had a big army, pound for pound the South's equal. They were not well equipped, but they had lots of spirit, and they were capable of one hell of a banzai attack, were willing to attack even when they knew they would die. What would happen if some fanatic, a corporal, captain, colonel, or general, decided to do an Oliver North and kick things off on his own? Once the first shots were fired, there would be no way to turn it off. Everyone would start shooting. *There were already reports of people in North Korea rioting over food shortages, and their desertion rates were up, way up. The natives were growing restless indeed.*

Let's not twist this old tiger's tail, I thought. *Better to let the beast die a natural death in his lair.*

Kim Il Sung had been running North Korea longer than Bill Clinton had been alive. He was the oldest established world leader, a Communist George Washington with sixty years to brainwash his people. If he and his son said hop, their people would hop right into hell for them. All I can say is, thank heaven for Jimmy Carter, at least for this one. It makes me laugh how the Beltway crowd groused at him for going over to Pyongyang and having tea with Kim Il Sung. All he did was make peace and spoil the biggest bloodbath of our blood-splattered century.

And then, just as the tension began to defuse, the North Koreans shot down an American helicopter. The chopper got lost. The North Koreans spotted it over their turf and shot the pilot through the heart. Bobby Hall, the copilot, crashed and was taken prisoner. I thought, *How can it be that here in the most dangerous place in the world we are flying a chopper so obsolete the pilot doesn't even know where he is?* Today you can buy a car, press a button, and the global positioning system will fix your location, telling you exactly where you are. We have plenty of military equipment that can locate itself within a few meters in just a few seconds. But not that chopper. With North and South Korea at the boiling point, why were we sending green pilots on patrol in obsolete airplanes that did not have the latest location-finding gear? I just could not comprehend the military logic of this development. The stupidity stood out for all to see.

After Hall was captured, he immediately spilled his guts, forgetting the soldier's sacred code: name, rank, and serial number. He was on television singing the same way Michael Durant did in Somalia. I could not believe it. Courtesy of CNN, he told the world he didn't know why the Americans were there and that they shouldn't be there. Then, when he got home, instead of being cashiered for spilling his guts after getting lost where he never should have been in the first place, he got a call from the President of the United States giving him a pat on the back, and his hometown decked itself in yellow ribbons.

It was nutty. But the administration was desperate for good news and the American people are desperate for heros. We are the only country in the world that offers an incentive package if you get captured, which even includes a little tin POW medal. *A medal, a fucking medal for going in the bag!* When I was a Raider operating deep behind enemy lines in Korea, every man in the outfit carried a pistol. Our mind-set was, if you got captured, the last round went between the eyes. *Now it's hop in bed with the enemy, sing like Pavarotti, and they make you a hero. Makes me want to puke.*

The day Kim Il Sung died, I knew I had one hell of a story and I wanted to get back to Seoul fast. So I finished briefing my ROK Army hosts quick smart. They stopped me for the usual presents:

belts, shoulder patches, badges, a watch. Then I climbed in my car and raced back to Seoul. I had been in the country for a month and by a stroke of luck I was the only reporter from a major American publication in town that weekend. I called my contacts and they all said everything was cool. The Eighth Army had been put on an increased level of alert, but that was not exactly a call for World War III. The restaurants and bars were filled with warriors from the American 2nd Infantry Division. If the 2nd wasn't getting ready to fight, no one was.

Our leaders had just treated us to a phony war.

When Admiral Crowe was Chairman of the Joint Chiefs of Staff he said, "South Korean forces are capable of defending themselves." Well I say, "Why the hell don't they?" While I was there, one serving Army colonel told me the answer to my question was that South Korea was a jobs program for the U.S. military and the U.S. defense industry. Amen. The Korea crisis was a buildup against an illusory threat. Even good guys like Senator John McCain of Arizona were sometimes frothing at the mouth. I have got a lot of time for Senator McCain. He was a POW during Vietnam. But to me it seemed crazy to go to war in Korea even if the North Koreans were getting the Bomb, which they were not. Pakistan has the Bomb. Pakistan is just as nuts as North Korea, but we do not bomb or threaten the Pakistanis. When the time and market are right, we take their money and sell them F-16s.

Speaking of connections, two former South Korean presidents—Roh Tae Woo and Chun Doo Hwan, both also former generals—have been put on trial for corruption, among many other charges. An American defense contractor who brokered F-16s is under investigation by the American and South Korean governments for allegedly bribing South Korean officials. Next to prostitution, the weapons' racket is the oldest game going—and it's more lucrative by far.

There is still a Yankee-go-home attitude in South Korea and Japan. So why don't we?

If the United States pulled out its Army and Air Force but did not cancel the understanding that the South Koreans are our allies, and if we warned the North Koreans not to move an inch across the 38th parallel or we'd return them to the fourteenth century, we could pull our ground forces back to Guam. Why not let the Navy provide the U.S. presence in the form of another Great White Fleet? That would cost less and make us a lot less vulnerable than we are now with the 2nd Division parked where it is.

Why do we station 34,000 Americans in South Korea when that country has built a crack army of 5 million men? Why are we spending $2 billion a year keeping our forces in South Korea? It in no way makes sense to me. Right now, the 2nd Division is deployed in the shadow of the DMZ well within artillery range and well within missile range and well within reach of the North Korean special forces. When the whistle blows, our soldiers sleeping in their barracks are supposed to run to their vehicles and drive to their assembly areas, get their act together, then go to their fighting positions. Based upon what I saw, I don't think they would even make it out of the barracks. If North Korea ever decides to march south, it's not going to say, "Ready or not, here I come."

The only warning we'll get is the sound of ten thousand cannons firing simultaneously.

If I were a North Korean commander, I would aim my guns directly at the American barracks. At 0-dark-hundred, I would open fire. Within a few minutes most of the Americans would be dead in their bunks. Then I would hit the motor pools to kill whoever survived. Finally, I would have my special forces block every road to the assembly areas. Those killers would set up ambushes to wipe out anyone left. Anyone with any military sense can see that the 2nd Division is on a suicide mission. Why should those warriors be there? They can't save the South Koreans, who are perfectly capable of saving themselves. The only reason they are there is to serve as a trip wire—policy geeks love trip wires—and as a pipeline for the military industrial complex.

Why not bring those 34,000 soldiers home? Within those two brigades we have about 6,000 warriors. The rest are support troops. Of these, 3 in 10 rotate out every few months. The turnover is 120 percent every year. Nobody stays long enough to become proficient and develop the teamwork needed for war fighting. Because every available inch of ground in South Korea is devoted to agriculture, there are few places to train. The rotation policy destroys unit cohesion. But even if we stopped the rotation and kept everyone in place for two years, they could not adequately train. So what's the point in putting them there?

The irony is that while we're stretched too thin on a suicidal mission, the South Koreans are going full tilt to build a stronger military. They're buying the latest Apache-type choppers, F-16 jet fighters, night vision devices, global positioning satellite systems, secured communications systems, radars, new artillery missile systems. While we beat the war drums, they're saying, *We better buy a lot more equipment.*

Whom do they buy it from? From the same guys who are beating the war drums. Which is why we had our little phony war.

CHAPTER 8

GREAT WHITE GOD: HAITI, 1994

"Colonel Hackworth. Come back to home. We have trouble. Come. Come. Come. Quick."

Mita the houseboy was frantic. He was calling on the two-way radio from my base at Plantation LeClerc up in the hills overlooking Port-au-Prince. Frenzy was the norm in Haiti, and Mita was a cokehead. But by his voice I could tell this time the problem wasn't just voodoo or blow. I had no choice. I jumped in the car and hauled ass up to my hideout in the jungle—which resembled a scene from *Apocalypse Now.*

When the Haitians around Plantation LeClerc first sized me up, they saw a lean old guy with white hair in a green camouflage cap. I always wear my Vietnam patrol cap when I go into a war zone. It has three diamonds on the back, which indicate a leader, and HACKWORTH, which says it's me—and on the front it has master jump wings and tabs from the Rangers and Special Forces. The cap helps when you're covering an invasion in shorts and a T-shirt. The Haitians spotted it and from the moment I turned up at Plantation

LeClerc in the late summer of 1994, they began treating me like a godly figure sent down on a spaceship to save them. It made for good company. But then, they expected magic from me, like Red Sea partings or conversations with a burning bush.

When I careened up to the villa, I found maybe eighteen people mobbing a woman and her two daughters. Everyone was wailing, shrieking, hysterical. Mita grabbed me and started translating the cries from high-decibel Creole to raggedy-ass English. It wasn't easy to follow, but the nub of it was that the cops had arrested and dragged away a little kid, maybe fourteen years old. He was the tear-streaked woman's son.

She pointed her finger at me.

"Go and get my boy."

Holy shit, what next? I know what this means. What this means is trouble. Big trouble.

I'd walked past the police station enough times to know I didn't want to drop in. It was a cinder block place no bigger than a good-sized closet and most days nine men were jammed into it. Nine bad dudes with guns. That's where she wanted me to go. The unknown always makes for the worst kind of combat. I'd walked into that kind of trap too many times on orders and it was nuts to volunteer.

I'm never going to see these people again. Their men aren't my men. I mean, we're talking a squad of Haitian Army deserters. I'm not responsible for them.

But I could not say no. Honor was at stake. The Great White God—and the first time I'm put to the test I chicken out?

No fucking way. Impossible.

"Get your weapon and let's go," I said to Sergeant LeClerc.

My loyal sergeant was standing there watching to see what I would do. I had hired him as my bodyguard and named him after the place where I'd holed up, the ruined villa of a famous American dancer. He'd deserted the minute he learned American paratroopers might be on the way to Haiti and it didn't take much time for eight of his friends to follow him. I enlisted them all in the LeClerc Division, lovers not warriors, the sweetest, sorriest motherfuckers I had ever had under my command.

Sergeant LeClerc grinned as he tucked his pistol in a back pocket under his red T-shirt and fell in. I told the rest of the crowd to stay behind. One of the boy's sisters was about fifteen—we took her with us—and Mita, feeling a lot bolder now, decided to tag along.

We went through a broken-down wall, trotted along a narrow alley, made a sharp turn and came out at the police station. When I walked in, the kid was sitting against one wall. I could see his white teeth through a bloody lip ripped open from the nose down.

If he doesn't get stitched up fast he's going to lose half his face.

I could relate to him. Almost fifty years before, an Italian lieutenant smacked me in the mouth while two of his goons pinned my arms, ripping my mouth in the same way. The morning after, my lower lip was as big as my fist. But the lieutenant didn't get away without paying a price—as he struck, I crotch-kicked that sucker so hard he's probably still choking on his balls.

The police sergeant was sitting there looking at me. He had a pistol stuck in his belt and a little Swedish submachine gun on his lap. Definitely in charge.

"I want this little boy," I said.

"Bad boy. Crazy boy. He says he going to kill police. Crazy. We take him away. We put him in jail."

"He's a little boy. He doesn't know what the fuck he's doing."

Mita was peering in the door watching me argue with the cop—and it blew him out. He was stoned on coke. He was with me. In his mind that gave him power, a license to carry on, and right there while I was getting down to negotiations, he started shooting his mouth off, threatening all those guys with guns. He was stoned all the time. He had guts and I loved him. But now wasn't the time.

"Mita, shut your fucking mouth and don't say anything," I shouted at him. "Say another fucking word and you're outside."

The police sergeant was watching all this. He smiled. So I leaned over and tried again.

"Now, I want the boy."

While I was leaning over the desk, I noticed the cop was staring at the silver pen Eilhys had given me for luck. I was wearing it on a chain around my neck where my dog tags used to be. I was watching his eyes.

Shit, that's what he wants. Eilhys will understand.

"Tell you what," I said. "I'll give you the pen, you give me the boy."

He nodded.

I took the pen off and handed it to him. Then I grabbed the kid's hand and in three steps we were at the front door. But when I looked out, I saw an enormous crowd milling around the police station. Everyone was chanting. Their eyes were actually bulging with fury, madness. And when I looked over my shoulder, the sergeant was standing up and leveling the machine gun at the door.

My God. We're going to have a massacre.

In my mind I raced through the possibilities. One goon was standing just inside the door with an M-1 rifle, my old weapon, and there were seven other dudes standing around, also with M-1s. I knew I could grab the weapon from the thug at the door in about one second, take out the sergeant first, then hose down the rest, BOOM, BOOM, BOOM. But wait a minute.

There's supposed to be eight rounds in an M-1—what if that fucker only has four? I've had it. The Great White God—DOA on his first Haitian mission.

That's what happens when you go out to save the world. I wish I could stop doing it. I wish the whole country could, but I guess it ain't gonna happen.

That summer, while President Clinton was belly bumping with Raoul Cedras and holding hands with Jean-Bertrand Aristide, I made two trips to Haiti, the first in August, then a second in mid-September. The situation had started to go to hell in a handbasket the previous fall only ten days after the disaster in Mogadishu. The President had sent 193 U.S. troops and 25 Canadians aboard the U.S.S. *Harlan County* to Haiti as part of a deal to retrain the local military. Aristide was to return from exile a few weeks later. But a hired crew of Cedras gunmen turned up at the docks growling and waving AK-47s and President Clinton meekly withdrew. The *Harlan County* turned tail and headed for Guantanamo Bay. America was humiliated. We had just gotten slammed in the teeth, a superpower getting chased away by a few two-bit thugs. What a message to their brothers around the world: Wave your weapon, watch the Yanks run. Teddy Roosevelt must have been spinning in his grave. Remember, at the time, we had plenty of Haitian spies on the CIA payroll. If our spooks had been on the ball, they would have sniffed out the demo in advance and killed it, or set up a counter demo.

The ship should have arrived with a Marine escort in one of our small, amphibious chopper carriers. Gunships should have been hovering above the dock, and gunners on the *Harlan County* should have been manning deck guns. A Marine platoon should have marched down the gangway, formed a riot wedge, and stomped forward with fixed bayonets. The bad guys would have taken to their heels like the listless drag-asses they really were. Instead, they kicked hell out of a few reporters, banged around some cars, and shot up a few clouds. It took nearly a year for the President to recover.

The following summer he tried to bluff Cedras off Haiti, keeping the 24th Marine Expeditionary Unit with two thousand warriors aboard cramped ships for almost two months while Cedras gave him the bird. This happened because Anthony Lake, the flake who heads the National Security Council, knows as little about military security as his nerd boss, and General John M. Shalikashvili, the Chairman of the Joint Chiefs of Staff, would rather trot along behind them wagging his tail than provide strong military leadership.

In August, I flew down on my first recon. I arrived thinking an invasion could come within a week. My sources were telling me the operation was on: The 24th MEU would be off the coast; the best weather conditions were coming up, moon and tide. I left World Headquarters and flew through Miami and a couple of Caribbean islands before I reached Port-au-Prince, where Spencer Reiss, *Newsweek*'s correspondent, a good man, met me and introduced me around. I was a complete island greenhorn. In hindsight, I see how naive I was about local realities. I didn't take any gear except shorts and T-shirts. That seemed like the right tropical uniform to me. But General Cedras was a very pompous prick and when I went to his headquarters to ask for an interview, the vibes I got from everyone were, *How can a beach bum like you see the Man?*

His chief of staff and his military advisers kept me cooling my heels for days. I was dumb—should have borrowed a pair of long pants. But then things began looking up. During a stop at the American embassy I bumped into a sergeant whose father had served with me in Vietnam. Then I ran into an old Special Forces buddy who was working for the Company. After that I had a good fix on what was going down.

All the maneuvering over the horizon showed how thinly we were stretching our warriors. The 24th Marine Expeditionary Unit had just returned from a six-month tour off Somalia and Europe. They had been on a very tight string, at sea for six months then home for only three days before they were recalled from leave. They hardly had time to unpack their seabags, kiss their wives, and play with their kids before Clinton rushed them off to Haitian waters. To throw those guys back just as show troops indicated a total disregard for their welfare. The Haitians were such limp dicks that we could have cut soldiers out of cardboard and glued them to the ships and it would have accomplished the purpose of the President's bluff. Instead, we had to take a force just back from a long mission and send them off again before they even had regained their land legs. That's how overstretched our fighting forces are now. And when the President's Tarzan noises didn't work, we had to pull back the 24th MEU.

During that first false start, the press pack had taken over the Hotel Montana in Port-au-Prince. The reporters spent a lot of time chasing gossip and feeding off one another. Among them were a lot of old hands who had been around Haiti for years and really knew the island. But they'd seen so much, all of it bad, that they had grown extremely cynical. Nothing much seemed to move them. I got caught up in that kind of reporting myself. The second time I went down, I knew I had to change my ways.

By the middle of September the signals out of Port-au-Prince and Washington began to look like the real thing. Having wasted a couple of weeks in Haiti the first trip, I waited until the last minute to hit the beach. My intel was good. At World Headquarters in Whitefish I was getting bulletins even before some of the units involved in the mission. All week long, calls from my guys came into Whitefish telling me this time the invasion was on. That Thursday I left World Headquarters and flew down to Miami.

I hooked up with Bill Gentile, the *Newsweek* photographer with whom I was to work. A compact guy in his early thirties with reddish hair, Bill was a first-rate shooter experienced in Latin America with a lot of combat behind him in Nicaragua and El Salvador. Bill could tell even more combat stories than me, I suspect, but he chose not to. He was very quiet, cool, with plenty of balls, and a close reckoner of risks. He would die to get a shot even when he knew no way was it worth dying for. He was a thinking man's shooter and he'd been around the track in Haiti more times than he could count. A good man who knew his occupation was a deadly game.

We flew from Florida into the Dominican Republic, getting in about midnight. Both Bill and I were bone tired.

As we were checking in, a couple of old mates from the press came rushing up.

"Let's have a beer," they said.

When I was young and dumb, I'd have gone drinking with them until dawn, but I don't do that anymore. So, I said, "No, I'll pass." On the way up in the elevator, they pressed and I had to tell them again I was not up for any drinking. I was going right to bed. I shut the door to my room and opened my bag to get out my toothbrush. And there was a knock on the door. I'm thinking it's Bill, who was on the same floor, saying, ah come on, have one drink, anyway with me. I opened the door expecting to see my sidekick.

Standing there was a beautiful, voluptuous Dominican about eighteen years old. She just pushed the door open, came in, and shut it behind her. As she was coming in she said, "Is anybody with you?"

"No, I'm all alone. What do you want?"

"Do you want to fuck?"

My first reaction was to check her body, and she did look great. *But after fifty years I had met a special woman and finally had the brains to make an unconditional commitment. I wasn't going to fuck it up just to fire a wad into a twenty-buck whore.*

"No, no," I said.

"You don't want to fuck?" She couldn't believe it.

"No, no."

"Come on."

"No. No. *Io sposato.*"

That bedroom Italian was the closest to Spanish I could get. I could only think to say I'm married.

"What's wrong?" she said. "You queer or something. You know. Look at me. Touch, touch, touch."

"No, get out of here. I want you out of here."

So, almost spitting on me, she stamped out and slammed the door.

Oh man, how things change.

There were no flights into Port-au-Prince. Invasion fever had hit the city and the airport was closed. I thought at first I would hire a boat and have it drop us off on the coast. But we discovered that buses were still leaving for Port-au-Prince, so the next morning, along with a platoon of reporters, we jumped one. We hit the road at 5:00 A.M.

Because the back door was the only way in, a trip that normally took two and a half hours took us more than twelve. All along the way, two-bit Haitian cops, customs officers, and soldiers stopped us on one pretext or another. Every one of them wanted a couple of bucks. Arrogant bullies, common thieves. The worst I had seen in the third world.

Around nine that night, we made it to Port-au-Prince and the Home-Sweet-Hotel Montana. I was starving. I dropped my bags and went down to the dining room and ordered a plate of spaghetti. I had just lifted the first fork when the evening rumor mill started grinding. There was a tremendous noise as every reporter in the hotel came pounding down the stairs, running to their vehicles, cranking them up, and screeching off down the hill.

"What's going on?" I asked a buddy I'd met on the last trip.

"President is having a press conference," he said.

So I put down my fork and hitched a ride to the presidential palace.

Everyone in Haiti loves rumor and gossip and the press corps in the palace had taken up the local taste for it. We were all leaning forward. Out walked one of Cedras's toadies, Emile Jonassaint, the junta's puppet president. As he took his place in front of the podium, a huge moth flew in and plunked down on the carpet right in front of him.

Voodoo.

Everyone was going like, "Ohhhhhh, omen, man."

The toady looked down at the moth and up at us. He announced Raoul Cedras would stay in power, no U.S. force or threat of force was going to drive him out. That left the next move up to President Clinton.

I knew I had to get out of the Hotel Montana. I didn't want to get trapped the way I had been on my first visit. When the invasion came down, it was possible the press would be sequestered, as they were in Desert Shield and Desert Storm. And I wanted to get to know the Haitians, face-to-face. The next day I met a Canadian named Cameron Brohman, who had come to the island as a student and stayed on. He had lived in Haiti for ten years and he was working as the general manager of Katherine Dunham's estate. The dancer was persona non grata to Cedras because she was supporting Aristide and she had not been able to come back; so Cameron was keeping an eye on her place. He also did some stringing. When he was not ripped, he was a very knowledgeable guy. I asked him if I could make the place my base of operations, offering to pay for my bed, food, and all the rest, and he said sure, why not, fine.

So I checked out of the Hotel Montana and moved up to Apocalypse Now. The estate was surrounded by twenty or thirty acres of triple-growth jungle. In the center behind a steel gate was a stone building with a kitchen and a series of master bedrooms. I set up in one of the bedrooms; Mike Sullivan, a reporter from National Public Radio, took another. The jungle was creeping in, the swimming pool was drained, everything was crumbling apart. Every night you would hear gunfire, and the next morning another corpse would be lying out in the road.

Mike set up his transmitter on the roof, so we had instant connections anywhere in the world through his satellite phone. Down by the kitchen there was a phone that worked intermittently. Three or four times a day I got a call from Eilhys in Whitefish relaying tips from World Headquarters. My guys in the States were calling her the Spymaster and she was passing on their hot poop. We had no lights because all the power was off. But we had candles. Two old women who had worked for Katherine Dunham cooked for us. So there in the middle of our jungle we had candlelight dinners with beautiful Haitian and French cuisine. Every night Cameron would invite an expert on Haiti as our guest, one night a professor of sociology from the university, the next a famous artist. They would regale us to the accompaniment of small arms fire and I would learn about this long-tormented land.

Just beyond the estate was Soray, a section of dilapidated shacks and broken walls. The ground was controlled by Cedras's gestapo. It was a dangerous place to hang out, gunfire all the time. I mentioned the neighbors to Cameron.

"Oh, don't worry about it," he said. "We've got friends around here with weapons."

When I asked if I could meet some of them, he brought in a big, strapping six-foot three-inch guy who had deserted from the Haitian Army. I hired Sergeant LeClerc on the spot as my military adviser and bodyguard. He brought to the game one old pistol and one new submachine gun. He also had some pals with a couple of rusty rifles. I hired them, too. Now I had my own squad. I always kept one guy full-time on security outside my room. I was the best-protected guy in town, except for the junta leaders.

Now that I had my safe house squared away, I did a recon looking for an observation post. I found a really good place up on a hill in the highest part of Port-au-Prince. On one side, I could see the airfield where the airborne drop was sure to come. On the other side, I could cover the bay, where I figured the amphibious operation would take place. Then I went down the hill and started to check out conditions on the ground.

What I saw didn't square with all the macho gas coming out of junta chief Cedras and his henchmen. The first thing I did was check out the airfield and the port. To my military eye, it looked clear no one was going to do much fighting for the Boss. I saw no sign they meant to defend anything. Zilch. No one was preparing fortifications, no one was filling sandbags, no one was digging in machine guns, no one was preparing a counterattack, no one was getting the equipment and weapons squared away and ready for battle. I could see in a minute if the Americans came in hard and fast, they would coldcock these swaggering assholes in less time than Mike Tyson could put down Boy George.

Outside the presidential palace, I saw an armored car. It had a flat tire and the machine gun on top was bright red from rust. It wasn't going anywhere. Yeah, the army was ready, ready to limbo. When I stuck my nose through the wire out at the airport, I could see their choppers were broken-down and most of their planes were parked on the side of the runway gathering cobwebs. Strike off their air force.

After that I took a look at their artillery. To size up any enemy you have to know where the artillery is, what shape it's in, how much ammo is available. I knew they had a fair number of 105-millimeter guns, but the artillery pieces I saw were all in very bad condition. I didn't want to be anywhere near them when they fired. Most of them would probably explode. The Haitians were not doing any crew drill either and zero maintenance. No wonder. Strike off the cannons. By the end of my recon, I could see the Haitian Army wasn't a serious military force. It was Keystone Kops commanded by comedy commandos.

Then Sergeant LeClerc gave me his rundown on troop strength and morale. At the time he had lit out for the hills three or four days earlier, his unit had 500 men in it. Each day I had him telephone his barracks, and each day the unit dropped: from 500 to 200 to 70. Finally, he called and nobody picked up the phone. The deserters were saying the army would never fight for Cedras.

All the time General Cedras was chanting like Churchill about how Haitians would fight on the beaches. He was bluffing, and the CIA could see everything I could. If our own spies were any good at all, they must have been reporting the bluff to President Clinton and that military moron Anthony Lake. Based on Haiti's manifest weakness, we should have been much more forceful in confronting Cedras diplomatically. I would have said, "Look, fella. You and I both know you can't fight your way out of a retirement home. Your military is going to bug out. We are going to land tomorrow at nine o'clock. We are taking over." Cedras was bluffing with only weak cards in his hand. If we raised, he had to fold. We should have read him better.

By now my own truth squad was really warming up. An intelligence type in the U.S. embassy was telling me what was happening with the diplomats. A colonel and a captain in the Marine Corps were feeding me intel from that end. A Special Forces friend working for the CIA knew everything about the master plan for the invasion, how the operation would go, who would move with the first Special Forces, who would knock off the key installations, arrest the key figures, secure the vital areas. The special ops guys had reconned everything to prepare the way so the Marines and the Airborne would have a softer landing. A member of invasion commander Lieutenant General Hugh Shelton's staff who went to almost every briefing kept me up-to-date with his doings. A major from the 10th Mountain Division and another major from Special Forces also contributed to the intel. A captain and a master sergeant in the Air Force and two Navy SEALs pitched in as well. The intel was excellent. With every operation, my network was growing larger.

These guys were not being disloyal to their service or to their country. They were not giving away military secrets. The higher cause that motivated them was concern for the welfare of their warriors. They didn't want to see good men wasted in idiotic operations. They knew they could trust me not to misuse their information. I would not expose them or put any of our troops at risk for the sake of a scoop.

After I heard we had sent in advance teams of Special Forces, I spent a lot of time trying to locate them. My guess was they were hidden in the embassy or staying in the quarters of American diplomats whose families had hightailed it home. One night at a restaurant I saw a big group of guys wearing high and tight haircuts. I knew they were warriors, though they looked too young to be Special Forces. I followed them home and discovered they were Marines, twelve of them, part of an embassy augmentation team that had infiltrated in-country as tourists. They were hidden at a huge safe house. I walked up and talked to the civilian guards, who told me that thirty Marines were quartered inside. But I could never find the Green Berets. Because they are physical fitness fanatics, I got up very early in the morning and went to places where I thought I might find them running. So I was out near the embassy, by a park, everywhere I could think of; but they had gone to ground. Deep. I couldn't spot them.

The night President Clinton talked to the nation, I was out in the garden listening on a portable radio. Aircraft were overflying the town; everyone was running around wetting his pants. I knew for sure the invasion was coming on Sunday because the son of one of my men from Vietnam was in the invasion force. He was feeding intel to his dad and his dad was relaying it to my darling Eilhys in Whitefish. After that operation I made her head of my own central intelligence agency.

"We're going to be sealed," the kid told his dad, who told Eilhys, who told me.

When you are running an airborne operation, you bring everybody to a closed area near an airfield and seal it. Nobody goes in. Nobody goes out. You get your ammo, your briefing, the battle plan. But you can't stay sealed for more than forty-eight hours with 5,000 or 6,000 guys. When you go to seal, the operation is on and everyone's locked and cocked for the mission. Since they had sealed on Friday, I could be sure they had to move by Sunday.

I found out the exact time of the invasion, about twenty-four hours before H-hour. Mike Sullivan and I climbed a hill, set up his satellite phone, and got ready to report. It was dark and we had an instruction book for the phone and a flashlight to scope out the booklet so we'd know which dial to turn and button to push to get the thing up and running. We were fiddling with the radio when I realized the moonlight was more powerful than my flashlight. It was almost like day.

My God, the President is sending our paratroopers down under a full moon.

The Chairman of the Joint Chiefs wore jump wings—at least he punched that ticket in training, though he was only a five-jump commando, not a seasoned paratrooper—but he should have known about the moon. Going down the chain of command, the commanding general of the XVIII Airborne Corps should have known. And the commander of the 82nd Airborne Division, a longtime paratrooper. And the brigade commanders, the battalion commanders, the company commanders, even the green platoon leaders. All of them should have known you don't jump under a full moon in a combat situation.

Why? Because you are a perfect target. With so many guys hanging there, swinging helplessly back and forth as they float down, even a single fanatic with a shotgun or automatic weapon could spray a huge area and hit dozens of warriors. I felt sure the Haitian Army and police—they numbered about seven thousand—would demonstrate their chronic cowardice. But I wasn't so sure about the Attachés, Cedras's goons. They were very well armed, frothing at the mouth, and they had everything to lose. They were scared, furious, very dangerous—and complete fruitcakes. If a bunch of those guys were holed up around the airfield, even a couple dozen of them could create havoc with the airborne drop. Afterward I double-checked my own weather report with the Pentagon's. I asked a friend at Army intelligence, "What were the conditions on the night for the invasion?" He said, "Full moon conditions, favorable for night operations." If you're crazy. Any paratrooper worth his chute will tell you, no way would you jump into combat on a full moon. Not unless you're hankering for a Purple Heart or body bag.

What really upset me was that no one had sounded off. Everybody just closed their eyes, said, "Wilco, sir," and the Pentagon went into max spin.

They're all just going along—politics over tactics as usual?

That was Sunday. Congress was reconvening on Monday. The temperature was running against using military force in Haiti. Was that why the President had decided he had to move on Sunday night? Was it a political decision that counted for more than the safety of our paratroopers? The same kind of corruption had put politics before tactical sense in Somalia, and our Rangers were sent into battle without tanks. With this President, I could see how such a bad military move could be made. He was so inexperienced he would have us grab a rattlesnake by the tail because that was where the noise was coming from. But where were the military leaders who should have said, "Hold on a minute, Mr. President"? Thank God the 82nd Airborne got only halfway there before they did a 180 and returned to base.

President Cedras and General Philippe Biamby would have been the only two men on the beach. The Haitians are normally gentle people, not out-and-out crazies like the Somalis or Bosnians, and when they thought the Americans were coming they fled Port-au-Prince by the thousands. The buses out were crammed full with riders standing in the aisles and sitting on the roofs. Anything on wheels had taken to the roads: bicycles, carriages, wagons, tiny French cars jammed with people, baggage on top; people running with their few possessions in battered suitcases or black plastic garbage bags hauling ass for the hills. It reminded me of Korea in 1950.

Jimmy Carter's plane was the first in weeks to touch down in Port-au-Prince. He came out with Sam Nunn and Colin Powell and improvised a press conference inside the terminal. The press pack waiting outside was sucked in like ants by a giant vacuum. Outside the terminal was a big tree. I sat underneath it where I could get some shade. I didn't give a damn about what the Three Wise Men of Peace said. I wanted to scope out the whole scene. About five minutes later two big semi trucks, the kind you load with cattle, pulled up, packed with people. The ramps dropped; the people jumped down and mustered up. Guys with bullhorns and cellular phones directed them to waiting pickup trucks. Everyone ran over to the pickups and grabbed signs reading NO INVASION, CEDRAS FOREVER, GO HOME

YANK, and that sort of thing. The guys with the bullhorns bellowed instructions, then moved everyone to the airport gate to intercept where Carter and his entourage would exit on their way to their meeting with chief thug Cedras.

By the time Carter and Nunn and Powell came out, the demonstration, choreographed like a ballet, was ready. The chants began.

GO HOME, MR. CARTER! GO HOME, MR. CARTER! GO HOME, MR. CARTER!

The motorcade could not get through the gate because the demonstrators were blocking it with their bodies.

By this time the media pack was running out of the terminal and setting up its cameras. Hot news. Images of Haitians demonstrating against the invasion. To the guy back home it had to look spontaneous, when in fact the whole thing was rigged. But the cameras had all glommed onto the celebrities inside the terminal. None had been in position to lock onto the cattle trucks and the guys with the bullhorns.

At first I was amused, but then something happened that made everything very unfunny. I was standing on the perimeter of the rent-a-crowd and suddenly I saw that there were guys in the middle of it with weapons and frag hand grenades.

What if they are fanatics, they are close enough, what if one of them rolls a grenade under Carter's limousine? We have one dead ex-President.

I don't think Carter or Nunn or Powell ever realized they were in real danger. There were guys in that crowd who could have poked a rifle at them or rolled a grenade right under them. Again, none of this turned up on the TV screens back home. A camera can only be as smart as the people directing it. By itself it doesn't give the full picture. We all know we shouldn't believe everything we read. We'd be a lot better off if we knew we shouldn't believe everything we see.

Finally, the motorcade punched through the crowd. The order had obviously gone out to let Carter pass. The motorcade picked up speed, five, then ten miles an hour, and they were gone. The minute the cars disappeared, on command the dial-a-ride group loaded back into the cattle cars and scooted on over to military headquarters. By the time I got there, the same demonstrators were back in business at the new stand. At the *Newsweek* bureau that night I saw the CNN coverage of the demonstrations. There was no way to tell the whole thing had been rigged.

The Haitian military headquarters was in a big white building with columns in the front built by the U.S. Marines in the 1920s during their last visit—a trip that lasted two decades. It faced the presidential palace, and the rear was all fenced in with a little parking lot. The Sunday when Carter, Nunn, and Powell came to town, their convoy of sleek limos provided by the State Department was parked in front. The security people were keeping reporters and everyone else twenty or thirty yards away behind a barricade.

As I came in from the airport, I caught sight of my CIA friend shooting the breeze with a guy in front of the building. So I went up to the security guard and flashed my retired Army ID card. He looked at it and let me pass. There were the trapped reporters looking on as I wandered up front and started talking to my friend the spook. I was finding out from him what I could about the talks, how they were going, what had been said in the cars on the way in from the airport.

"It's really hot," he said. "Why don't we sit in the car?"

His car was a big bulletproof limo, air conditioning running. We sat in the cool interior with an Army Intelligence type, the three of us, bullshitting about what was going down. When I stepped out, all the press guys were glaring at me. Giving me that you-son-of-a-bitch look I've grown to know and love. You could see them thinking, *He's got contacts and we don't. Now he's used them right in front of us. He's finding out what's going on. Sonuvabitch.* The CNN guys and gals were frantic, especially the one the troops call "Miss Death," Christiane Amanpour. After I finished, I walked to the edge of the herd, the guard passed me through, I got in my vehicle and drove away to the sound of gnashing teeth.

I had a Deep Throat who was an adviser to Cedras. I mean this dude was in his back pocket all the time. The CIA and most of the Americans hated him because he was working for the wrong general. He was one of Spencer Reiss's Deep Throat contacts and he was kind enough to share. We clicked like an accuratized pistol.

"I've got your book," he said. "I bought it in Santa Monica."

"Fuck, man, that's my hometown." I thumped him on the back. He ran into his office, got it, and asked me to sign it, and we kind of became coconspirators. He was a Canadian and he had served in the Canadian paratroops. Even though he was technically the enemy, I liked him—after all, he had pretty good taste in reading. He was always dropping stuff in my ear about what was going down. Sure, he was double-dealing, looking for some *Newsweek* insurance, and considering the murderous bastards he slept with, he was playing a smart hand. One slip and he was finely sliced machete meat.

When Jimmy Carter, Sam Nunn, and Colin Powell came down to negotiate, they brought with them a guy from the National Security Council. He asked for a private room where he could call Washington on the scrambler. The CIA techs swept the room for bugs and pronounced it secure. So the NSC guy was in there talking constantly to Washington, directly to National Security Adviser Tony Lake and others, reporting on how the negotiations were going and not going. For a while they were swimming deep in the toilet.

The Pentagon was reaching the fish-or-cut-bait moment when they would have to kick off the airborne armada out of Fort Bragg, North Carolina. The timeline reality was, the airborne assault force had to launch by 1900 hours so they could drop, consolidate, brush away the opposition, and secure their objectives before first light. The schedule was tight.

Cedras was holding his ground. At one point my spy was upstairs in his office with Clinton's dream team while the NSC guy was downstairs in his "secure" little cubbyhole. People were running back and forth with messages. Finally, in the "secure" room, Washington said, "We're launching the 82nd Airborne now." And the NSC guy relayed the word to a runner who beat feet to whisper the info to Carter.

What Tony Lake's man didn't know was my guy was up in the room directly above him. The NSC guy also did not know the Canadian had drilled a hole into the inner sanctum. He had his ear to a glass pressed to the hole and he could hear everything that went on in that room. And everything he heard, he told to me the day Cedras walked.

The dream team had all their high-tech stuff. Every expensive spy gadget in the inventory. But they were being screwed by the most time-worn spy gimmick in the book: the old ear-to-the-hole-in-the-wall trick.

"I was like a fly on the wall," my spy said. "I heard every word." He flashed the signal that paratroopers were on the way. General Biamby went busting into the meeting with Carter, Powell, and Cedras, shouting Washington was tricking them, the paratroopers were coming, the time had come to quit talking and get to battle stations.

At that moment, Cedras knew President Clinton finally meant business. We were going to force his ass out even if we had to shoot our way in to do it. But he didn't get the truth from the dream team. He got it from that guy with his ear to the hole in the wall. At that moment General Biamby was the only warrior left in Haiti. But the instant he came in and said the time to fight had come, Cedras started thinking Miami was looking better and better all the time. He was a good poker player with a weak hand, but he was able to win a big pot from Clinton, who was holding all the cards. The take was an instant membership in America's Ex-Dictators Millionaire Club—some reports said the CIA had given the Cedras mob up to $12 million in getaway goodies. Cedras came away with a nice life, total amnesty from all the lives he'd snuffed and all the crimes he'd committed. Now he's leading the good life at U.S. taxpayers' expense. That's what the State Department policy crowd calls realpolitik.

So Cedras took a powder and the next day the outward migration reversed and ecstasy flowed from the Haitians returning to Port-au-Prince. They were never warlike people like the North Koreans or the worst guys in the Balkans. Most were lovely people who'd had been screwed by two hundred years of bad politics and worse politicians. Around the airfield where the Americans first came, tens of thousand of Haitians turned up to watch the friendly invasion drop from the sky. They were cheering and waving and throwing kisses and shouting *"Vive l'Amérique"* at the top of their lungs.

The performance of the U.S. military in Haiti that day was brilliant. They did everything right. In the beginning they had to be prepared to fight; then they had to turn on a dime and supervise an instant peace deal. They had to change their mind-set, scramble, break away from the detailed fighting plan they had been working for months. The only problem I could see was with the initial rules of engagement. In the beginning, American soldiers were down on the ground securing key installations. But at first they were watching the Haitian police beat Haitians and doing nothing. The rules of engagement prevented them from stepping in to stop it. You can't really fault the brass on what was such a rapidly changing situation; but it is true the invasion was a humanitarian mission and they could have been better prepared to stop the thugs.

It wasn't easy. The Haitians were high on democracy. They were thinking, *We're free. We can say what we want to. We can do what we want. The American sheriff will protect us.* They didn't realize the bad guys hadn't been blown out of the saddle yet and the police and Attachés would come down mean and hard. The atmosphere up around Plantation LeClerc reflected this dangerous innocence. Things began to get weird. And I had put myself in the middle of the rain forest with people who were very superstitious. They started calling me "Father," as if they saw me as a white god who was going to save them, a protector God had sent down, to watch out for them, to fend off all evil.

It was eerie. "Oh, Father, here's your breakfast. Father."

I would be writing in my room and someone would pad in and say, "Father, would you like this coconut juice we've made specially for you?" Wild trip.

As the days passed, I spent a lot of time walking around the plantation and in the Soray slums. I was right in the center of things and I began to get to know the people. There was still gunfire every night. The next morning I would find bodies parked up against our rock wall, the people killed the night before by the Attachés. The thing that really blew me out was all this was happening only twenty minutes from downtown Port-au-Prince where there were twenty thousand American soldiers. In the center of town, American soldiers were building their camp, lining up their tent pegs, getting everything all squared away. That is how you always do it. I see no fault there. But out in the boonies, nothing had changed. Go twenty minutes or four hours out and the Attachés were still in charge.

At first the Americans had to secure themselves and they were not able to guard the Haitians. The purpose of their coming was to liberate the people and guarantee their security, but that couldn't be done straightaway and it wasn't. I suppose that's why the people around Plantation LeClerc treated me like a god. They wanted me to set them free from the police, the Attachés, and the rest of the leftovers from the gestapo-like junta government. During my first trip I had thought an invasion did not make good sense, but now I had seen their fear up close and I changed my mind. I began to think what had been very wrongheaded in Somalia was a good idea in Haiti.

There was no way around the fact that most of the country's current problems could be traced in one way or another to the United States. For over a hundred years American fruit companies and industry had gone down there and ripped the Haitians off. Cheap labor. Slave labor, practically. You could hire a Haitian for a dollar a day to manufacture anything. Pin little wings on toys, pick fruit, make baseballs, whatever you wanted.

We had also polluted their morals. The jet set had used Haiti as its own personal sex shop. You could buy young girls or young boys for practically nothing. Port-au-Prince was a big homosexual holiday scene. Sex with a ten-year-old was easy, fast, and cheap with no questions asked. I started thinking, *Wait a minute. This country is really screwed up, but we're the ones who did the screwing. We've left the place a basket case, and the basket is filled with AIDS. For the price of one B-2 bomber every ten years we could set it right. And maybe if we set the example here, the Japanese could do the same in Myanmar or the Germans in Rwanda. Not cheap charity, not two bucks to the Red Cross and forget it, but something genuinely humanitarian, a rich nation helping a poor nation it's fucked over.*

My military judgment was not whispering these thoughts; it was just my feelings as a human being. My job was to go down there and analyze the place as a battlefield, check the feasibility of an invasion, estimate the casualties, and shoot the word to *Newsweek*. But you couldn't spend time among the Haitians without understanding the bad shit they had been through.

The day Mita dragged me off to save the little kid from the Soray slammer convinced me.

While I was standing in the doorway of the police station wondering how many people were going to be killed, a pickup truck carrying a dozen Haitian cops with M-1 rifles and billy clubs came screeching up. The stormtroopers piled out and started beating the hell out of the crowd. For a moment, they dispersed like geese unlucky enough to be flying over an NRA firing range. Then they regrouped and started throwing rocks. *Brave devils,* I thought. *Stones against M-1 rifles.* The senior noncom in command of the shock troops came stomping into the police station. The first thing he saw was me and my cap.

"You military? You colonel?" he said.

"No," I told him. "I'm a reporter."

That made it even worse. I had seen the brutality of the attack. And those things were not supposed to be happening with twenty thousand Americans securing Haiti.

The little boy with the split lip, the one I had paid for, was pressed up against the wall. They grabbed him and put him in the back of the pickup.

"No, no, no," I started yelling. "We had a deal."

I started to lift the kid out of the truck. He was a little guy, maybe eighty–ninety pounds. He froze, put his feet stiff. He was too scared to get out.

"Come on, come on, come with us," his sister wailed.

But the kid would not budge. The truck drove off with him. God knows what happened to him. I walked back down the hill feeling sick. *Is this really the way it's got to be?*

Scraping out the rot was going to take a lot of work. One day not long after the showdown at the police station, I was out with Bill Gentile working on a story about the Haitian boat people. We'd gone to a little fishing village on the coast. At the first checkpoint out of Port-au-Prince a goon demanded to see our identification, our passes, our press cards, our passports. He had the gun pointed right at us and looked just mean enough to let off a burst. At this point, in perfect English, the guy said to me, "You will not pass unless you give me twenty dollars."

And I said, "I'm not going to give you twenty dollars or twenty cents. I'm not going to give you shit."

He didn't understand what I had told him. He had memorized only the English sentence he had spoken, the greed words. *Here's a guy who speaks perfect English except that's the only thing he can say.* He gave me a hard stare. I said a little prayer and told the driver to press on.

We went to the fishing village and spent some time with the boat makers, chasing our story. Haitians would sell their home, their goats, everything they had, for a ticket on a little rickety rowboat that hopefully would take them far enough out to be picked up by the Coast Guard. Most of them would then be sent back. But they would blow everything they had trying to escape, that's how bad Haiti was under Cedras and company.

As we were sitting there over huge lobsters plucked by one of the fishermen out of his trap for our lunches, another soldier came up and asked us what we thought we were doing. He took us at gunpoint to a small police station where a very arrogant sergeant began to abuse us. He said we were under arrest. Then, again at gunpoint, they drove us another ten or fifteen miles to a major police station. This time an even more arrogant captain accused us of being spies.

Enough of this bullshit.

"You're speaking to a U.S. colonel," I said. I pulled out my ID and slammed it down on his desk. "When you speak to me, you stand up and say, 'Sir.' You got that?"

He came to attention and locked his heels. His whole attitude changed. Up to that moment he had abused us, made us wait while he finished his lunch and played with his whores, but when I started pounding on his desk, he didn't know how to react.

I completely lost my cool—saw I had him, I was on a roll.

"You dare arrest us? Do you think we are here in this country to allow you to treat us this way?" I just tore into his ass backward and forward.

Our translator-driver was a sweet little man about sixty years old named Pancho. He had lived all his life under these horrendous conditions. He was sweating and his hands were shaking. He kept saying, "Oh no, boss. Don't talk this way, boss. Boss, we in a bad fix."

But the captain was cowed and now he did a complete 180. I concluded right there most of the Haitian authorities were bullies. Sure, they had guns, but once they saw they were up against a greater force, they tended to cut and run like all bullies do.

"You may leave," the captain said. "We have made a mistake."

He practically carried us back to our car, and we drove off.

The ordinary people were completely different. One day Bill and I went out on patrol with a unit from the 10th Mountain Division. A couple of infantry squads went through a village near the airfield to check it out. They started out in tight military formation, guns deployed, soldiers alert. The more we moved into the village, the more kids came and crowded around us. By the time the patrol reached its objective, the soldiers were covered with flowers and the kids were holding their hands. The Haitians were awed, the grown-ups as well as the kids. And that feeling didn't last just for a day. The Happiness Circus had come to the island, bringing to the Haitian people a never-ending carnival of joy and wonder at their liberation from their oppression. The Haitians' awe at our military's paraphernalia was touching and probably similar to what the original natives felt when they met up with Christopher Columbus and crew in 1492. The gods in armor with fire sticks and big ships with wings and now a sky full of planes and helicopters with more big whirlybirds landing in an hour than that savaged country had seen in its lifetime.

I stayed for three weeks watching our troops settle in. Along the way I met a really good man, Air Force Master Sergeant Brian Sunday. He had read *About Face* and he said, "Boy, anytime you want to know what is going on, come to the field and I'll tell you."

One morning I met him at the front gate of the airfield. When I asked where he wanted to talk, he suggested the tower above the gate. There was an M-60 machine gun up there and it was cooler because the wind blew through. So we climbed the ladder and I was sitting with my notebook out interviewing him.

"Uh-oh," he said. "Here he comes again."

"What do you mean?" I asked him.

Looking down I could see two Humvees pulled up at the gate.

"That's George Patton," the sergeant said.

He pointed down to Brigadier General George Close. He was sticking his fist out of the Humvee punching the air. A gate guard didn't have his helmet on exactly right and the guard next to him was also slightly out of uniform. General Close was drilling them each a new asshole over the egregious crime. I couldn't believe what I was seeing.

This is a general doing a corporal's work. So what do corporals do?

General Close was so totally absorbed he didn't notice a Haitian standing right next to him with a grenade in his pocket. Earlier that morning when I'd hooked up with Sergeant Sunday, the first thing he'd said to me was "We got a breakdown." The Americans had started a program to buy back grenades and guns from the Haitians. But so many grenades had come in that the policy was changed and the word hadn't gotten out to the people. That was the breakdown.

He had then pointed to a guy by the gate and told me to check out his right front pants pocket, where I saw the pineapple imprint of a frag grenade. He wanted to sell it but the buy-a-grenade special was over.

"Shit," I said. "Let's buy it." I thought, *Christ, if you find one grenade you might find your way to a case. Not a bad deal.*

I had checked my wallet. All I had was twenties, no small bills, and the going rate for a grenade was five bucks. Sunday was in the same shape. He had said the Air Force explosives guy was on his way and would deal with it.

Now I looked down from the tower and saw the same dude standing there while General Close was steaming and reaming. While the general worried about the dress code, the Haitian could have rolled that frag grenade under his Humvee and blown his ass halfway to Puerto Rico along with a half-dozen other soldiers. About an hour later, someone did throw a hand grenade in downtown Port-au-Prince. It blew up, killing and wounding a dozen people.

The grunts were also steaming, all suited up in Kevlar pots, Kevlar vests, and all their other gear. They were not even allowed to roll up their sleeves. The Army and the Marine Corps seem particularly susceptible to the almost military necessity of looking good on the tube at the expense of the welfare of the troops. The image always has to be recruiting poster gung ho. Appearance over substance, show over go. Our guys had to look razor sharp all the time although they were out in the boiling sun melting every day. Even rifle company commanders would say privately, "I can't believe what the CG is making my guys do." But the generals were into micromanagement and spinning and if they turned on CNN and saw a soldier out of uniform, they went ballistic.

I had a lot of respect for General Shelton, but I didn't think as highly of Major General David C. Meade and General Close, the two men who were running the 10th Mountain Division show in Haiti. I later wrote a piece in my syndicated newspaper column, "Defending America," criticizing them. *Why weren't our troops dressed for the tropics? The fatigue uniform didn't breathe. It wasn't all that different from the one their grandfathers wore on Guadalcanal. It should have been made of lightweight material. Why didn't they have tropical shorts like the Australians? Why did they have to dress as if they were on parade when the temperature was 110 degrees in the shade? Meanwhile, of course, the Wild Blue Yonder Gang from the U.S. Air Force was laid back and cool in T-shirts, shorts, and soft caps. At least the Air Force has the good sense to look after their people.*

As I said after Desert Storm, I do invasions, not occupations. When occupation duty clicked in, I headed back to Whitefish. I left Plantation LeClerc at midday on October 20, hung around the airfield watching General Close do his act, then hopped on a C-141.

A few days later, at five o'clock in the morning, the 10th Mountain Division staged a predawn, precision raid on Plantation LeClerc. Troops came pounding in to roll Mita and Cameron out of bed at gunpoint. The raiders rounded everyone up, searched the whole place. They had been told they were hitting a guerrilla training center. Some spooky U.S. Special Forces rogue was supposed to be there teaching Haitians how to conduct underground operations; they were going to find platoons of trained "Gs," weapons caches, and a lot of other bad stuff. They had orders to capture a white godfather who always wore a hat with Special Forces and Ranger markings. My hat all right. Everything else they got wrong. *God, our multibillion-dollar-a-year intel machine from Pearl Harbor to Haiti never gets it right!*

When I got back to World Headquarters in Whitefish, no sooner did I walk through the door and say hello to Heidi, my Army-trained executive officer, than the phone rang. I picked it up.

"Dave? This is Dave Meade. Haiti. Tenth Division. I didn't appreciate your column, Dave."

"That so?"

"Yeah. It's not true. We're looking after the troops. You damn well know it."

He was giving me the party line, but he was as hot as a two-bit pistol that had just plunked a hundred rounds downrange.

What's going on? He's chewing out the wrong guy—I'm not a soldier boy anymore! Tell him to go fuck himself.

"Look, that's a bunch of bullshit," I said. "The troops were melting in the sun and you guys didn't care and you were running around micromanaging over chickenshit. I simply tell it like it is. My concern is the soldiers, not you and not your image. If you don't like the story, that's your problem."

"Well," he said, backing off, "we didn't have air conditioning."

"Bullshit. Don't tell me that. I went through your command post at the airport. Remember, I went everywhere. I saw that air conditioning going, so don't tell me you didn't have it."

"Well, I mean later."

"I don't care about later. You guys were sitting there under air conditioning while those troops with all their shit on were being cooked well-done in the sun."

Then, snap, snap, he softened up and I knew I had him. I wasn't going to let him bully me or say my reporting hadn't been accurate. He changed his whole approach.

"I am concerned about the troops," he said. "I make them wear their vests and all of that."

"To protect them from what? You weren't in the middle of the Battle of the Bulge. There was no enemy opposition whatsoever."

Then, smoother than ever, he said, "You must come down, Dave, and visit us again, and spend a little more time with us, and you'll be able to really appreciate just how good we are." *The old bait and switch.*

"Look," I said, "I'm going to be on your case until you get the job done right and you look after the troops. I'm going to tell the American people when you don't."

Click.

Meade was a fast-tracker wonder boy. Punched all the right tickets: skippered a battalion; commanded a division artillery, which is equal to brigade command; then he was aide to the Army Chief of Staff. Really a connected dude. But how did a Perfumed Prince with not one lick of troop smarts get command of a light infantry division? Something is really screwed with the Army's Division Command selection system. More proof that nothing's been learned since Vietnam, where Westmoreland, also an artillery man and like Meade, a micromanager supreme, was in charge of a light infantry war, which he never understood and we consequently lost.

Meade was not a people person. In Haiti, the rate of his division courts-martial and Article 15s, which are courts-martial in miniature, was well above the Army norm. This was a general trying too hard to be perfect, and this was his chance to lead an invasion force.

It would have been great for him had he been cool. But he was out there micromanaging so there would not be a mistake on his watch. The inside word on the E-ring was he had been headed for four stars. Now he lost it; he plain self-destructed.

Word came through later that General Close hated me even more than the troops hated him after my attack came out in "Defending America." He was so pissed off he started bumping into his own fenders. Whenever people asked him about the column he would say, "Hackworth? I don't care about Hackworth. He's a bum anyway." Reports like that kept coming back to me. After Haiti, when the 10th was back at Fort Drum, General Close left. They gave him the standard hail-and-farewell party. A young captain rang me and said, "We really want to stick it to him." So I said, "Hey, man, anything for the troops." And he said, "Send us a photograph autographed: Dear Georgie, with warmest regards, your pal, Hack."

So I did, and they framed it and presented it to him at the farewell. A captain said he turned about nine different colors of purple while the troops glowed in the heat.

POSTSCRIPT

In one particularly bad case, General Meade's responsibility for not doing his sworn duty resulted in the court-martial of Captain Lawrence Rockwood, a scandal that embarrassed the Army. Rockwood was a military intelligence officer. Before the invasion, he had concluded that the prisoners held in Haitian prisons were being abused, even murdered. After the invasion, he felt strongly they should receive top-priority treatment, not unlike the priority American forces have always given prisoners—in World War II, the concentration camps, the Japanese and German POW camps. He knew in many cases the United States ran special operations to prison camps to secure prisoners, to keep their guards from wasting them. It had happened during World War II, it had happened in Korea, it had happened in Grenada in 1983, and in Panama in 1989. So there was nothing out of line in his thinking.

Even before the invasion, Captain Rockwood had made a number of recommendations to General Meade on what needed to be done about the prisons. The general and his staff knew about the situation; it just wasn't one of their urgent priorities. They were into "sleeves down and pots on." Several weeks passed after the invasion and still nothing had been done about the prisons, even though the Marines had checked one out and had found conditions exactly as Rockwood predicted. Captain Rockwood told his boss, went through the chain of command. Still, General Meade and his staff were slow to react. Rockwood exhausted everyone in the division. He even went to the division chaplain. Rockwood raised a lot of hell, but he was only a captain. No one from Meade on down paid any attention to what he was saying.

Finally, he went to the division's inspector general and filed a career-killer complaint that General Meade was not following the Hague Convention, the Geneva Convention, the Articles of War, not doing his sworn duty, not following his officer's oath. That captain stood in the door and pissed off everyone in the chain of command to bring the truth to the general in order to get some action on the prisoners. The inspector general worked for General Meade. Catch-22. Nothing was done.

Finally, in desperation, Rockwood went down to the prison himself, taking his M-16 and not driving in a proper military vehicle with a proper vehicle escort or following a bookful of other petty regulations. He had become Eliot Ness obsessed with the prisoners' plight. He scaled walls, banged on doors, pointed his weapon at people, demanded to look into some of the cells. There he found guys on litters in the hospital dispensary who were in such bad shape their skin peeled off when they were moved. A few hours were enough to confirm all his worst suspicions. The prisons were cesspools filled with people who were dying.

The response of the military was to send a major from the embassy, who trotted in, disarmed Captain Rockwood, and returned him at gunpoint to his unit. When he got back, they yelled at him for not being present for duty, charged him with being AWOL, violating four or five chickenshit regulations. Still, even though he'd been to the prison, even though he'd proven what was wrong inside, no one would listen to him. By that time he was so frustrated, so upset, he got into a screaming match with his colonel, the division intelligence officer who had sat on his earlier reports. So he was slapped with a bunch of charges, including insubordination, using profanity to a senior officer, being AWOL, violating orders, and conduct unbecoming an officer.

They packed Rockwood back to the United States, saying he had lost his marbles, he was nuts, he had flipped. At the same time, a colonel went into the prison and found everything Rockwood said was true. The place was a concentration camp without ovens. But nothing happened until December 1994, when a congressman who had talked to Rockwood went to the prison. He found five hundred prisoners in one small room standing in three or four inches of shit, covered with sores and infections, people very badly sick. I suppose that's where the cops in Soray wanted to put the little kid they dragged away from me. Only after the congressman raised hell did General Meade and the Army take action.

The one-week court-martial the following March was a kangaroo court from the presiding judge to the jury panel to the legal eagles. A star prosecution witness, William H. Parks, was found on two occasions inside the jury deliberation room. When challenged for this improper conduct, a display of incredibly bad judgment on the part of a senior Army lawyer, he told the judge he had been just "socializing" over a cup of coffee with the jury and had not discussed the case. One visit to this "off limits room" was bad enough; two set the tone for the entire proceeding. Throughout the marathon trial, which averaged twelve hours a day, the body language of the Army judge, Lieutenant Colonel John Newberry, said it all. He slouched down in his chair, yawned as if bored out of his brain, and constantly eyeballed the courtroom clock like a ten-year-old waiting for the school bell to ring. His attitude seemed to be "Get on with it. I've got a golf game at noon."

The captain was found guilty on three of five counts. He could have gotten ten years. Instead, he was dishonorably discharged. He had to forfeit sixteen years of exemplary service. He was only four years from his gold watch and a paycheck for life. They booted him out.

Rockwood has appealed. There is plenty of reason for the court-martial to be thrown out. General Meade has faded away. After Haiti he retired. *Thump*. His career was over. What he did, I do not know. But my guess is part of it was that the morale of the division had plummeted to whale-shit depth. Meade took a great division, which Bill Carpenter had reformed and good leaders like Mike Plummer and Johnny Howard and Steve Arnold had honed to a fine edge, and he gutted it, reinforcing the timeless Army adage: There are no bad units, only bad commanders. Another source, a former 10th Mountain Division assistant division commander, told me the grunts universally hated him as did his boss, Lieutenant General Shelton, who "saw right through him."

The Pentagon should review the whole sorry affair. Anyone with any sense would throw out the court-martial, give Rockwood a slap on the wrist for violating those chickenshit regulations, then give him a medal for doing what was right and decent and very American. We haven't heard the last of this case—the appeal could go all the way up through channels to the Secretary of the Army and to the Supreme Court. I received hundreds of letters from Americans who were truly horrified at how Captain Rockwood was treated.

The scary thing is having an all-volunteer regular military—which our forefathers warned us against—so removed from the larger American society that it went through the entire Rockwood spectacle in lockstep. It's a classic story of how the chain of command so often ignores what someone down at the bottom sees. No one was interested in listening to any drum beats from below; all anyone cared about was going along to get along.

Within the Army, the institutional view of Rockwood is that he was emotionally unstable and completely out of tune with how tough things can get in the third world. "He went on his Rambo mission armed like a postal clerk after a bad day at the stamp window," one outstanding officer told me. But even this guy, who had closely examined the case, says the Army mishandled the case. "A court-martial, for God's sake," he said. "Meade was consistent. I'll give him that, dumb from beginning to end. It was letter of reprimand stuff. Send him home, give him a job counting basketballs at the gym, and zap him on his efficiency report. He could have retired early and found a lucrative position in some animal rights group and spent his spare time wearing a tinfoil hat and talking to Venus."

The Army is on the horns of a real dilemma here. If Rockwood was nuts, why did no one find out until Haiti? He had served sixteen years. Either the institutional view is dead wrong. Or, if Rockwood's critics are right, the whole personnel system doesn't work. Either way, the institutional view misses the real point. Whether Rockwood was a "flake" or not, the prison he exposed was a true cesspool, exactly the sort of shit the invasion was supposed to clean up—not on D-day plus sixty but immediately, the way Ike did with German concentration camps at the end of World War II.

Why weren't our commanders in Haiti as sensitive as Eisenhower? They were so locked into the execution of their neat little timetables they were blind to an ugly problem staring them in the face and screaming to be corrected. Some of our very best professional soldiers now respond to third world conditions with numbness. I know one superb warrior who says, "After five months in Somalia, I probably would have shrugged off Christ on the Cross." This ain't the way it's supposed to be. Our founding fathers opposed a standing army because they were afraid of the Roman mind-set, its coldness, its brutality, its indifference to common humanity. An American warrior should never be callous or uncaring. Besides, don't forget what happened to the Romans.

What was really on trial back at Fort Drum was not Captain Rockwood, but General Meade, the all-volunteer military's mind-set, and the leadership of the U.S. Army. When you think about it, the sad thing is that superb soldiers did a brilliant job in Haiti executing a tough mission. But what will be remembered was how the 10th Mountain Division court-martialed Captain Rockwood because a few higher-ranking officers who should have known better chose to act like stormtroopers rather than thinking, caring leaders in the Army of a republic that stands for freedom, justice, and equality for all.

CHAPTER 9

"KING OF BOSNIA": THE BALKANS, 1996

The Admiral was in Naples. Earlier that day he had heard something about the President of the United States inviting Captain Scott O'Grady to lunch at the White House. This set him thinking. What about the young, hard-charging Marines who had saved Captain O'Grady's ass from the Serbs who wanted to fry it. Hell, we have a big house, he thought. Why not do something special for those heroes?

There was one problem. The right day happened to be his thirty-third wedding anniversary. So he picked up the phone and called his wife Dorothy.

"Dottie," he said. "What about it?"

What about having those young guys over for a beer?

"Yes," she told him. "I won't stop you. Go ahead."

After that, the Admiral called one of his commanders and asked for a favor.

"Look, you pose the question, because I don't want the folks to think this is a command performance."

So the commander relayed the invitation to Colonel Marty Berndt, of the 24th Marine Expeditionary Unit, the guy who had led forty-one Marines to O'Grady's rescue.

"You mean all of them?" the colonel said in disbelief.

"Yeah, all of them."

Then the Admiral called in his chef.

"Bernie," he said. "How many cases of beer you got?"

"I have plenty here, Admiral. Don't worry about it."

"Bernie, how many cases you got?"

"I have five."

"Go get twelve more cases."

"Twelve cases?"

"They're going to drink beer faster than you can put it out on the table. Go get twelve cases and put them in the fridge and get them cold."

Come to think of it, twelve cases were nowhere near enough. So the Admiral sent Bernie out again.

"Bernie," he said, "we are not going to run out of beer."

Admiral Leighton W. Smith is not a Perfumed Prince. The Commander of all NATO forces in Bosnia is a stud of the old school, a warrior, one of the finest of the few we have left. The beer-run story reminded me of a general I knew in Vietnam, the other kind. He once told a company of paratroopers if they snatched a prisoner for him, he would spring for a case of beer. Let's see, that was twenty-four cans for about two hundred guys. And any one of those hard-asses could have killed a whole case on his lonesome.

I know a NATO officer who calls Admiral Smith the "King of Bosnia," because he has the most extensive powers ever granted a U.S. military commander outside of war. He has the authority to use military muscle to carry out every major element of the Dayton negotiations and Paris Peace Treaty. His NATO peace implementation force (IFOR), sixty thousand strong, is the biggest military operation in Europe since World War II. Not since Admiral Chester Nimitz commanded Allied forces in the Pacific during World War II has an American naval officer commanded such a large ground force.

A few days before Christmas in 1995, the Admiral flew to Sarajevo to take over the job of running Operation Joint Endeavor. His fixed-wing aircraft could not land because the airport was socked in. Douglas MacArthur would have told the commanders on the ground, "Stand by. This will lift someday. I'm going back to Tokyo. Call me when it's lifted; two, three days; two, three months; I shall return."

But not Snuffy Smith. He called down to General Sir Michael Walker, his British deputy, and said, "Take command, buddy, I can't get in. I'm flying to Split and I'll come back by chopper. Take command and kick ass and get things rolling."

No photo op. No bullshit.

I knew he had won a chestful of medals when he flew 280 combat missions over Vietnam. In 1972, after we had lost 128 planes in seven years trying to take out the Thanh Hoa Bridge in Hanoi, he dropped it with a two-thousand-pound bomb. With Generals Powell and Schwarzkopf, he belongs to the Never Again Club of Viet vets who have sworn never to let another Vietnam happen on their watch. In the spring of 1995, when the Balkans started lighting up again like Dante's Inferno, I decided to check in with Colonel Dave Hunt, a true gunfighter who was working closely with the Admiral.

"This guy's not Westmoreland," Colonel Hunt told me. He wouldn't bend with the slightest political wind. He wouldn't go along to get along and then wail the politicians had tied his hands. Colonel Hunt said, "Smith is the ultimate truth teller. His greatest strength is his leadership ability and his talent to cut through the BS in microseconds. There will be no mission creep, no Somalias, no Vietnam disasters on his watch."

This was a guy who had started out as a man of action. He spent most of his service in the cockpit of fighters or on the deck of a carrier where he rose from fighter jock to squadron commander to skipper of the U.S.S. *America*, a giant flattop. But along the way he also turned himself into one of the Navy's finest thinkers. He spent as little time as he could in the Pentagon, but while he was there, as a three-star, he developed the current blueprint for the post–Cold-War Navy. The strategy, called "From the Sea," moves the Navy away from five decades of blue-water operations combating the Soviet Union into a global course for the twenty-first century. Stanley Arthur, a retired admiral and former Vice Chief of Naval Operations, once told me, "Even though a lot of people take credit for this brilliant concept, Smith put it together and was the principal author."

During the eighteen months he kept watch on the Balkans long-range from Naples, Snuffy Smith directed the 3,515-sortie air campaign that helped blast the Bosnian Serbs to the Dayton peace table. He enforced no-fly zones over Bosnia. He directed the humanitarian supply airlift into the country. He also policed the arms embargo against the combatants. "Here's the real question," Colonel Hunt said. "Can the rest of the high brass put up with such a pure fighter?"

I knew I had to meet this guy.

On New Year's Day, 1996, I flew from New York to Munich and on to Zagreb where I picked up IFOR credentials and wangled seats on a military flight down to Sarajevo the next day for me and Leif Skoogfors, *Newsweek*'s shooter. Leif was a tall guy, six-feet-three and fifty-six, with a hippie beard and haircut; a sweet guy, very laid back on the surface with a volcano burning within. He was Michelangelo-brilliant with his camera, a complete pro, at ease with people and shit hot smart about his specialty, the military. He had covered Northern Ireland and El Salvador and Nicaragua so he knew how to handle himself in tight corners and he knew the new weapons and other stuff much better than I did. A good man.

I got my gear ready and caught some shut-eye. The next morning when I came down to the lobby to check out, I saw a Special Forces major standing by the desk. He had master jumper's wings, and a Vietnam combat patch, meaning he had to be an old airborne guy, so I introduced myself. We were both waiting for transport out to the airport and we started shooting the breeze.

"What the hell are you doing here?" I asked him.

"Civil affairs."

"What's that?"

"I'm here to do the information war. We have a radio station, a newspaper, a big budget, a lotta resources."

A Green Beret for information. Never heard that one before. I took his phone number, hooked up with Leif, and flew south.

Three years earlier the pilot had corkscrewed us down, hot-landed, and taken off again with less than a minute on the ground. As I ran from the bird, mortar rounds were thudding all around the airfield and I had to dive into a bunker before the Serbs could blow me away. This time we glided in and softly touched down. The airfield now belonged to NATO. It was military efficient, our own little O'Hare in Bosnia, humming with incoming and outbound aircraft, forklifts moving equipment, trucks shuttling personnel. The old bunker was still there, but they'd turned it into a telephone station.

Two guys from Snuffy Smith's liaison office were there to meet me. I had met one in Washington and the other, a Marine Corps colonel, I had known at Quantico. When they heard I was coming in, they volunteered to pick me up. They had two armored Humvees. So we all piled in and headed directly for Admiral Smith's headquarters.

"How are things going?" I asked them.

"Problems," one of them said.

The NATO staff had been sitting in Naples since 1945, fat,

dumb, and happy. Now they had to operate from the saddle of a horse and they'd forgotten how to ride. They were having a hard time just coordinating and communicating. The same thing had happened with the units deploying out of Germany. They were used to having their autobahns and their airfields and training centers and everything else on a rigid military grid organized around a German clock. Suddenly they were out of their playpen and they couldn't cope: V Corps was taking so long to bridge the Sava River that General Dennis Reimer, the Army Chief of Staff, was telling friends it was the biggest embarrassment he could remember.

As soon as I got to Admiral Smith's headquarters, I got a little taste of the confusion in miniature. I showed my IFOR pass at the front gate and went in to meet his info officer.

"I have an appointment with the admiral this afternoon."

"No, it's tomorrow."

"I don't think you know what you're talking about, so here's what I'm going to do. I'm going to check into a hotel and get squared away and I'll be back here around two-thirty. Why don't you go see the executive officer and tell him I'm here for my two-thirty appointment."

The info captain said, yeah, right, yawn, yawn, and I beat feet over to my hotel. About five minutes after I checked in, there was a phone call from the captain.

"Be here at two-thirty. The front gate."

"Roger on that."

The headquarters was about fifteen minutes away, so that afternoon I zoomed over. The captain was out front.

He took me into Smith's offices and I spent some time hanging around the anteroom to the Admiral's office. There were clues all over that this was a warrior, not a desk pounder. Two young Marines were pulling security for him. A Perfumed Prince would have had them flanking his door at parade rest in dress blues, shoes gleaming, eyes locked, coming to a rigid Achtung every time the Admiral or anyone else with a star farted. These two guys were wearing cammies and sitting in lounge chairs.

"What do you think of the Old Man?" I asked them.

One of the kids checked me out.

"I'd take a bullet for him," he said.

About that time, one of the Navy SEALs who had done security for General Shelton in Haiti came walking by. I had met him when I went busting into General Meade's airport office one day on the tail of General Cedras. "Hey, man," he had said. "You can't go in there."

"Like shit I can't. Watch me," I had told him. And he had said, "Who are you?" "David Hackworth, who are you?" "A SEAL," he said. That was all he had to say. I pulled up. But instead of breaking my neck with one hand, which he could easily have done, he and I had gotten together later for a beer. Now he was walking by Snuffy Smith's office and he spotted me.

"Hey, Hack," he said. "Come into the bunker."

I followed him into another room where four more SEALs were holed up. It turned out they were Smith's personal security detail. All studs. In a fight, barroom or battlefield, they would be their opponent's worst nightmare. We shot the breeze for a few minutes and arranged to get together that night for a Good Evening Bosnia brew. There is no better company than these rare and unique fighting men. Being in the vicinity of their testosterone levels would even recharge the batteries of a eunuch.

When I got back, the Admiral was waiting for me at the door. No full colonels to escort the visitor into the royal throne room.

"Leighton Smith," he said, and he stuck out his hand.

I was looking at a lean guy in his middle fifties, not tall or imposing, but in good shape and very relaxed. The thing I liked first about him was he got his own coffee. No mess stewards fluttering around in white jackets. Just a coffee bar over to the side with a contraband bottle of Jack Daniel's standing tall despite General Order Number 1—no alcohol—from General George Joulwan, NATO's Perfumed Prince Supreme Commander.

Most offices belonging to high brass have all the trappings of their imperial station: flags and a Glory Wall covered with plaques and pictures with President X or Prime Minister Y and Joint Chairman B or S, mementoes of past commands given by "loving subordinates" who were more than likely damn glad to see them leave. Smith's office was different, like the man: unadorned and plain, functional. You could smell the fresh paint that made everything trim, shipshape.

I was just setting up my tape recorder when a Navy captain named Rusty Petrea walked briskly in and handed the Admiral a spot report. It had to be urgent. An aide would never break into an appointment just to chew the rag.

"Uh-oh, here comes the bad news," Snuffy said.

He scanned the report quickly.

"Agghhh," he said, looking like somebody had punched him in the gut. "Mines. Goddamn mines."

Three British soldiers had been wounded and he seemed to take the news as a personal loss, as if two of his sons had been hurt in an automobile accident.

The captain left and we started to talk. The Admiral is the son of a pig farmer from Mobile, Alabama, and his homespun delivery is full of barnyard analogies and southern slang. The act is attractive, not phony. It disguises a mind honed as sharp as a Ranger's trench knife. He laughed when I told him I thought he had really hit the ground running, once he could land, and we spent a little time talking about the party he had thrown for the Marines who rescued O'Grady.

"I meant to take nothing away from Scott O'Grady," he said. "He's a wonderful young American who deserves the attention he got, and he, by the way, has consistently made the point: 'These are the real heroes.' " For a while the admiral had thought about scouring the countryside for every young lady in southern Italy, but there hadn't been enough time. So he had put all his energy into the chow and the beer. The young Marines walked in at first ramrod stiff; then they took a look around. They were wearing short-sleeve shirts with open collars.

"Sharp-looking guys. They loosened up when they found out there was beer and a hell of a lot of food. And they were just as polite as they could be. I walked around and Dottie walked around with them. The fact that it occurred on our thirty-third wedding anniversary made it very special to us. Boy, we kept the food going: pizza, shrimp, turkey."

He thought for a second, then said he thought it was rare in life you could share something so special with anyone. He wasn't talking about the buffet.

"Admirals are real people," he said. "They care about real people. That's the important point."

From any other guy this would have sounded like crap straight from the back end of the bull. But Snuffy Smith actually meant it. You could see it in his eyes. If I had the power, I would have made him Chairman of the Joint Chiefs of Staff right there. The perfect man to blow away all the politically correct Perfumed Princes who since before Vietnam had threatened warriors' lives with fuzzy-headed social experiments. He's a warrior who cares about the troops, a swabby who worries about groundpounders. There ain't all that many around.

Snuffy Smith had the balls to snatch O'Grady right out of the middle of the Serb Army in broad daylight. But that was only half the story. The truth was if the Chairman of the Joint Chiefs and the President had been listening to the Admiral all along, O'Grady would never have been shot down and the Marines would not have had to risk their leathernecks to save him. Admiral Smith didn't tell me this story. I got it from a soldier for the truth elsewhere in the NATO command.

General John Shalikashvili, the Chairman of the Joint Chiefs, told Congress that we didn't know any SAM-6 missiles were in residence the day the Serbs brought down O'Grady.

This defies belief.

The intelligence community knew about the SAM site. Our pilots knew they were up against SAMs. Our satellites, our intelligence aircraft, were all linked through the intelligence community to the Chairman of the Joint Chiefs. In Naples, Admiral Smith knew about them. Month after month, he tried to persuade NATO, the diplomats, the Pentagon and the National Security Council people to do something about them. "We knew where all the SAMs were," my source told me. "But we didn't have the balls to take them out."

The day O'Grady went down, an intelligence report from a U-2 spy plane identifying the exact site of the SAM that got him reached the Pentagon—about six hours before he took off. Everyone knew about that SAM except O'Grady. Word didn't reach him before he climbed into the cockpit. This was a spectacular example of how we spend ourselves broke on intelligence and then do not use it in time to do any good.

O'Grady wasn't supposed to be doing anything particularly dangerous the day he was shot down. He was just flying air cap on an operation called Deny Flight. The job was not to attack targets on the ground but to destroy any aircraft that might wander in, specifically Serbian aircraft flying ground support. He was flying calmly along when the SAM-6 painted him. Before he could do anything, the missile exploded and ten days later he was on *Larry King*. Why did this happen? It happened for the same reason the Rangers were sent to capture Aidid without armor or an air cap. President Clinton would not let Admiral Smith take out the SAMs for diplomatic reasons. He was afraid it might escalate the war.

When O'Grady nearly bought the farm, the President, of course, said he was very upset our airman had been shot down. He directed the Chairman of the Joint Chiefs of Staff to appoint an investigation team to fly over to Europe to find out what had gone wrong. Can you believe it? The SAM that blew O'Grady out of the sky had been inside the Banja Luka Air Defense envelope for two years. The Pentagon knew it was active. In sending out the investigating team, the Chairman of the Joint Chiefs was trying to misdirect responsibility, to pin it on someone other than those who were really to blame. We did the same thing in Vietnam. Our pilots saw the Vietnamese putting up SAM sites. They were told, "Don't touch them. You will enlarge the war." So the Vietnamese hardened their air defense. Not until then did the authorization come through to take them out. Too late. A lot of guys wound up in the Hanoi Hilton because of this stupidity, a lot of guys who knew Snuffy Smith. O'Grady was a hell of a lot luckier.

The men ultimately responsible were the Chairman of the Joint Chiefs of Staff, the President, and National Security Adviser Tony Lake. They had been given full warning. This was not a real investigation. The President's people were just playing damage control. Luckily for them, O'Grady turned out to be such an attractive young guy the image manipulators were able to lionize him and divert attention from the stupidity that had nearly gotten him killed. No wonder he quit the Air Force.

At the time I thought the whole sorry episode had only pointed up once again our chronic intelligence problem, the same one that hurt us from the first shots of World War II to Desert Storm. We have an expensive, high-tech apparatus that collects intelligence rapidly but cannot get it quickly enough to the user. At Pearl Harbor, primitive radar picked up the Japanese, but we thought we were looking at our own planes, not raiders from the Japanese fleet. During the Battle of the Bulge, the kids in the foxholes knew the Germans were preparing for battle but the word didn't reach Ike until too late. Again in Korea, we grunts knew China was in the war because we were capturing Chinese. But the intelligence folks didn't get the flash until we were surrounded by a huge Chinese army. In 1968 the intelligence guys responsible for the Viet Cong order of battle knew the Tet Offensive was coming; they had the clue in their in-boxes, but it was buried three days deep in other material. They had so much intel coming in, they couldn't read fast enough to keep up with it.

We now spend about $30 billion a year on intelligence that arrives too late. Who warned us the Soviet Union was about to fall apart? Might have been useful information. Might have saved us a lot of money. During Desert Storm a general I know took me aside and said, "I'd swap all this highfalutin intelligence information for one spy who has dinner regularly with Saddam Hussein." Afterward, General Norman Schwarzkopf went in front of Congress and blistered the intelligence community. He said his intel had been so overwhelming there'd been no way to read it all.

We need to reform and consolidate the intelligence community, get rid of its multitiered bureaucracy. We can't have a system where the Defense Intelligence Agency won't trade a secret with the Central Intelligence Agency, and the CIA won't deal with the National Security Agency because all three want and need scoops. The guy with the most scoops ends up with the biggest budget next year. That is what Washington intelligence is all about. A few forward thinkers around the CIA and the Pentagon have been talking about these very ideas. Talk is easy, but follow-through is what's needed.

Admiral Smith must know all this, though he would never talk to me or any other reporter about it. But as we sat there going through the small talk that seems to begin every media interview, I couldn't help thinking back to the previous summer and fall and the bulletins I was getting on Bosnia from my own intelligence operatives:

"CINC wants to use massive NATO air. Anything on the road, shoot it. NATO has gone to the UN and said, 'Here's our battle plan.' UN won't approve."

"Smith is terribly frustrated. No human intelligence. Running everything through what they are picking up on satellites."

"No guidance from National Security Council. Nothing but questions, questions, questions. Then no replies but more questions."

"We've got a terribly weak Chairman of the Joint Chiefs."

"Air strikes less than 50 percent effective. Incredibly disappointing."

"Bomb Damage Assessment still 50 percent, which is very, very terrible. All kinds of excuses, especially from U.S. Air Force. Best plane is the F-18D, which is all-weather, and the old A-10 Warthog. Rest is junk. Flies too fast, can't loiter, won't do job."

The big difference between Bosnia and Desert Storm was that the Iraqis had no place to hide and they didn't fight back. A spy told me, "The SA-7 shoulder-fired missile, fired by determined Serbs, is a pee burner. It's keeping us above fourteen thousand feet where the guy in the airplane can't see worth shit. We're just using lasers, which is wonderful if you can see the target. But you can't see through the clouds, making the target impossible to hit."

At one point we were using cruise missiles to get around the weather. I learned at the time that when Admiral Smith found out about it, he smoked a few asses—you don't use a weapon that big and that expensive to knock out a stinking little radar. By the end of the air campaign, a source was saying, "Now we're just plowing up old ground. I think we're spending a cool fortune. We're snapping at a billion dollars."

The air attacks were terrifying. But their most significant effect was to make it possible for the Croats and Muslims to advance on the ground. In effect, NATO supplied the Bosnian Muslims and the Croats their own air force. Big-scale action passed after late September and October because the Bosnian Serbs had their ass handed to them by NATO air. Two things then happened. First, they realized they could no longer fight the kind of war they had been fighting with someone else hammering them from the sky. Second, Serbia cut them off from their logistics.

At that point, they could still say, "Okay, we can try to fight with this guy coming down from the sky, we can go into the dense woods, we'll be like the Viet Cong and perhaps we can survive the beating from the air." But at the same time, they had to be saying to themselves, "We are not going to get the fuel and the ammo, we are not going to get the financial support every month to pay the soldiers." Insurgents need a sanctuary to supply all the goodies. Without one they are dead meat. Those were the forces that prodded them into seeing that for the time being at least they were in a no-win situation. Only then did it work for Richard Holbrooke, the President's negotiator, to lean so heavily on them. Airpower alone did not force them to go to Ohio and do the deal at Dayton. No matter what the Air Force says, life in the New World Disorder is never that simple.

All of this had gone into shaping the hand Snuffy Smith was holding the day I talked to him. He had plenty of problems. He had to work with a 120-mile supply lifeline that stretched over bad roads, wide rivers, and banged up or downed bridges. Thick ground fog and heavy cloud cover could shut down his air resupply traffic or aerial reinforcements if a unit on the ground got in trouble. Thousands of wild card rogues who didn't give a damn what their country's leaders agreed to on a scrap of paper in Dayton could set ambushes, snipe, lay mines, and launch terrorist attacks. When winter lifted and the crazies left the warmth of their stoves, anything could happen.

The Admiral knew all of this, but he radiated confidence. He began by telling me he thought that IFOR's first accomplishment had been to recycle the NATO troops who had been serving under the United Nations into IFOR soldiers with real muscle. He said, "We were concerned that taking off a blue beret and putting on a green one was going to be a mind-set change that the guys on the ground would not be able to adapt to. We were dead wrong. Our guys put on the hats. 'We are different,' that's what it showed. 'We are different—get the message now. We are not going to be humiliated. We have a mission. It's up to you to help us.' The word got out. People welcomed them. I heard the stories. About the second or third day, this Brit was somewhere south of Banja Luka, and he saw an old man out there cooking a pig. The guy walks up and says 'You know in my country we always cut the thing up and put it in an oven.' The old guy says, 'I thought you boys were here to tell us how to make peace, not how to cook our pigs.' "

I could see the story had tickled the pig farmer's boy in him and I couldn't imagine it coming from any Perfumed Prince. For a while after Snuffy got his fourth star, there was a campaign around the Pentagon E-ring to polish him up. "Leighton" was in and "Snuffy" was out. The name came from a cartoon character and to every grunt who uses it, it carries the highest respect as well as true affection. No one I know talks that way about Stormin' Norman. Admiral Arthur says, "When Snuffy puts on his country boy act, it fits perfectly." One minute he is delivering a stag joke, the next he is inspiring men to go into the teeth of hell for their country. There is no guile about the man, just the opposite, but believe me, he's a master salesman. He could sell the local Ford dealer a Cadillac.

The Dayton agreement gave him more kick than an Alabama mule. U.S. negotiators wanted to avoid any repeat of Lebanon, where 241 Marines were killed in a terrorist attack in 1983, or the Rangers' firefight in Mogadishu ten years later. The military annex to the Dayton peace agreement put it this way: "The parties understand and agree that the IFOR commander shall have the authority, without interference or permission of any party, to do all that the commander judges necessary and proper, including the use of military force to protect the IFOR and to carry out its responsibilities, and they shall comply in all respects with the IFOR requirement."

Translating for me before the interview, Colonel Hunt had said, "The first guy that violates the agreement will get it hard, right in the teeth, and I doubt if there will be a second incident."

The Admiral's own version was a little more diplomatic. He said, "We got some real tough crunch times coming. Sarajevo's going to be one. There's still a powder keg with all kinds of fuses lit. I think there are fewer fuses, and fewer people have matches. But they're still out there and we have to be very cautious that we don't underestimate the seriousness of the emotion that surrounds us."

The Bosnians were clearly expecting miracles from Snuffy and this bothered him. It also bothered him the media was perched like vultures on a dead cactus waiting for the first sign that IFOR was screwing things up as badly as the United Nations.

"I can't make this a crimeless country any more than I can make Houston, Texas, or Chicago, Illinois, crimeless," he said. "No matter how many people we put on the street, we can't stop lawless activity in a country where people are armed to the teeth and emotions are running so high. This is a police problem, not an IFOR problem. What I have said is, 'Here, look, we want to try to establish a secure environment in which the people of this country can begin to live life in a more normal way. This doesn't mean I can allow everybody to go anywhere they want anytime they want to do it without some thug taking advantage of them—and we've got lots of thugs in this country.'"

When I asked him how this was going over, he said he had been talking about it the night before to the Bosnian foreign minister. He told him, "I don't want to dot *i*'s or cross *t*'s about what's in the damn peace agreement," which seemed to me to be a pragmatic working attitude. But Bosnia is not exactly the land of pragmatism. "Let me tell you," he said. "He did not want to hear that."

The first thirty to forty-five days were sure to be critical to the chances for peace because they would put to the acid test the zones of separation around the former lines of confrontation, particularly in Sarajevo. Within the first thirty days all parties were to withdraw four kilometers from their old positions, somewhat less in places around Sarajevo. After that he had the right to put forces in and provide military security.

"Big word, 'military security,' " he said.

"What we are hearing from the Bosnian Serb side is that at D plus thirty, when IFOR comes in, they're gone. Because right now they're scared to death of IFOR."

So where the United Nations had been seen as a toothless old hound, no one was eager to mess with the new boys in town. But echoing Secretary of Defense William Perry, Snuffy said he was not there to play cowboy. He was going out of his way to reassure the Bosnian Serbs, who had a lot on their conscience and on their nerves. One of the first ideas he had was to do a *Larry King Live* on a Serbian talk show. Someone called in and said:

"Admiral, is it true that IFOR is going to arrest all of the Serbs in Serb Sarajevo [the Serb suburbs of Sarajevo]?"

"I said: 'Absolutely not. I don't have the authority to arrest anybody. The only thing I will do is this, if in the normal course of our duties we come in contact with an indicted war criminal, we would detain him and turn him over to the International Tribunal.' "

In my mind the really tough question was whether the diplomats would try to hogtie him so tightly he couldn't do the job. If that thought worried him, he didn't show it. When I asked him if he was satisfied with the rules of engagement for Operation Joint Endeavor, he said, "The NSC gave us what we asked for—and it was not easy. There were a lot of people who wanted to limit it. Our position kept coming back. They'd say 'This is a compendium. What contingencies do you want?' and I said, 'Everything. I put the goddamned thing together and that's what I totally wanted. Now give it to me.' "

"And you got it?" I asked him.

"Yeah," he said.

I thought back to the image of President Clinton and Anthony Lake, the National Security Adviser, smoking victory cigars after Snuffy rescued O'Grady—as if the accomplishment were theirs. If they're smart they'll give him a loose rein and they might get a few more of those cigars. If they try to flimflam him, I think I know where he will tell them to cram their stogies. He is not going to let his troops be treated the way the Rangers were in Mogadishu.

"Let me tell you," he said, "when we say the senior soldier present has the authority to order the use of deadly force if he believes that he or the soldiers in his charge are in danger, that's the order."

He didn't give any orders to shoot to wound.

"That doesn't mean I'm asking anybody to shoot to kill," he said.

But any grunt in the field now knows that Snuffy will back to the hilt anyone who has to straighten out combatants who cross over the line. The more I talked to him, the more it seemed to me that for all his aw-shucks style, his political sense was a lot better than anything I had seen among the policy crowd. He was not just a military man. He had a lot of finesse. He made a point of observing that the main military thrust of Operation Joint Endeavor would not come in the first six months of 1996, that the true test would come after that.

"The big event this year will hopefully be the elections. And so, if we have this environment in which the elections can take place, where people can feel free to go to the polls and vote, that's going to be a key. My function is to keep the warring factions apart. Let's get the mines pulled up, get the weapons under cantonment areas, get back from the zones of separation, do the transfer of territory, stop the fighting.

"Prime Minister Silajdzic told me, 'I've been asking you guys to come here for four years. We can't make peace ourselves. You've got to help us.' I can do that. I can understand that. The people of this country want peace. I can testify to that. There are people here who are going to hate each other the rest of their lives, and that won't change. And there are going to be people who are going to be distrustful about what is said to them. They have been conditioned to not believe it until they can see it, touch it, feel it, smell it, and in some cases eat it.

"Let me tell you what the Serbs are telling me. This is a legitimate expression of fear on their part; I think it's overstated, but here's what they say. 'You've got one hundred forty-two thousand Bosnian Serbs who will not live under Muslim rule, and they're going to leave. Now what are we going to do with one hundred forty-two thousand refugees in the middle of the winter?' And my answer to them is 'They don't have to leave. I can do something.'

"They said two things: that they won't live under Muslim rule—they said they're scared to death and they want me to buy some time for them. And I didn't say yes, and I didn't say no. I accept the responsibility for creating, at least in the minds of some, very difficult thoughts.

"What I told them in the room was this: 'I can do something about the fear. I can try to help you with that. I can put forces in and if we see atrocities occur, we can stop it. But I can't do anything about the psychological impact or the psychological block about living under Muslim control. That's something you're going to have to work on.'

"Fear is a natural by-product of the environment in which they live. And I don't blame them for being afraid to come. And I don't blame them for saying, 'Who is this guy Smith? I mean a Navy guy sitting in a landlocked country running an Army business. This doesn't make any sense to me, so why should I believe him?' "

If Colonel Hunt was right about what you see is what you get with Snuffy Smith, what we were getting was a natural leader with a gut fighter's experience and a diplomat's subtlety, a combination of the best traits of three great generals: U.S. Grant, George Marshall, and Dwight Eisenhower. Under the good ol' country boy act there was a damn sharp brain, a sledgehammer-simple style—and an iron will. I left his office that day wishing to hell we could clone him. It would drive the Perfumed Princes nuts.

I drove back to my hotel that afternoon seeing plenty of clues Sarajevo was changing for the better. The first time I'd come, the city was a free fire zone, definitely a hot, dangerous place where it was very easy to leave your front door and walk into a body bag. Back then the streets were mostly deserted. When you did see people, they were moving fast to avoid the incoming mortar rounds or the buzzing lead in sniper alleys. The only vehicles using the streets were armored cars. I could smell death everywhere. The locals wore the same vacant, thousand-yard stare you see on the faces of scared and exhausted combat infantrymen.

Now the fear was gone in downtown Sarajevo and the city was buzzing with energy, not slugs. All is relative, of course. Over two million rounds of killing stuff had pummeled the city, gutting parts of it like Berlin after World War II. The Muslim section had suffered badly. Before the fighting it had been a magnificent place with towering high-rises and charming old neighborhoods. Now hardly a building was free of scars; everything was still blackened and broken. But it amazed me to see how high spirits were running among the people—I was impressed with the dignity of the inhabitants: 30,000 Croatians, 100,000 Serbians, the rest Muslims.

Considering they had come out of four years of war, they looked good, dressed well. It made me think back to Italy at the end of World War II. The Italians in the middle of similar ruins were gloomy, really down. That was because they had fought on the wrong side, I suppose. But in Sarajevo, these people had fought well and you could tell they felt they'd won their battle. There was a certain pride in their walk, like the quick stride of a GI on leave in Rome after a long hitch up at the front. Not a swagger. Just a walk with a lot of pride, almost dignity.

I got back to the hotel to find the rate had doubled overnight, from 100 marks to 200 marks. Inflation was ripping through the country, but this was too deep a gouge for me. Everyone was trying to soak the rich Americans. Leif and I shopped around and found a guy who wanted to rent his apartment for 700 marks a month. So we took it, and made it our base. It wasn't much; the electricity was rationed and you only had water in the morning from five to seven—so if you wanted to shower you had to get up early. Every day, one of us got up, filled all the buckets, and we had water to flush the toilets throughout the day.

That night I met the SEALs and we had an old-fashioned reunion, shot the shit about Haiti and other exotic places, talked about the new mission, put away $90 worth of beer. They all felt the same way about Snuffy that the Marine posted outside his office had felt. Remember, you can't bullshit these guys. When a general or an admiral is an asshole they'll guard him with their lives but they won't cover up for him.

I spent the next few days checking out our psy-ops information war. The Admiral was putting a lot of faith in it. When I told him I thought it might be the key to the mission, he had said, "You've hit the nail on the head. If this were a perfect world I would have had those guys in here a month ago." The problem was to win the hearts, and minds of the Bosnians, not to piss them off with an arrogant display of who had the biggest dick.

This hasn't exactly been one of our natural strong points. In Vietnam, one of our stated military objectives was also winning the hearts and minds of the Vietnamese people, but that was done by grabbing them by the balls in the belief their hearts and minds would follow. Several good men in the psy-ops told me too many of the U.S. high brass, including General William Nash and General James P. O'Neill, who cut their teeth in Vietnam, still didn't quite get it. When O'Neill crossed over into the country, a militiaman asked him for his passport. "That's my passport," the general said, pointing to a grunt's M-16. "This cowboy attitude won't change until the top brass who fought in Vietnam change their balls and hearts' mind-set," one of the info-war Green Berets told me. The Brits, French, and Bosnians were already talking about the American elements of NATO as John Waynes, too quick on the draw. The Admiral knew it. I believed he would retool or relieve the jackboot types. I had no doubt he could turn off the charm and cut nuts with the best of them.

Psy-war was a game the other side could play. One day I was talking to the mayor of Sarajevo, who was eagerly organizing a visit by President Clinton. Right after I left his office I happened to talk to a colonel who was following the arrangements.

"Visit's scratched," he said. "Ain't gonna be one. We can't guarantee his security."

That afternoon at five-thirty a rocket fired from the Serb side of town across the river struck the Sarajevo tram killing a woman—it took off half her head—and wounding nineteen other people. I immediately beat feet to the scene. The rocket hit about two hundred meters from the battered Holiday Inn where the JIB was located and reporters often stayed. The shot was not random. It was a little calling card.

Everything was pretty well cleaned up by the time I got there. So the next morning I got up early and visited the three IFOR armored cars that had been stationed right where the rocket had come in. They were there every day. I talked to the French paratroopers manning them. It turned out no other reporters had talked to the soldiers—everybody had been over at the Five O'clock Follies, the same story in Bosnia as in Desert Storm. Few reporters talk seriously to the grunts. Instead, they go to the conference show-and-tell, listen to the horseshit, and argue back and forth with the official minders. There are always a few press springbutts trying to impress everyone with their cleverness. The countries may change, but the acts that make up the Five O'clock Follies never do.

I bypassed all that.

"What happened?" I asked the paratroopers. It turned out I was talking to the gunner who had responded to the rocket.

"At five-thirty I was in the turret in my armored car," he said. "I saw a flash from a window of this apartment across the river, like two hundred meters away, and I heard an explosion—the tram—and I immediately started shooting, putting heavy fire from my cannon into where I saw the flash, also my machine gun, as did the other two armored cars."

This was fascinating because at that very moment a lot of reporters were yammering that IFOR had been sluggish in its response. The implication was they'd been no better than the United Nations. In fact, the French gunner had cut loose exactly the way Snuffy Smith had authorized his troops to cut loose; he had been aggressive, not tame, just as the Admiral had predicted. I was convinced the French boy was telling the truth because he was only a private and he had no reason to lie to me. In fact, he had very specifically said, "Now listen, if my captain comes you must go away because I am not authorized to talk to anyone." But I had kept on talking parachute stories, so he probably thought I was from military intelligence. The real point is that you can usually get the straight skinny if you know how to talk to a soldier.

Unfortunately, I discovered our psy-ops people were even slower than the regular media in getting out the truth. Our information war was a great idea but badly executed: The radio station was twenty-four hours behind everyone else with the story; the mayor of Sarajevo was angry because no one from psy-ops had checked in with him; the colonel in charge thought the mayor had it backward, that he should be checking in with us, and there weren't enough staffers fluent in the languages needed to get broadcasts out smoothly. In short, we had a long way to go before anyone was going to win anyone's hearts, minds, or anything else. Everyone was still trying to get everyone else by the balls.

Under the circumstances, President Clinton decided to skip Sarajevo. He flew into Tuzla at the end of the week, keeping a promise he had made to the troops at Baumholder to visit them in-country. The President chose to do it Douglas MacArthur's way.

The President was due at eight so everyone turned out at six.

Because of the weather he couldn't land; his plane remained in orbit. Finally, it flew into Hungary and didn't get back until after lunch. The whole time the troops were virtual prisoners at the airfield, little cardboard cutouts the President needed for his photo op. They had to stand at the airfield from six in the morning until two in the afternoon waiting for him. If I had been General Nash, I would have dismissed my troops and let them go sit by the fire. Then I would have met the President and explained to him it was just too damn cold to keep good men and women out there freezing, but I did have five or six men by the stove in my quarters waiting to talk to him. Instead, it was the arrival of Caesar, and all the frozen legions had to wait at attention.

I am sure a lot of old paratroopers felt like throwing up when they saw the pictures of the President, that Vietnam draft dodger, wearing their coveted red beret and wading in among the grunts. That photograph was the real reason for the visit and every editor who ran it fell for a cheap media manipulation. This surprised me. Ronald Reagan overworked this game so flagrantly that editors got onto him and started looking more skeptically at these set-up situations. They did the same thing when George Bush got too clumsy with them during his losing campaign against Bill Clinton. But this time they were complete suckers for that red beret.

Afterward, I talked to a lot of soldiers, and those who had actually shot the breeze with the President were very generous in their reaction. They said he was charming, electrifying. A lot of this was because he is a supercelebrity and these were ordinary guys who don't get many chances to rub elbows with the great and the near great. Obviously, the President is a magnetic personality; no one can take that away from him.

But once again, as in Baumholder the month before, a lot of grunts felt they'd been used. Still, they were big about it.

"It wasn't such a bad deal," one of them told me. "At least I got a shower."

It was the first shower he'd had since leaving Germany a month earlier.

Later, I got a letter from the mother of one of the kids who'd been in the President's stage show that day. She told me her son had been wearing the same shorts he had on thirty days before when he arrived in Bosnia. He'd been shipped from Hungary with two-hundred men on boxcars with no heat or sanitation facilities, a scene right out of World War I or World War II. The railroad cars had been so cold the men got down on the floors only to burn their faces from the heat of the wheels. When they were sent onto the tarmac to meet the President's plane, they were told to cheer on command. The security guys stripped them of all their ammo, leaving them exposed to snipers or any other Bosnian crazies in range.

The mother was angry. She called the Pentagon to complain. Finally, an Army major told her she was right, but couldn't tell her why it had happened or what was being done about it.

Nothing was being done about it. When the hell did a photo-op Caesar ever care?

CHAPTER 10

DESOLATION GULCH:

THE BALKANS, 1996

Route Pepsi ran straight as a bayonet through the center of no-man's-land. The road cut through the middle of a field, an ancient cornfield turned into a killing zone. Weeds three feet tall poked up through cracks in the mine-covered surface; the asphalt was littered with shrapnel and spent ammo. For more than three years no one walked this road. From one line of trenchworks a Muslim sniper would waste you, if the Serb sharpshooter across the field didn't draw a bead on you first. Or if a mine didn't convert your body into purple mush.

Now the sun was just beginning to come up over the American peacemakers.

On the muddy track in front of the now-vacant Serb fighting positions, Captain Kevin Volk, commander A Company, 3rd Battalion, 5th Cav—the Black Knights—was standing alongside a band of unarmed Serb officers and soldiers. His mission: to complete the withdrawal of the Serb forces from the Zone of Separation and link up with his battalion S-3, Major Ray Castillo, who was coming north with a delegation of Muslims.

Mists swirled up over the field. Through the gray morning fog, you could see the Bosnians still tramping around their forward positions, as if coming over to Route Pepsi didn't seem like such a good idea.

"They can't be trusted," muttered Captain Vidakovic Milorad, a Serb operations officer.

The captain and his men were edgy, angry. You could see standing out in the open scared them. After four years of battles, all their senses were screaming, "Get down. Take cover. A sniper has you in his sights."

Captain Volk got on his radio to prod the other side. Through the mist the Bosnians finally began to move out of their trenchworks. They took their time. A lot of time.

"We're happy to see American soldiers here. They are bringing peace," said Captain Milorad.

The little commercial sounded as hollow as an empty coffin.

It was past high noon when the ZOS was finally empty. Down Route Pepsi came the Bosnian delegation, led by Major Castillo. The Bosnians and Serbs met and shook hands as woodenly as marionettes.

Over to the side, Specialist Dewayne Stoner, pulling security for Captain Volk, swept the scene with his alert, sniper's eyes.

"Didn't look like too much happiness out there," he said.

I turned and talked to Mumanovic Semsudin, the Bosnian battalion commander. He was a big unsmiling dude with bitterness and hatred seeping out of every pore. I asked him how many casualties he had taken.

"We kill more Serbs than they kill Muslims. They can't be trusted."

Later, a hard-bitten A Company sergeant summed it up for me.

"All sides have a lot of hate left in them," he said. "I don't trust any of these bastards."

As quickly as I could, I had gotten out of Sarajevo and into the Bosnian countryside. I had contacted a brigade commander, Colonel Gregory Fontenot, a pal of Dave Hunt, who set me up with one of his companies where I could see a little bit of everything. He sent me to Kevin Volk's A Company, a mechanized infantry outfit of Bradley infantry fighting vehicles and attached Abrams tanks. A Company had two infantry platoons and one tank platoon, which is one hell of a steel punch.

I found the outfit in Ulice about ten miles to the south of Zupanja. The Serbs had ethnically cleansed this Croat village. Every house was bullet riddled, blown apart, trashed. The Serbs had liberated roofing timbers and almost every other scrap of lumber to fortify the bunkers and trench works they'd dug several hundred yards from the center of the village. What timber not used for fortifications had gone into firewood to make it through the subzero winters.

The landscape looked like a battlefield from World War I. Very eerie. It reminded me of Korea in the closing phases of that war. The enemies had hunkered down in opposing trench lines, bunkers, and fighting positions. Mines, booby traps, and debris were scattered everywhere. Barbed wire snaked through the fields. You could smell the stench of rotting garbage, piss, and shit oozing from the ground. Rats big as your arm slithered for cover as you approached.

The troopers called the village Desolation Gulch. The road that ran through it they named Desolation Boulevard.

Leif and I drove up from Tuzla in our UN-white Japanese four-wheeler rental. We stopped first at battalion headquarters. Lieutenant Colonel Anthony A. Cucolo III was the commander, a tall, good-looking West Pointer. We shook hands.

"My dad was in your regiment in Italy," he said, and from that moment on Leif and I were honorary Black Knights.

After the introductions we piled back in the four-wheeler and beat it over to Desolation Gulch. The guy who met us was First Sergeant Brent A. Stoneberger. He had a million-dollar scar running from the top of his eyebrow to the center of his cheek, and he could swear like a mule skinner. A very good man. "I intend to bring this company home with everyone we brought here," he told me. "If I piss off a lot of people along the way for being tough, so be it." I'd want my son to be in his outfit.

Sergeant Stoneberger was Burt Lancaster in *From Here to Eternity*—a great guy who ran a first-class outfit, one of the sharpest and most disciplined I've ever seen in the U.S. Army. Total pros. Back in Germany they'd fired the top score in the Bradley and won a bet with Colonel Fontenot, who had to shave his head. During the time I spent with them I never saw a soldier out of uniform, a dirty weapon, an unalert warrior. And I never heard a leader raise his voice.

On the surface, their morale was sky-high despite the conditions they were up against: boot-top mud and walk-in-freezer cold and being treated by the top brass no better than their great-grandfathers had been in the miserable trenches of France during World War I. On watch and on patrol, with the wind chill turning them blue, they were out there standing tall in the turrets of their Bradleys, always alert. What I was seeing represented a triumph of moral fiber and warrior spirit over logistics. They hadn't had a shower since New Year's Day, when they left Croatia. They were eating MREs or gluck from a can. When not standing guard or out on patrol—they were putting in eighteen-hour days—they slept in large, crowded tents or sacked out in their combat vehicles. Meanwhile, back in Tuzla, the home of Task Force Eagle, the Army was building fancy, heated tents for the REMFs along with PXs, mobile shower units, and almost all the comforts of home: hot chow, telephones, mail, newspapers, you get the deal.

Every soldier I talked to had a horror story to tell about the company's deployment from Germany. Nothing went right.

"A royal rat fuck," snorted one sergeant.

For three years the division knew the Bosnia mission was on the front burner. It rehearsed the tactical side of the job right down to a gnat's ass. But the top brass paid little enough attention to the logistics other than writing up their neat little movement plan.

Normally when a unit—from a 100-man rifle company to a 50,000-strong corps—deploys into an operational area, it always remains ready to fight. It stays together so it will not lose its cohesion and punch. A small advance party goes ahead to square away everything for the arrival of the main body. Everything is supposed to be organized in advance: food, lodging, latrines, security. The advance team then meets the main force at a prearranged spot and guides it to the new position.

Since the U.S. Army has been doing this trick for more than 220 years, believe me, it has the drill down to a science. But what happened to A Company and almost every unit in the twenty-thousand-strong U.S. Army slice of IFOR was the total fulfillment of Murphy's Law. Everything that could go wrong did go wrong. Spectacularly wrong. From the top NATO generals and their huge staffs to the division staff to the Air Force transport people, all the military logisticians blew it big-time. Those responsible for one of the biggest logistical fuck-ups in U.S. Army history should be sentenced to the line troopers' shit-burning detail for the rest of their natural lives.

First, the Air Force forgot to tell Captain Volk and the other commanders when the flight for Hungary was taking off with their command vehicles. They had to scramble for a later flight. Then the Army hired commercial buses to ferry the grunts. In Company A's case, the German bus drivers refused to take them to the rendezvous point in Hungary. Instead, they dropped Company A in Croatia. Then the train with the Bradleys and tanks took three days to reach the river crossover point into Croatia. The snafu left Captain Volk in one place, his command vehicle in another, the rest of the vehicles and crews in yet another, and the grunts in a fourth with no commo linking them or any support in place. They were scattered over fifty kilometers apart in a foreign country at the dull end of a long war. Using the wisdom common to street-smart sergeants, Stoneberger at least had the sense to check his unit into a Croatian hotel.

Finally, Captain Volk managed to bum a ride on an Air Force aircraft, find his command vehicle and driver, and, after three days' searching, round up the rest of A Company. Their ammo didn't arrive for three weeks. If at any point they'd had to fight, they would have been in deep shit.

"Someone with stars should be shot at dawn," said Division Command Sergeant Major Jack L. Tilley. "Next time we'll have a division deployment plan that works."

Their first job in Desolation Gulch was to make sure the Serbs and Muslims were hanging up their guns and pulling back from the ZOS, and to supervise the clearing of Serb and Muslim mines while building a permanent base camp. In the early days, Serbs and Muslims continued to snipe at each other across the lines; but they didn't take potshots at the Americans.

"They know we've got too much firepower," Sergeant Stillman Maxwell told me.

Night after night Captain Volk met with the Serb battalion commander, Captain Mladen Gajic, in a smoke-filled room at Gajic's headquarters. Over a gutful of stomach-wrenching Turkish coffee, they worked out the day's wrinkles. West Point–trained, Captain Volk politely refused the Slivovica fruit brandy the Serbs offered him. IFOR soldiers, he told the Serbs, "are not authorized to drink alcohol."

General Joulwon's candy-ass, no-beer policy drew the only bitching I heard from the grunts. "It sucks," said one of them.

In January the sun drops out of the sky like a mortar round. By 5:30 P.M. it was pitch dark at the Desolation Gulch corral. One moonless night Sergeant Thomas E. Tomlin got word over his tank radio to take his combat engineer vehicle—a plow on the chassis of an M-60 tank—and link up with A Company. He had to go down a road that crossed a minefield but the Serbs had probed it, the Army had rolled it, and word had come the lane was good to go.

"Kick it in gear," the sergeant told his driver, PFC Andrew Ringbauer. "Move out."

The CEV was moving ahead cautiously in the tracks of a roller tank that had gone before it.

BOOOOOOOOOOOOOOOOOOOOOOMMMMMMMMMMMMM.

The sound every IFOR member dreads.

The sixty-ton vehicle shook like a toy in the hands of a malevolent child. The loader, PFC Bradley J. Smith, was peering into the darkness from his position in the open hatch when he saw a blinding flash. The concussion banged him against the turret ring, covering him with dirt and gunpowder.

"It smelled like cordite, as if our main gun had fired," he told me later.

His first thought was his buddies had bought it. Down below, Specialist Alex Toirkens, the gunner, was thinking about the thirty rounds of bunker-busting ammo he had in the turret. If it cooked off, all of them were finely chopped liver.

The explosion tore off a one-ton steel track and sheered away the road wheels. Miraculously, no one was hurt. A four-vehicle convoy had just passed over the same mine without touching it off.

An hour or two earlier that afternoon when the sun was beginning to go down, Leif and I had gone out on the same road with Tony Stoneberger. While there was still light we were hoping to get a few pictures of the unit that was sweeping for mines on a strip of road intersecting the Muslim and Serb positions. The sergeant got in his Bradley and we tagged along behind in our tin can four-wheel drive.

The road was seriously muddy. I am talking mud in places up to your knees. We ran into the oncoming convoy and got into a little traffic jam with a tank and a couple of Bradleys that had just come through the "cleared fields" in no-man's-land. The Bradley with Sergeant Stoneberger just popped off the road. I guess he considered it relatively safe. The Serbs had probed it; the Americans had cleared it. And those vehicles had just passed through.

But by then the sun was almost down.

"Look, it's dark," said the sarge. "Why don't we do this in the morning?"

"Fine, sounds good to me."

It took fifteen minutes to wheel the Bradley and our four-wheeler around on the narrow road. And while the drivers were doing it, I heard the boom up the road. No one who has ever heard a mine go off could miss the sound. We had stopped about one hundred meters short of where the buried trouble was waiting. I believe my guardian angel again said, "Whoops, there's a little problem up ahead, so I've got something for you. It's called a traffic jam." So instead of beating the CEV down to the mine, we were safely stuck in the mud.

Mines sneak up on you that way. Especially when they are not supposed to be there.

Theoretically, after the Serbs had done their clearing drill, the Americans should have swept the road with mine detectors. I found out later that American engineers had been swamped with readings from their detectors because the road was filled with shrapnel and spent bullets. You could not turn on a detector without getting ring-a-ding-ding and then ding ding ding. But then you would just recover a slug or a chunk of steel. After a dozen or so such false alarms in below-zero weather, it's only human nature to say "Fuck it."

But then nothing is more freaky than a mine. We have special seventy-ton Abrams tanks outfitted with one-ton rollers. One of these rigs had rolled the road three times.

At first light Leif and I and the First Sergeant went zooming back. We found the CEV and a crater big enough to stash a washer-dryer. I was standing there with two Serb officers and a Serb grunt they called Hawkeye, an ordinary soldier, with incredible vision. Hawkeye took a little walk around the crater. Then he got down on his knees and started digging carefully with his fingers. In a few seconds he had three big tits from a second mine sticking up from the ground. He looked around, repeated the digging, and uncovered a third mine a few feet away from the hole. Antitank mines are ordinarily planted in a triangle of three and the CEV had set off only one. I had been tromping all around them. All of them had been probed, swept, rolled over, walked over, and still not detected. Sergeant Stoneberger looked down and shook his head.

"This kind of drill only shows you just how goddamn dangerous it is regardless of how careful you are," he said. "It's gonna be a long year."

Soldiers understood the dangers they faced; the policy boys didn't. In early January I received a smart-ass letter on White House stationery from Anthony Lake at the NSC. I had written several stories for *Newsweek* pointing up the threat that mines presented to our troops on the ground in Bosnia. *One thing I know is mines. The first*

KIA I ever saw was in Italy a few hundred miles from where Company A was rooting in a minefield. An Army captain tripped a Yugoslav mine, which ripped off both his legs. In Korea, it was three years of wall-to-wall mines, and in Vietnam, five years. I was wounded once by mines in Vietnam, and my Hardcore Battalion of 600 warriors took over 1,500 mine casualties in a year and a half of tripping through the minefields in the Mekong Delta. Lake wanted to correct me. He wrote: "You assert that because the mines used in Bosnia are made mostly of plastic . . . [that] they can only be found by probing with a bayonet. It is true that most of the mines laid in Bosnia are plastic; however, they do have metal components and thus are detectable by our highly sensitive metal detectors and explosive-sniffing dog teams."

If the head of the National Security Council knows as much about the rest of the world as he knows about mines it's no wonder we get into the kind of trouble we do. I have taken the Special Forces mines and demolitions course. I have walked over more minefields than I care to remember. So I think I know a little bit about them. I had just come out of a field in Bosnia where I had seen mines that in spite of all the best techniques, including plows and rollers, damned near blew a tank to the scrapyard. And I am getting a letter from this asshole who has never worn a uniform except maybe one from the Boy Scouts telling me about explosive-sniffing dogs.

President Clinton has surrounded himself with too many such dork advisers whose knowledge of the military is similar to Lake's. When he was elected, he assured us he was going to make the White House staff representative of the country. A fair share of women, blacks, Hispanics, you name it. But I don't think you will find in a senior position of responsibility a fair—or intelligent—representative of real veterans, people with combat experience, people with some sense of what it actually means when Clinton sends Rangers into battle in Mogadishu or infantrymen to keep the peace in the Balkans.

The President gets his advice from General John Shalikashvili, a Perfumed Prince who has as much combat experience as a GI Joe doll. He also has his own personal four-star, General Wesley Clark, another Perfumed one, coincidentally an Oxford and Little Rock pal, who has rocketed up in the Army as fast as Alexander Haig and is just about as popular and phony. No one around the White House seems to have a clue about war or war-fighting.

In one sense I suppose it's unfair to criticize the President this way. He is totally naive about anything military. When he visited Korea's DMZ in 1993, for example, he stared intently through binoculars at northern positions. After about five minutes of this "scoping," a soldier standing next to the President noticed he had not removed the binocular caps. I have often wondered what Clinton was thinking during those long five minutes of darkness. What is he supposed to think when some general marches up in uniform with row after row of ribbons on his chest? Most are having-been-there or keep-your-mouth-shut medals and in the middle there's usually that pseudo Silver Star, the Distinguished Clerk's Politically Correct Service Medal, blinking away. But what does Clinton know about medals? Right away this kind of President is going to be in trouble.

What is the official rationale for the Bosnian operation? One: We have to show leadership. Two: We have to preserve NATO. If NATO does not go into the Balkans and do the job, the possibility exists that NATO might disappear.

If I were Chairman of the Joint Chiefs of Staff, I would argue that NATO became obsolete when the Berlin Wall came down. The Warsaw Pact is no longer lurking over there on the other side. The purpose of NATO was to stop the Soviet Union. Now the Sovs have cold-conked themselves and the purpose has disappeared. What is really keeping NATO alive are the arms makers who sell the tanks and the missiles and the airplanes that NATO uses and the politicians and top brass who are in on the game. Any other way of trying to explain why we still have to have American forces in a European defense agreement where even most Europeans no longer want us, does not add up.

NATO is just one more trip wire. Whenever the next Bosnia blows up, the wire will get tripped again and we will once more have to send troops.

So I would like to propose a new approach. Before any more wires get tripped and forces are rushed to a killing field like Bosnia or Somalia in another futile and costly mission:

- The President should approve the plan.
- The Congress, both the House and the Senate, should vote on it.

Then, for the first thirty days of the operation, the President and the representatives and senators who have signed off on it should go as the first wave to the field of battle. They should dig the trenches, patrol the roads, and clear the mines. After they get a taste of the dangers they otherwise ignore, then the troops can come in and finish the job. And since the President's National Security Adviser, Anthony

Lake, knows so much about mines, I think we would all feel a lot safer if he were the first guy down the road.

Of course, I would rather trust Snuffy Smith. Right now, the Admiral has the brutal job of writing the book on peace enforcement. He sums up his biggest problem in a single word: "expectations."

"Every single day something new comes up on the horizon," he told me. "This is a brand-new operation. I must tell you there's some nervousness about this. Obviously NATO is going to see this through and it is going to succeed. Now at least we've got the right start. Where we go from now, we'll have to see."

My own guess is we have just dipped our toe into the Balkan swamp and we are going to be getting in a lot deeper before we get out. If Tito could not teach these crazies manners in forty years, how can Clinton expect to accomplish much in one?

The bombing campaign caused the Serbs, Croatians, and Muslims to have profound respect for the Big Stinger in the Sky. They are going to be very reluctant to take on Uncle Sam's airpower. Before the year is up (November 1996, election time), there will probably be an incident or two ranging beyond sniper, terrorist, and mine level. But with the firepower Snuffy Smith has at hand to throw at them, I'd hate to be the first guy to cause a shootout. He is going to have everything thrown at him. And when that happens, every Bosnian is going to say, "Holy shit, did you hear what happened to the three little raiders? The Americans dropped twelve B-52 runs on them and twelve million rounds of artillery." It is going to be overkill like you will not believe.

Right now the principal players are thinking this way: The Yanks are coming. They are going to throw billions of dollars at us to shut us up and be peaceful. So let us grab the money and stuff it away and when they leave in one year, we will go back to it.

Because this civil war is not over yet.

To understand the level of hatred in the Balkans you do not have to be a historian. You do not have to go back one thousand years. Forget about all the atrocities in World War II, World War I, Kosovo in 1104. All you have to do is consider the past four years. These people have seen their fathers killed, their brothers killed, their mothers and sisters raped, their relatives disappear. The wounds are too deep to heal in a year. The only thing sure to happen within that year is the American election. Politics, not reality, has wound the clock.

So they will take the money, hide their weapons, wait until we leave, and then resume the war.

That is what ordinary people say: Croats, Serbs, and Muslims, the little guys.

The big guys—the mayors, the hustlers—say, "Everything is going to be great. Boy, we'll put a big hotel over here and another one over there." They are busy looking at it from the point of view of how full they are going to stuff their pockets. But that's not the attitude of the people, of the average little guy.

I see two similarities between Bosnia and Haiti: the clock and the gun. According to the clock, we will have spent eighteen months in Haiti. We were to come out just as we plunged into Bosnia. The President assured us he has set a one-year alarm clock on the Bosnian mission. The last President who said that was Harry Truman. He said my unit would be in the same neighborhood for only a year or two and it wound up staying nine, trying to keep the Yugoslavs and Italians from killing each other. The point is all too clear. There is no way you can set clocks on missions so complicated and dangerous.

The other point of similarity involves guns.

In Haiti the bad guys went to ground after we invaded the country. They've been hiding while the good guys patrolled the streets. The first thing they did when we came in was hide their guns. When we leave, the guns will come back out. In Bosnia the same thing is going to happen but on a much larger scale. The CIA has satellite pictures showing the Serbs responded to the Paris Peace Treaty by hiding tanks and cannons in barns, haystacks, caves, and holes in the ground.

Tanks, not AK-47s. And why do you think they were hiding them?

I asked Snuffy Smith whether he thought announcing a one-year clock was wise. He was too loyal an officer to contradict the Commander in Chief. But he did say, "I believe there are definite drawbacks to laying a time limit down and saying, 'OK we're going to be in one year and we're out of here.' But on the other hand, if you're willing to stick to it, then it has some advantages." He put the best possible spin on a bad situation: "Let me tell you how I used that last night on Serb TV. I said, 'Look this is an opportunity for you. Peace is better than the alternative and we don't have very much time. Help us, help yourselves make this work.' "

I pray that Snuffy is right. But I also think it quite possible we will spend at least a year and at least $4 billion or $5 billion on an exercise that might be totally in vain.

If you told me we were going to Bosnia for a generation—let's say for twenty years—I would think that might be enough time for a healing period, enough time to forget what happened to your mother and sister, your father and brother. But to set a reasonable time limit, we would have to decide whether we as a nation can afford to outpost countries like Bosnia for twenty years at say $5 billion a year; whether we are willing to take hundreds of casualties. One big chopper goes down and we lose thirty guys. One transport aircraft goes down and we lose one hundred. A sergeant trips a mine and we lose one.

You'd have to be a pretty damn good fortune-teller to predict the final outcome in Bosnia, but if we look over our shoulder at Vietnam, Somalia, and Haiti, we can get a pretty good fix on how the situation is likely to sort itself out. My guess is that we are going to see more of the same: We are either going to cut and run or throw more good money after bad. We got into this swamp because of Dayton, where all the crooked players had a gun pointed to their head. From the very beginning, the Dayton deal was flawed and it immediately began to unravel like a wool sweater teased apart by a cat pulling at every loose thread. The guy who held the pistol at Dayton, Richard Holbrooke, was the first to grab his medal and run, leaving our troops hanging out there to police no-man's-land.

All the players on that ancient battlefield are pros at dealing from the bottom of the deck. Little more than a month after the Paris Peace Treaty, Colonel Hunt was involved in catching the Bosnian government, supposedly the victimized good guys, in a major deception. Through signal intercepts, Admiral Smith's headquarters found itself reading mail between terrorist groups and offices of the Bosnian government including the president's. Terrorists were working out of Bosnian government safe houses.

Twenty miles west of Sarajevo was a very beautiful ski chalet—really a terrorist training school and bomb factory. Colonel Hunt watched it for weeks, putting all the pieces together, then tasking French commandos to conduct a raid. They rehearsed, conducted dry runs, perfected the operation. While they were practicing, intel showed that the terrorists were about to strike. Colonel Hunt struck first.

A U.S. Special Forces Black Hawk helicopter landed next to the chalet and ten commandos unassed the bird. When the point man rushed to the front door, out popped a terrorist to see what the noise was. The commando used a quick karate blow to take out terrorist number one. Meanwhile, the rest of the squad stormed the chalet and found two terrorists by the fire, their AK-47s against the opposite wall. The raiders smacked the two terrorists against the wall, disarmed them, and tied them up.

While this was going on, the Black Hawk lifted off and disappeared. From outside the chalet, it looked as if the safe house was still terrorist cozy. When the raiders checked out the place, they found sniper rifles, rocket launchers, and children's toys wired with plastic explosives, including one booby-trapped ice cream cone. They also found sand table mock-ups of the U.S. embassy in Sarajevo, Admiral Smith's headquarters, and French headquarters in the Post Telephone and Telegraph Building.

At that moment a car drove up and unloaded three more terrorists, who walked through the front door into very deep shit. They joined their brothers in handcuffs. A few minutes later, a third vehicle arrived—this one with Iranian diplomatic plates—carrying five more little bomb throwers. The raiders seized them, too, along with boxes of documents showing how to build bombs, other terrorist plans, and a forty-six-page operations order for blowing up the PTT Building, which quartered French troops.

Under interrogation, the terrorists squealed like little pigs on the way to the slaughterhouse. This wasn't just a rogue operation. As the intel guys waded through the documents, they discovered an incredible paper trail with the Bosnian government's fingerprints all over it: bills, blank passports, and a lot of other spooky documents. The idea was to blow up the PTT Building and blame the attack on the Serbs, following the earlier pattern of mortaring their own people to win sympathy. The trail led directly to President Alija Izetbegovic's cabinet and to five other terrorist bases in Bosnia.

The terrorists who attacked the World Trade Center in New York got life in prison for their operation. It doesn't work that way in Bosnia. These murderers were held for twenty-four hours, then released into the custody of the very government that had recruited them and paid their bills. Talk about a catch-22. This is the sort of rat fuck we can expect from now on.

It worries me when I see Bosnia soaking up 10 percent of the U.S. Army's total combat and support power. It worries me even more when I see how overextended we are in a world where we have made commitments to so many places. There will be somewhere between fifteen and twenty battalions in Bosnia. We have had up to twenty battalions in Haiti. We have a battalion out in the Sinai on another peacekeeping mission. We have battalions in Kuwait and

Macedonia and the equivalent in Green Berets with the Kurds in northern Iraq and troops outposted in dozens of other hot spots. We are talking about putting others up on the Golan Heights.

The problem is even worse when you consider the multiplying factor of these deployments. Consider our battalion in Macedonia, for example. We don't just have a single battalion tied up. We also have a battalion getting ready to go to Macedonia and a battalion that is retraining to become hard-edged war fighters again after a tour there. It's not that you lose only one battalion that is deployed, but you lose the battalion that just came out of there and you lose the battalion that's preparing to go. So for every battalion on these scattered missions, you can multiply the effect by three.

The Pentagon calls what we are seeing OOTW—operations other than war—missions. They are drawing away our combat power, dissipating it. Everybody is staying busy, but everybody is stretched to the limit doing it. It is in the nature of these missions to dull the mightiest heavyweight sluggers. Here we are taking a well-trained prizefighter and putting him on a line to hand out chow at the Salvation Army. To be a good fighter you have got to work the speed bag, work the heavy bag, you have got to spar. To be a soldier you have got to train, have live-fire exercises, use your main guns, fire and maneuver, and you have to do this over and over until you do it right. Mistakes on the battlefields end up in stretchers and body bags. We have got our warrior built up to fight wars, but he is doing missions where he has to pull his punch—and this pulls all his muscles.

When that battalion comes out of Macedonia, it is not ready to fight. It has been on a peacekeeping mission, so its marksmanship is lousy. Its cohesion is loose, its ability to maneuver, to employ artillery, helicopters, air simultaneously—all that gets rusty. It needs to be sprayed with WD-40 in a heavy way.

We have a limited number of battalions. Given the number we are deploying so far and wide, who is going to guard the candy store when a real threat comes? Who is going to be able to react when some crazy like North Korea explodes? Who knew in August of 1990 that Iraq, a country we were providing millions if not billions of dollars, worth of military supplies, would turn around and attack our ally Kuwait? I really worry about what is going to happen in China. The Chinese are snapping at Taiwan now, saying they might just have to take back their island. When you look at Russia's instabilty, it's obvious we need a military that is not fragmented and spread all over the world on missions that do not directly strengthen our national security. We have got to get our act together before it's too late.

Charity begins at home. We need to take care of America first, deal with our own domestic problems, before we continue rushing off to try to solve every global disaster because bleeding hearts, unmindful of the consequences, watch the tube and decide "America should do something."

The purpose of American armed forces should be to defend America.

CHAPTER 11

ADMIRAL BOORDA: WASHINGTON, 1996

Did I kill Admiral Boorda?

Is a news story worth a man's life?

The answer to both questions is no. But that's the easy way out. Admiral Boorda's tragic suicide deserves harder thought and a full accounting.

On May 16, 1996, the admiral drove to his quarters in the Washington Navy Yard, where he went into the backyard, drew out his son-in-law's .38 revolver, and shot himself in the chest.

A year earlier, during the summer of 1995, I had gotten a tip from an old and trusted friend, Roger Charles, a retired Marine lieutenant colonel who works for the National Security News Service, a watchdog outfit in Washington. Roger told me he suspected that Admiral Boorda, the Chief of Naval Operations, the highest-ranking officer in the U.S. Navy, the man responsible for restoring the Navy's integrity and effectiveness after Tailhook and a terrible run of other scandals, might be wearing medals for valor he didn't rate.

This sounded odd. Soldiers and sailors, airmen and Marines prize awards for heroism even more than Olympic competitors cherish their gold medals. A lot of civilians don't seem to understand that within the military, outside of cowardice under fire, there is no greater disgrace than to wear an undeserved valor award. To warriors, medals are not just pieces of ribbon. They are sacred, the ultimate symbol. They say you've been there, you've stood tall. At a glance, warriors can look at one another and determine exactly where and how well they have done their duty and how much they've bled. Medals are the military's DNA chart. They command instant recognition and respect. Men and women die for valor awards. Wearing one that's unauthorized is absolutely despicable to military people. And every serving warrior knows that to fake your ribbons is not just a dishonor. It is also a court-martial offense.

What, then, was Admiral Boorda's explanation? *Newsweek* arranged to interview him at the Pentagon. "We will tell the truth," the admiral told one of his aides. When he killed himself, the full truth probably died with him.

I felt absolutely sick that Admiral Boorda had committed suicide. I spent my own life as a soldier being responsible for my actions. I have seen young men go out on my orders and die. As I grow older, those memories are always in mind. They do not fade. As a fighter and now as a writer, I certainly understood that the consequences of this story could be grave. It is tragic that Admiral Boorda apparently chose suicide over answering questions about those two valor awards that were not in his records. President Clinton said afterward that as Commander in Chief he wished he could have intervened in time to tell the admiral the issue was not worth his life. Every decent person feels the same way.

Yet I pursued the story. Why?

To me integrity is the very foundation of the U.S. military. Midshipmen at Annapolis, cadets at West Point and the Air Force Academy, ROTC students, those who become commissioned officers, are all imbued with "special trust and confidence." Those are the words that appear on an officer's certificate of commission. The military is held to a far higher standard than anyone else in our society, a standard independent of rank, not binding just upon ensigns, lieutenants, and captains, but all the way up to our top generals—and admirals. All of them are supposed to exhibit these virtues equally. An officer's integrity is the basis and justification for that officer's authority to lead.

These leaders are responsible for the security of the United States. If they lie about readiness, if they lie about the enemy's capabilities and strengths, if they lie about the way they conduct operations or wars, then the nation is in danger. In my lifetime I've seen our country fall into terrible trouble because generals and admirals lied about Vietnam. Their lies doomed a generation of young Americans, ripping this country apart, nearly destroying it. So I pay attention to lies in the military. I know firsthand how dangerous they are, how they can multiply, how one can feed another. I think they should be brought into public view so the citizens can judge them. And I think to ignore them or to gloss over them or to wish them away can be fatal for the country.

My job as a correspondent is to check things out. I've got the military beat. I receive half a dozen complaints a week from my sources about the brass—generals and admirals joyflying around in expensive aircraft at the taxpayer's expense, living like kings in castles, misappropriating huge amounts of taxpayers' dollars. I'm a watchdog. I have to check out each one of those reports because my job is to tell the American people how ready their defense forces are—and how well their tax dollars are being spent. Without such scrutiny from people like me, there would be more Vietnams, more Tailhooks, and billions more dollars wasted.

At least half the time when I check out a lead, I find it's not true. When that happens it goes in the trash can where it belongs. But if allegations are true, if everything checks out, they become stories.

What kind of leader was Admiral Boorda? On the basis of my interviews, I found he was colorful and competent, but also controversial. He came up through the ranks the same way I did, so I could identify with him about many things. He cared about sailors exactly the same way I care about soldiers. He wanted to make changes, to get the Navy turned around at the very moment it is in a terrible funk. He was reported to be damned proficient in his trade. He set the example in everything he did—except what he wore on his chest. That's what I wanted to know about. Why was an officer of such distinction, who certainly knew better, wearing medals his records didn't support? What did those now-you-see-them-now-you-don't Vs say about his integrity, his character, his ability to live by the code of honor—and his emotional stability?

I expected a strong defense from Admiral Boorda and from the Navy. Not suicide. If the allegations had been false, there would have been no story. Now it is even more vital that the story be told.

Here is what I know.

It all began in February of 1995 when Charlie Thompson, a former producer for CBS's *60 Minutes* and a Navy Viet vet who knows his medals, mentioned to Roger Charles that a number of Navy officers seemed to be wearing decorations that looked suspicious. He said the scuttlebutt within the Navy was that something wasn't quite right with the fruit salad on Admiral Boorda's chest.

To understand what happened next, you have to understand Roger. He is one of the most squared-away guys I know, an Annapolis graduate, a meticulous investigator, a right-down-the-line detail man, a guy who checks and double-checks everything, who dots every *i* and crosses every *t* before he makes a move. He is careful. Very careful.

Roger mentioned the tip to me. In March 1995, more than a year before the story broke, he put in a Freedom of Information Act request asking to see the documents covering Admiral Boorda's decorations. The Navy was slow to answer, but there was nothing out of the ordinary in the delay. At the end of July, Roger received a file with nineteen enclosures. Among them were a 1965 Commendation for Achievement Medal to Lieutenant Jeremy Michael Boorda and a 1973 Navy Commendation Medal. The first was awarded for "meticulous training procedures, diligent organization effort, exceptional leadership characteristics, and zealous devotion to duty" while aboard the U.S.S. *Craig* off Vietnam between April and August of 1965. The second was for "peerless managerial competence and unsurpassed technical knowledge, tireless energy, patience and professional knowledge" aboard the U.S.S. *Brooke*, during a second tour off Vietnam.

During the Vietnam War, the Navy issued such awards—"being there medals"—by the boatload. Neither of Boorda's was for valor in combat. Roger thought this confirmed Thompson's hunch, but the citations alone were not enough to justify a story. He wanted to see if there were photographs showing the admiral wearing the unauthorized V devices. He called the Navy's public affairs office to ask that an official photograph of Boorda be mailed to the National Security News Service. The Navy refused, saying Roger had to come to the Pentagon to pick up the photograph. At the time, Roger was working on several other stories that had higher immediate priorities, so he put the Boorda file into his "inactive stories" drawer until he could get the photograph. There it remained until early April 1996.

Then, on page 3 of the April 8–14 issue of *Defense News*, a weekly trade newspaper, Roger noticed a color photograph of Admiral

Boorda with his fruit salad in full view, including Combat Vs on both the Navy Commendation and on the Navy Achievement ribbons. Roger reopened his inactive-stories drawer, cross-checked the documents, and saw the citations did not state that Admiral Boorda was authorized to wear the Combat Distinguishing Device—the Combat V—on either medal. The implication was inescapable. The admiral was improperly, perhaps fraudulently, wearing the two combat devices.

Roger still needed to double-check regulations to make sure no subsequent changes in the rules had authorized Boorda to wear them. Fine reporter that he is, he checked the regs from 1963 to 1996. Nothing was there to authorize Admiral Boorda's Vs. The language governing the medals came from a 1969 Navy Marine manual. It read as follows: "Personnel aboard a ship are eligible when the safety of the ship and the crew were endangered by enemy attack, such as a ship hit by a mine or a ship engaged by shore, surface, air or subsurface elements."

"You can't just be in the theater," Roger thought. "That ain't good enough."

So far, Admiral Boorda clearly had some explaining to do. But Roger was not satisfied he had all the right evidence. He is a very fair-minded guy. He went on to research the histories of the U.S.S. *Craig* and the U.S.S. *Brooke* to triple-check on whether or not they had come under attack during the young officer's hitch. They had not.

Only then did my phone ring. It was Roger.

"Hack, we have an admiral claiming to be something he ain't."

And the admiral turned out to be the Chief of Naval Operations.

We still didn't think one photograph was enough. So we waited. At the beginning of May other photos came in. A shot from 1977 showed Admiral Boorda without the Vs. The first time they appeared was in 1985, after he won his first star. In 1993, there they were again when he became a vice admiral with three stars.

I was in Whitefish at World Headquarters. It was a Saturday and I phoned Maynard Parker at *Newsweek* in New York.

"I think we may have the Chief of Naval Operations wearing phony decorations. Commendation medals with V devices."

Parker had been a U.S. Army lieutenant during the early stages of the Vietnam War. He instantly saw the significance of the story.

"My God."

"I want some space in the magazine."

"I'll talk to the national editor. If the story checks out, we'll do it."

I pushed ahead on gathering more information. I spoke to one Pentagon insider familiar with Admiral Boorda and his record. The first thing he said was there was no way he could ever speak on the record. Then he said, "You're on the right track."

And he gave me a number of Navy sources to call.

Roger was also at work. He had been very quiet in filing his Freedom of Information Act requests. He didn't want the Navy to sniff him out and cover up the facts. The Navy would later claim it had done an independent investigation. A Navy captain in the Judge Advocate General's office checked things out and told Admiral Boorda he was not entitled to wear the Vs. But so far as Roger could find out, no one in the Navy had pressed the issue of why he had put them on in the first place, no one had raised the issue of accountability or disciplinary action. At the same time, apparently no one wondered about the stability or emotional state of the admiral.

To make sure I wasn't being unfair, I spent a few days interviewing former military officers to get their views on the seriousness of Boorda's Vs.

"I've never encountered anything like it," Lieutenant General James Hollingsworth, a retired U.S. Army officer, told me. General Hollingsworth is an authentic hero and combat veteran from World War II, Korea, and Vietnam. He holds three Distinguished Service Crosses (the country's second highest award), four Silver Stars (the third highest), four Bronze Stars for Valor and six Purple Hearts. Among the military, his very name conjures up valor. "It's hard to believe that an honest man would do it," he said. "It doesn't make us feel good about our Navy, which is such a great, honorable service."

"There can be no explanation for it," Major General Jarvis Lynch, a retired Marine, told me. "It's mind-boggling, absolutely astonishing, distressing. I only encountered this on one occasion in my thirty-five years of service and the officer was quickly drummed out of the Corps."

I called General Lynch because in February 1995 he had written an article for *Naval Institute Proceedings* on the problems of morality and ethics within the Navy. Among other things, he had written that the time had come for the senior Navy brass "to set a living example of professional standards and moral courage of the highest order."

Another of the men I interviewed, Senior Chief Glenn Maiers, U.S.

Navy retired, put it in plain sailor's language: "Boorda's the guy who was hired to be the rudder of the ship. The ship's been taking on water for a long time and now the rudder's fallen off and the ship is spinning in circles. Everyone's going to have to look up that rusty chain of command and wonder who's providing the leadership, who's providing the integrity."

The general and the chief were not exaggerating. They were hitting the bull's-eye. Consider the record:

- May 1987: The U.S.S. *Stark*, a guided-missile frigate on patrol in the Persian Gulf, is hit by two Exocet antiship missiles fired from an Iraqi warplane. Thirty-seven of her crew are killed. The ship's high-tech electronic warfare system either failed to detect the emissions from the Exocet's homing radar, or the system was turned off because of its high propensity for false alarms.
- July 1988: The U.S.S. *Vincennes*, an Aegis cruiser on patrol in the Persian Gulf, mistakenly shoots down an Iranian Airbus killing all 290 innocent civilians aboard. The Navy says publicly the ship's skipper thought he was under attack by an Iranian F-14, although the information is known to be false.
- April 1989: An explosion aboard the U.S.S. *Iowa*, a battleship, kills forty-seven.
- December 1990: Naval Academy classmates chain Gwen Dreyer, a female midshipman, to a urinal. The first classman involved is subsequently commissioned as an ensign. Dreyer's parents complain the investigation was botched. She resigns from the Academy.
- December 1990: Three top naval officers are relieved of duty and reprimanded for concealing cost and schedule problems in connection with the Navy's highly secret A-12 stealth aircraft program.
- October 1991: Admiral Kelso, Chief of Naval Operations, publicly apologizes for the Navy's flawed investigation of the explosion aboard the *Iowa* in which a seaman was originally blamed for the blast.
- July 1992: Admiral Crowe, Chairman of the Joint Chiefs of Staff, admits on national television the Navy knew all along that the cruiser *Vincennes* was inside Iranian territorial waters when she shot down the Iranian airliner.

Later, asked by Congress why this information had not been presented earlier, he says, essentially, no one asked.

- April 1993: The Pentagon's Inspector General issues a devastating indictment of the naval service in a report that says dozens of senior officers knowingly lied to investigators about their knowledge of drunken and lewd misconduct at the 1991 Tailhook Convention in Las Vegas where more than eighty female Navy personnel were manhandled.
- February 1994: A Navy judge issues a finding that Admiral Kelso committed perjury regarding his whereabouts during the Tailhook Convention. Kelso moves directly to retire. The Senate endorses his retirement at four stars without challenging the judge's finding.
- April 1996: In a major speech at the annual meeting of the U.S. Naval Institute, former Navy Secretary James Webb, an Annapolis graduate, bluntly criticizes the Navy's leadership for an excess of political correctness and a lack of moral courage.

Admiral Boorda wasn't responsible for all of these disasters. None of them happened on his watch. But he was the leader responsible for cleaning them up. So the issue of his own medals inevitably raised the question whether he might have been part of the problem rather than the solution.

On Tuesday, May 13, after my weekend call to Maynard, I phoned an editor in New York to check on the week's story list. It surprised me to discover the editors appeared to be considering postponing the story. Then Evan Thomas, *Newsweek*'s Washington bureau chief, got on the phone. Evan was supportive, enthusiastic. He said it was a "big story" and offered to help in any way he could.

At the last minute, I needed some help. Three days before, I had had oral surgery and my jaw was acting up. So Evan agreed to interview Admiral Boorda. The meeting was set for Thursday at one o'clock. I filed all my material to Evan Thursday morning so he would have it before the interview. Then Roger briefed Evan and two of his correspondents, John Barry, the regular Pentagon hand, and Gregory Vistica, a reporter who had outraged the Navy with a book on the Navy's faulty leadership.

"We are going to go out of our way to be fair," Evan said. *Newsweek* was going to present the evidence to Admiral Boorda and give the Navy two full days to reply. Evan, John, and Greg grilled Roger for an hour and a half, playing devil's advocate, looking for weaknesses in the reporting. At the end of the session, Evan said to Roger, "I think you have a pretty good case." They decided to press ahead with the interview, though Evan told Roger he wanted to take Barry, his own guy, and not Roger, over to the Pentagon for the meeting with Admiral Boorda. Roger briefed Barry for an additional two hours.

About this time, Rear Admiral Kendall Pease, the Navy's chief spokesman, called to say that since I was not going to be at the interview, Admiral Boorda wanted to postpone it. I've thought about this. I had been looking into another case concerning Admiral Donald Pilling, head of the U.S. Sixth Fleet based in Naples. An excellent source reported he was living high off the hog in a villa that cost the taxpayers $78,937 a year. According to my source, almost $50,000 in Navy funds had been spent to furnish the place with items that included Chippendale-style furniture and expensive imported rugs, all of this managed by a mess specialist earning $38,000 a year. I filed a FOIA request to verify this account. In response, the Navy furnished documentation showing furniture expenditures of $14,841; but my source says the remaining $35,000 was billed as "local purchases." I think now Admiral Boorda may have believed the interview was about the other admiral; that he hoped to schmooze me and kill that story, that when I couldn't come, he decided to cancel out.

Whatever the case, Evan told Admiral Pease the interview was extremely important and when Admiral Pease asked why, Evan told him about the two valor awards. That was about noon. Admiral Pease then went to Admiral Boorda to explain the situation. They locked on a new interview for two-thirty that afternoon. Evan was waiting in the Pentagon when word came through that there had been gunshots at Admiral Boorda's quarters.

Out in Whitefish, I was sitting in my office when the phone rang. It was Roger.

"Hack, Boorda shot himself."

"You're kidding me."

"No, I'm serious. He drove to his quarters and shot himself. He's in the hospital now. I don't know whether he's dead or alive."

The day before I had a momentary thought that something like this might happen. It wasn't based on anything my reporting had told me about the admiral, anything he had said, or anything anyone had said about him. It was just an old soldier's passing suspicion. I tried to put myself in his place. What was he going to do if the story came out? What would he feel when he went down to the Tank at the Pentagon on Monday and the Chiefs of the Marine Corps, the Army, and the Air Force looked at his chest? What would he tell his children?

Even so, I felt no hesitation about proceeding with the investigation. We had assembled substantial evidence, but we had printed nothing, leaked nothing. We were going to give the admiral every chance to explain. If the explanation was satisfactory, if we'd gotten it wrong, we would have stopped right there.

Now, instead of killing the story, Admiral Boorda had killed himself. I felt terrible for him and his family. As I was trying to absorb the news and think things through, the phone rang. The caller was Donald Graham, the chairman of the board and chief executive officer of the Washington Post Corporation. Don, an Army Vietnam vet, told me not to overreact.

"Hack," he said. "Don't feel guilty. You were doing your duty."

He said it was the third time something like this had happened on his watch at the *Post*. He was enormously kind.

"Don," I said. "Let me tell you something. You're a good general. You're looking out for the troops. But for many years I had to order fine young men into battle and many died. I feel really bad, shocked, about what's happened, but it seems to me it comes with the job." He called again a few days later to check on me. "I'm not calling as your boss," he said. "I'm calling you as your friend." What a lovely man.

What happened elsewhere around the media world was not so pretty. Over the next few days it surprised me to see a small number of my colleagues trying to distance themselves from me and the story. One day I picked up the *Washington Post* to see a false story by Howard Kurtz quoting an unnamed Pentagon officer who said that several weeks earlier I had been "bragging" about bringing "down a Navy admiral." This was a damned lie but Kurtz went with it anyway. He let the Pentagon spin him.

John Dalton, the Secretary of the Navy, John Lehman, a former Secretary of the Navy, and Admiral Pease went into maximum damage control. One of their partners was Admiral Elmo Zumwalt, a former Chief of Naval Operations and Vietnam vet who said he encouraged his men to wear the V devices. "You just assumed a ribbon authorized in a combat area carries with it the V," Admiral Zumwalt said, and the Navy cited a 1965 manual purporting to make everything all right. The *Washington Post*, led by the nose, then ran a front-page story under a headline that read BOORDA MAY HAVE WORN

RIGHT MEDALS. That was precisely what the Navy spinmeisters wanted everyone to think and the *Post* fell for it.

The truth was altogether different. During Vietnam, Zumwalt commanded what was called the "Brown Water Navy," the sailors who fought from small craft in the rivers and canals, brave warriors who came under fire and were entitled to their valor awards when they won them for engaging with the enemy. But Lieutenant Boorda served with the Blue Water Navy off the coast of Vietnam and under the command of the Commander in Chief of the Pacific. Even for the Brown Water boys, the regulations very clearly show Admiral Zumwalt had no authority to pass out Vs unless the combat conditions were met. Theater service alone didn't win you Vs for valor. And he had no authority to authorize anything for the Blue Water boys offshore under CINCPAC command, any more than Ike in Europe had authority to give medals to MacArthur's boys in the Pacific during World War II. Admiral Zumwalt didn't know what he was talking about.

But the truth was the last thing the Navy flaks were eager to embrace. The objective was to convey the impression that Admiral Boorda had made "an honest mistake." That's the whole problem with the Navy right now. Instead of looking right into a black hole of terrible difficulties and figuring out how to solve them, the Navy's first impulse in the scandals I've mentioned has been to cover up or wail that it's been stabbed in the back or spin to reporters that the latest scandal is "old news" and that corrective action is under way. The Navy is afflicted with arrogance and self-denial. That's the cycle that must be broken before the Navy can be brought out of sick bay and returned to health. It doesn't help when reporters let themselves be deceived by the damage-control artists.

At times, some Washington reporters seem to wallow in the same clubbiness that afflicts the people they cover. Outsiders offend and threaten them. The same day Kurtz took his swipe at me, he referred to Roger as "a Marine lieutenant colonel who had no journalistic experience" before joining the National Security News Service. All I can say is that Roger knew enough to find the facts and handle them accurately and fairly while the *Post* was being taken for a ride.

Even my own side exhibited Establishment Syndrome. Jon Alter, *Newsweek*'s media critic, was assigned to do an analysis of how *Newsweek* covered the story. When I saw it, I was stunned to see myself referred to as a "part-time contributing editor." Evan was quoted as saying I made him "nervous" because he thought of me "as more of a soldier than a journalist." He also said he was "wary" of Roger because he thought Roger had been "too conspiracy minded" about an earlier story, an investigation of the *Vincennes* disaster and cover-up. Alter didn't allow Roger to defend himself. Like Kurtz, he also failed to mention that Roger had won the Gold Medal for Investigative Excellence from IRE—Investigative Reporters and Editors, an association of professional journalists located at the University of Missouri Graduate School of Journalism—for his reporting on the *Vincennes* story. The whole business was particularly puzzling to me because Evan had been so helpful and supportive—until Admiral Boorda shot himself and the first incoming rounds started landing on *Newsweek*. For a few days I felt the way I did in Vietnam when the Viet Cong started slamming in rounds and the brass took to their heels—and their bunkers in the rear.

The worst moment came when Alter quoted John Barry as saying, "If the story had come to me, I would have been inclined to confide to [Boorda's] aide, 'For God's sake, get those Vs off of him,' rather than do a story that smacked of 'gotcha' journalism." Here was *Newsweek*'s own defense correspondent sounding just like a Navy flak. Was this the kind of journalist Evan trusted over the two old soldiers who brought him the story?

I felt disgusted. Sure, Roger and I have spent a bunch of our professional lives in the military, not the press pool. But maybe that's not such a bad thing. A reporter's job is to find the truth, not to make it disappear, to confront sources trying to conceal the truth, not to offer them sage advice on spin and damage control. I thought back to that time after Desert Storm when Mike Stone, the Secretary of the Army, wanted to invite me to lunch and I turned him down because I knew it wouldn't be ethical to break bread with him then tell a story about the Army that was true but would cause him problems. The truth had to come first. John Barry had been so busy practicing hey-let's-do-lunch journalism that it is small wonder he didn't discover or develop the Boorda story himself.

But it is trivial and ugly to indulge in professional backbiting over a story that led to a man's death. I think the real story of the story—how Roger discovered it and checked it and relayed it through me to *Newsweek*—shows that during the entire course of the inquiry there was never a moment of "gotcha" journalism. I believe the story was legitimate, the reporting professional and fair.

The minute you press beyond Boorda's personal tragedy, you run into institutional questions that are also deeply troubling. How is it that the military system lets someone so vulnerable, someone with such a deep, tragic flaw rise to the top and become responsible for the heaviest matters of war and peace? Admiral Boorda was the first man to deploy NATO in a war role when it dropped the first bombs on Bosnia. He could reach into his sea jacket and find nukes. He had that kind of power. A senior naval officer presides over a fiefdom all his own and commands an awesome force. The weapons are all there. At any moment he can issue orders with devastating international consequences. He must be impeccably honest, unflinchingly brave, and stable beyond the limits of ordinary civilians. This all along was the real issue, not simply the propriety of Admiral Boorda's errant Vs.

That said, his suicide was a tragedy. His career needed no embellishment. He should have been as proud of it as the Navy was of him.

CHAPTER 12

IT'S THE SYSTEM, STUPID:

WASHINGTON, 1996

Sometimes a friend from the services—past or present—asks what's eating me. When this happens, I reply, "I keep wondering if it's possible we're about to close out a hundred years of total war with nothing to show for it but better body bags."

I have spent half of the twentieth century, a century of horrendous violence, chasing wars. In my lifetime, the military has always reflected America at its best, the American tradition of stand-up-and-be-counted, shoot-straight values. Down at the fighting level, in many ways, the military is the last bastion of the finest American beliefs: Thou shall not lie, cheat, or steal still means something there. Today, it inspires me to go into the field with our young Marines, Rangers, grunts, fighter jocks. They still burn with the warrior spirit, the sense of selfless service I first saw during World War II. And I can feel it.

But when I scope out our top military leaders, too often I see political animals obsessed with their careers and bringing home the pork. This all-pervasive new breed of top dogs is made up of politically correct operators, from the President right on down—bloated, self-serving, and morally corrupt. All talk the good talk, but few have ever walked the hard walk. Because of such sorry leaders, I have had to watch the U.S. military go from a fine, duty-first organization of good men and women to an outfit fragged by the "system." For too many members of our armed forces, serving in the military has become merely a job, not a calling or a passion. Dedication, the vital glue that holds a military team together, has been grossly corrupted by the same sickness that is destroying America: *me, me, me.* The real tragedy is that this sick system is eliminating or driving out our finest young warriors. We are losing our very best, the ones who stand tall and win wars.

The essence of leadership is integrity, loyalty, caring for your people, doing the honorable thing. Over and over since Vietnam, I have seen political expediency killing these values. When slickness and cheap compromise run the show, people who refuse to cave in and play the game get zapped. And when that happens, the ultimate loser is our country.

This system mows down its victims in all the services. Truth tellers are not wanted. Consider the case of Colonel David Hunt, one of the finest serving warriors I know. As a young lieutenant in Korea, he threw himself on a live frag grenade to save a soldier. He's the only guy I know who can do more damage to a grenade than a grenade can do to him. He commanded two battalions, one for two years, in Korea. One day his division commander, Major General Jack Woodall, told him he had the best battalion in Korea—the hardest charging, the most efficient, the best trained, the most spirited. It could, it would, be the point battalion in war. But Woodall wasn't going to recommend Hunt for brigade because he didn't want to use up that vital slot. He wanted to save it for a new-breed corporate general, a young Prince. He told Hunt he'd never make general because he was too outspoken, too abrasive, too apt to piss off the wrong people. Here's what he really meant: No way ever would Hunt win any prizes as a salesman for the Military Industrial Congressional Complex. He wouldn't be a General Smoothie. He might not use the right fork. He'd have his mouth full of snuff and be looking for someplace to spit. Worst of all, he'd always tell the truth.

Here was a guy who could be another Grant, another Patton, another Abrams; here was a real fighter who could win battles and provide his warriors with genuine leadership. But he was doomed because he called 'em as he saw 'em. When I heard the story, I asked to see him. I hadn't met him, I just knew him by reputation. And he lived up to it completely. By chance, a few days later, I met with Jim Morrison, then the vice president of Business Executives for National Security, a now terribly lost and troubled reform group, and Senator Sam Nunn. When the senator asked me what was wrong with the military, I told him the Hunt story.

"How did this happen?" he snapped.

"The dancers and prancers go to the top," I said. "They don't make waves. They hustle bucks for their services regardless of the consequences."

"What can we do about it?" he asked me.

"Your committee confirms all senior officer appointments," I said. "You put the stamp of approval on anyone recommended for brigadier general and above. You've got to make sure that the Pentagon sends you war fighters, not Perfumed Princes."

Nunn said he would do it. He wrote up a goodie and sent me a copy. It basically said: When you are considering people for senior grades, don't just consider somebody with a perfect record who has punched all the right tickets. We need war fighters, we need people who tell it like it is. He introduced it in the 1991 Defense Authorization Report. The report stressed that leaders should spend substantial time with troops to learn and to bond. It was a giant step toward killing ticket punching, which cost so many lives in Vietnam and has done so much damage since. Senator Nunn felt good about writing the document. So did I. When he sent me a copy, I got a rush. *Oh, hallelujah,* I thought. *We finally cleaned up the system. Now we will get war fighters at the top.*

I called a buddy of mine, very smart about politics, an aide to a senior senator. And he said, "Ah, forget it, Hack. Nothing's gonna happen."

So then I went to Lieutenant General Bill Carpenter, who was by chance the top combat commander in Korea and Hunt's ultimate boss man.

"You've got to save this guy," I said. "We're losing an Abrams here."

Carpenter was one of our finest, fightingest studs and damn smart, too. We went back a long way. Bill commanded a rifle company attached to my battalion in Vietnam where he was recommended for the Medal of Honor.

"Okay, Hack," he said. "I'll look into it."

Carpenter spent one day with Hunt's battalion, one full day and night during a live-fire exercise. He stayed in the trenches with the soldiers while the bullets were whistling and explosions were crashing all around them. It was so real he even cautioned Hunt about how far he was bringing his boys in training.

"That's how you train soldiers for warfare," Hunt said. "You don't train them by watching a video. You train them on a field of battle, getting them as close to the real thing as you can."

"Yeah," General Carpenter said. "But you're on the edge of overdoing it."

Carpenter was impressed enough to put in a special, glowing report on Hunt, recommending he be promoted and sent to war college. That saved a true warrior's ass from being shuffled aside and put out to pasture as a lieutenant colonel. He made bird colonel; they sent him to the war college and Harvard to get his master's, all the right system punches to prepare him for a star. But he has never gotten a regular brigade, so he is finished in the Army. Because he's not a salesman. Just a great leader who knows his job and leads his warriors by follow-me example.

Recently I went down to the Pentagon to visit Dennis Reimer, the Army Chief of Staff. He's a good man, movie star handsome with short-cropped silver hair, a six-foot-three Paul Newman in Army green trying hard to change the system.

I mentioned Hunt.

"I love him," he said.

"Why don't you make him a general?" I said.

He looked at me and changed the subject to the threat of mines in Bosnia. I could see the pain in his cool blue eyes. It's the System. The fucking System. The deadly System we have to shoot between the eyes.

Right now, no matter what position you hold within that system, to buck it means death. James H. Mukoyama, a valiant captain in my Hardcore Battalion in Vietnam, rose to major general in the U.S. Army Reserves. In 1994, he went before Congress and testified that federal Reserve units were being converted to state Guard units at the cost of combat efficiency, risking soldiers, lives just to preserve pork. A Government Accounting Office study confirmed that the damage to the taxpayer was more than $180 million. Mukoyama's reward? The Perfumed Princes in the E-ring forced him to retire. The first Asian American in our country's history to command a division, he had served America for over three decades. The top brass treated him like a guy with the Ebola virus. No gold watch. Just a gaping, bleeding belly wound where he fell on his sword, doing the right thing for his country.

When it comes to big bucks, the Perfumed Princes don't hesitate to cheat, lie, and steal. They trashed Sanford Mangold, a brilliant Air Force officer who was sure to be a general. The Air Force had rated him in the top 1 percent of its colonels and he was a boy wonder at the Space Command. Then he blew the whistle on MILSTAR to his boss, a lieutenant general who agreed with him. MILSTAR was a Cold War dinosaur, a satellite system that could maintain communications in spite of radiation within a "nuclear environment." Sandy felt the system was no longer necessary since the Soviet Union was dead. He showed how we could save $640 million immediately and nearly $5 billion over five years by paring it down. The Army, Navy, and Marines all bought his idea.

Then Les Aspin gave the MILSTAR a death row reprieve even as the Defense Secretary was conducting a phony "bottom-up" review. The delay gave Sandy's enemies at Space Command an opportunity to save their Crown Jewel. Because they could not argue the case on the merits, they went after Sandy personally, stacking the deck, lying, and doing everything they could to diminish his credibility. It worked. MILSTAR was back on the books and Sandy was out on the streets. It took him 210 days and $50,000 in legal fees to clear his name. Later, I went to the Air Force Chief, General Ronald Fogleman, and told him that Sandy was a good man, a truth teller, that he should be part of the Chief's personal staff. Fogleman subsequently tried to right a wrong by awarding Sandy the Meritorious Service Medal for "outstanding leadership, integrity, and intellect." "I missed the chance for general," he told me. "But I did not join to be a general. I joined to serve my country." Again, America was the loser.

If you are not a Perfumed Prince or a courtier, you get killed. If you know how to work the system, it will work for you. Slickness is all. If you know the game, you can break all the rules. The trick is never to get caught—and if you do get caught, be sure to have friends in high places.

Not long ago in *Newsweek*, I nailed Air Force General Joe Ashy, who has four stars, for spending nearly $250,000 to fly himself and a blond bombshell from Naples, Italy, to Colorado Springs. When he was reassigned from NATO to head the Space Command in Colorado, he used a C-141 Starlifter, capable of carrying two hundred passengers for himself, a twenty-one-year-old female "enlisted aide," and his cat Nellie. The aircraft had a crew of thirteen, including a chef, along with a plush VIP compartment, the equivalent of a presidential suite—all for the use of the general and his twenty-one-year-old. To support this costly junket aloft, the Air Force had to arrange two hazardous air-to-air night refuelings. This took two KC-135 tankers, each with a crew of five, one from England for a rendezvous over the Azores, one from Delaware to refuel over the East Coast, to bring the general home. His trip cost the American taxpayer nearly a quarter of a million dollars. To fly commercial would have cost $1,465. In Colorado Springs, General Ashy then splurged another $100,000 for a new conference room—because the old one didn't have a view of Pikes Peak. The money wasted on that quarter-of-a-million-dollar flight would have bought sixty-two shells for an M-1 tank, more than enough for the crew's annual gunnery training. It could have paid for fifty hours of flight time to train fighter pilots.

I'll say this for General Ashy, he knows how to look after the troops. According to a retired Air Force sergeant major, he gave his "aide," a gorgeous Air Force enlisted woman, an equally gorgeous black negligee. To boost morale, I suppose. Generals are not supposed to have enlisted aides except as cooks or stewards. Their aides are supposed to be junior officers. A lackey on his staff coming to his defense over the flight told me, "All the senior officers do it, it's an entitlement." There was an investigation. The general got a slap on the wrist. He had to reimburse the government about $5,000. The official story was that he wrote and mailed his check the day the judgment came out.

My gut feeling was *This is a check that is gonna get lost.* I called Charlie Murphy, a former Marine who works for Senator Chuck Grassley, a Republican reformer from Iowa. Charlie is a bulldog, the Ralph Nader of high-ranking military corruption.

"Hey, Charlie," I said. "I got a feeling this check's never going to hit the bank."

"I do, too," he said. "But I got a plan."

His plan was to call the Defense Department's Inspector General in ninety days, long enough for the check to clear the Bank of Siberia, and ask for a copy of the paperwork. Three months later, I phoned Charlie again. He got in touch with the IG and discovered that the check had never been banked. The Air Force went to DEFCON 1 on damage control: It was lost in a drawer, locked in a safe, left behind when the Secretary of the Air Force flew overseas—everything but the tooth fairy ate it. Finally, the check turned up in the safe of Sheila Widnall, the Secretary of the Air Force, a good pal of General Ashy. Only then was it deposited. Later I discovered that the Air Force was going to pay $300,000 to buy a VIP pallet like the one Ashy used for the secretary so she could also fly high. I confronted the Air Force flacks with the story. A general and colonel huddled three hours concocting a cover. I was on deadline, so I rang an Air Force captain who was baby-sitting the action.

"What the hell's taking so long?" I asked her.

She was as frustrated as I was. She said, "They are crafting a reply."

"Out of what—granite?"

They finally came back and said the whole project was on hold. But I knew the order was already at the factory. Several weeks later, after the story ran in *Newsweek*, the Air Force quietly canceled the secretary's VIP flying palace.

While chasing the Ashy story, I was lied to more than I have ever been lied to in my life by officers, majors and colonels who should have known better, but were protecting their bosses. The experience left me with the view that if I could, I would eliminate every flack position in the U.S. military. All the flacks do is spin, deceive, and promote their service. This takes thousands of people and costs the taxpayer hundreds of millions of dollars every year.

The Ashy story was just one small example of the horror stories that now come my way every week from frustrated warriors. Only the big ones make fleeting headlines: Tailhook; the U.S.S. *Vincennes* cover-up—from Captain Will Rogers who shot down the Iranian Airbus to Admiral William Crowe, Chairman of the Joint Chiefs of Staff, who, according to Roger Charles, lied about what really happened—the Air Force sweetheart deal with Lockheed, a $500 million ripoff; the Army's attempts to hide friendly fire deaths during Desert Storm and now Gulf syndrome. These are only the tip of the iceberg.

Why is this happening? Too many generals and admirals no longer ask what they can do for their country but what they can do for their individual services and for themselves. They've forgotten about defending America and their sacred obligation to look out for the troops. They have also forgotten the oath they took as young ensigns and lieutenants.

It blows me out how badly the media cover the defense story. Half of the discretionary spending in the federal budget goes to the military, which eats up one sixth of the total budget. When the military screws up, the facts—and the stories—are earth-shaking. The media latch on to stories like Tailhook, and rightly so, but it is always off covering O. J. or finding out who Timothy McVeigh's girlfriend was in kindergarten when billions are being blown and lives are being lost. You'd think editors would devote more of their own scarce resources to bulldogging stories like MILSTAR; or the early Bradley, that flaming coffin; or the $39 billion B-1, a military albatross; or the Sergeant York antiaircraft system that cost $13 billion and couldn't shoot down a fat pigeon sitting on the end of its gun.

How different the picture is today from the way we Americans started out. My ancestors had muskets and hand axes and they carried their blankets and rations on their backs. Back then they were led by a few good dedicated men. If an order was bullshit, the troops sounded off: "Look, Captain, this doesn't make a damn bit of sense." We had a democratic army, not a Prussian military machine that clicks heels to all orders right or wrong. Back then the tooth was long and sharp and could bite like hell and inflict great pain on our enemies. There was little tail. But from 1776 to now we've grown an incredible bureaucratic tail and the teeth have been getting smaller and smaller. If we keep going this way, we will end up trying to gum our future enemies to death.

Our military setup is essentially the same organization George Washington had at the Delaware, except that in 1948 we threw in the Air Force. If IBM did not change with the times, it would belly up, which it almost did. But the U.S. military machine has locked itself in concrete.

Featherbedding is worse than stupid. It weakens our national defense. During the peak of the Cold War, the United States contributed almost 400,000 troops to NATO. After the Iron Curtain collapsed, that force was reduced to 100,000 troops. Our cutting edge in Europe today is made up of only four U.S. fighting brigades with a total of 12,000 fighters. Half of them are now in Bosnia enforcing the peace. But if you look around the assorted command posts in Europe, you will find 12,000 clerks, jerks, and generals—as many desk men as warriors. Even though the force level has been reduced by almost 300,000 people, the top brass has hardly been touched. Not one chalet or villa has been closed down or one VIP aircraft mothballed. A battalion commander in Europe told me, "All we do is suffer VIPs. In the American component of NATO, there are 4 four-stars, 6 three-stars, 27 two- and one-stars, and their staffs all oversupervising twelve combat maneuver battalions." *Never have so few been commanded by so many to so little good purpose.*

Since 1946 the United States has spent about $14 trillion (in 1996 dollars) on military toys and boys supposedly defending America. We tend to forget how much that really represents. If you lined up a fleet of bulldozers on the East Coast and pushed everything into the drink on the West Coast—every building, every car, every bridge, and every brick and stick made by human beings—and replaced them with brand-new stuff, you would still have a pocketful of change left over from $14 trillion.

Insane spending has turned our military machine into a gadget-loaded truck with a hundred more gold-plated cylinders than it needs. In 1996, defense spending continued at 80 percent of Cold War levels. This does not track when you figure 60 percent of all defense spending during the Cold War was to stop the Soviets.

Since the end of the Cold War we have cut troops and mothballed or retired thousands of ships, guns, and airplanes, but the United States still spends on average $300 billion a year defending the globe, more than the rest of the world combined. This massive post–Cold-War spending places an unacceptable burden on the American economy and saddles the nation with a military force that is far too powerful in some respects and too weak in others. The Japanese and the Germans spend a fraction of their GNP defending themselves. Why shouldn't they? We do it for them while they clean our clocks on the economic front. Meanwhile, the United States spends $30 billion per year on education—a mere 10 percent of the defense budget—on our youth, the hope and future of this country. Where are our priorities, especially when you consider that for the moment we have no serious enemy threatening us?

Look at our main adversaries for the foreseeable future: Cuba, Libya, Iran, Iraq, North Korea, Syria, China. The Islamic Confederation. All lack a strong industrial base. None of these countries, besides China, has the Bomb yet or an effective army, navy, or air force. Except for China their major capability is terrorism and chemical-biological sneak attacks. North Korea remains a serious threat, but it looks like the North Koreans may be only a few years away from bellying up.

Yet we are continuing to buy more and more Cold War relics designed to fight an "Evil Empire" that died when the Berlin Wall came tumbling down. Paradoxically, as the Pentagon budget increases, the size of our armed forces decreases. In 1991, during Desert Storm there were 2.1 million people on active duty. Today we have about 1.5 million.

Our troop strength has been reduced by almost 30 percent, yet we have seen no corresponding decline in defense spending, and the fat cats in the Pentagon and other high headquarters have not really felt the retrenchment knife. Ordinary Americans remain ignorant of the waste and the rip-offs until they hear the horror stories: the $7,000 coffeepots; the $12,000 hooks for the C-17 transport that cost a couple of hundred bucks at the local hardware store; and $250,000 airplane rides for generals. And that news seems to stick in their heads about as long as a Jay Leno joke.

The Perfumed Princes have it down to a science. Even when they retire, they pass through a golden revolving door and go to work for the same guys from whom they used to buy weapons. They get big salaries for ringing up their former subordinates and saying, My company's gear is just right for you. The newest twist is for foreign countries to rent our retired generals. For example, General Carl Vuono, who used to be the Army Chief of Staff, runs a company called Military Professional Resources, with two thousand retired American generals and admirals and other officers for rent. His boys were training the Croatians long before President Clinton sent in the regulars.

It has been nearly forty years since President Eisenhower warned us about the Military Industrial Congressional Complex, but we still haven't brought it under control. Military leaders, politicians, and defense contractors form what has been called the Iron Triangle, a dollar-gobbling, three-legged monster with the Pentagon, Congress, and the White House as the main players. They work in concert to keep defense money in play, many of the hoodwinkers moving from one leg to another as opportunities arise. The MICC has a strong and supportive constituency, the first leg being the politicians and their pork. Defense pork means jobs for the lawmakers' districts and states and jobs mean votes. Votes mean reelection. Building a bigger pig trough to wallow in is the endgame of the Iron Triangle.

Who needs more B-2 bombers, more Sea Wolf submarines? Who needs another Nimitz-class carrier? We could save hundreds of billions of defense dollars if we had leaders with enough common sense and courage to do the job right. What's a good citizen supposed to think when Congress approves buying a new fleet of bombers? The people who build the bombers have worked it all out. They know how many congressional districts there are in the United States. In district after district, someone is building a part of the new bomber. So when a vote comes up in the House of Representatives, the bomber gang can say to the politicians in Washington, "Look, you're going to lose jobs in your district if you don't vote for this weapon." The tail assembly is made in one guy's district, the rudder is made in another. And so it goes. Then the bomber cabal fans out into those districts and whispers to the people, "Your Representative wants to vote against our bomber, and that vote is going to cost you your job. Your local economy is going to go down the tubes. You gotta put the squeeze on him." So the faxes and phone calls and letters start raining down on Washington, and even the politician smart enough to know America's national security wouldn't miss a beat if no more bombers were produced will still vote for it. And every time he plays along with the bomber boys, he's sticking a knife in your back and keeping the guys who end up face-to-face with the enemy from having the right stuff.

That's how the game is played today and the tragedy is that too many of our top military leaders are playing right along with the politicians and the defense contractors. Among the Perfumed Princes in the E-ring, knowing how to sell a weapons system to Congress is prized far more highly than knowing how to fight a unit in battle. They are spending us broke on wonder toys while shortchanging the boys who do the fighting and dying. We have wonderful young soldiers, high school and college graduates full of idealism who would fall on a grenade in a second to save their buddies or their country. But as they rise through the ranks, they become corrupted. Somewhere around lieutenant colonel all too many become so obsessed with making it to the Pentagon's E-ring, they'd poison their mother if that's what it took to get them there.

The higher these officers rise, the more they lose their nerve—because they have more to lose. These are not leaders who would take the point, stand tall to correct a wrong, or fall on their sword for their men. Politically correct, they go along to get along. They are not risk takers: To take risks, you have to have balls; what they have is bureaucratic cunning. So this democracy of ours, which is the best form of government going, is hurting. That's what's causing the anger and frustration of ordinary citizens today. But few within the Beltway are listening to the drumbeat. Washington has become as bad as London during the American Revolution, when the king and lords could not understand why all that tea was floating in Boston Harbor.

The MICC is winning big-time. Let's consider a few examples. The Navy needs sixty attack submarines to sink a Soviet submarine fleet that has already sunk. During the Cold War, we had about one hundred attack submarines to do that job. We retired forty. We retired the U.S.S. *Los Angeles* even though it had twenty years of shelf life left. But now we are building one Sea Wolf sub per year at $4 billion a pop because they provide ten thousand jobs. They are certainly not a military necessity. My solution: Send those ten thousand sub-making employees to Hawaii. Even if we pay their salaries, all room and board, and throw in all the luaus they can hula to and grog they can drink, we will still save about $3.4 billion a year by not building, crewing, and maintaining the unneeded subs.

We are currently building two new aircraft carriers and the MICC is pressing for a third. With planes, goodies, and crew, the total life-cycle cost for these three will be about $60 billion. This will give us eleven carriers. Building them means a lot of jobs in Virginia, one place where the Republicans and Democrats do not have a gridlock. Both Senator John Warner (R-Va.) and Senator Charles Robb (D-Va.) pull the same rope at the same time in the same direction. Carriers in Virginia mean jobs, and jobs mean Warner and Robb will return to their MICC seats of power.

We now have more flattops than we need. Even in Desert Storm, we only used six, four in the Red Sea, two in the Gulf. Many naval experts say five big carriers and six mid-size carriers will do the job nicely. The U.S. Navy has almost twice the number of carriers of the combined forces of the rest of the world: One of our big carriers has more strike power than all other foreign carriers combined. Russia has only one clunky supercarrier, and it's on its ass for lack of maintenance, spare parts, and serviceable aircraft with only eight qualified pilots able to launch off its rusted deck.

We also have a new fleet of fighters in the works. The F-22 stealth fighter will take the place of the F-15. The total package for four hundred F-22s will run over $72 billion. The purpose of this new gold-plated wonder weapon is to shoot down the Soviet Union, even though the Evil Empire crashed in 1990. Some strategists like former naval officer and Assistant Secretary of Defense Lawrence J. Korb, now of the Brookings Institution, say we could maintain our technological edge and keep our stealth capability as sharp as it was in the Gulf War with fifty to seventy-five of these aircraft. The Pentagon has dozens of other costly wonder weapons—many are relics from the Cold War—fighters, missiles, and whiz-bang stuff. The MICC says we urgently need these weapons even though they are aimed at an enemy that has ceased to exist. On top of that the individual services want even more.

Budgetary sleight-of-hand jeopardizes our warriors and steals defense dollars from where they are needed to sharpen combat readiness. Grunts don't have many guardians or much political clout.

Grunts don't contribute to political action committees like the people who make stealth aircraft or $60 billion worth of aircraft carriers. Between 1991 and 1993, seventy members of the House Armed Services Committee scooped up almost $3.5 million from PACs. The two biggest beneficiaries were Pennsylvania's John Murtha, who gobbled up $331,200, and Texas's Charles Wilson, who raked in $285,150. Both are big cheerleaders for a costly, bug-ridden, half airplane–half helicopter that does two things well: crash and waste heavy-duty money. It's no coincidence the V-22 Osprey is built in their districts.

Defense contractors make up the second leg of the MICC. They have had a license to steal since just before World War II, and with forty-seven years of Cold War, this abuse has become a virtual science. Until only recently, most defense contractors operated on a cost-plus basis and the taxpayers paid dearly. Flat-out cheating, heavy padding, and out-and-out corruption have become endemic. Not long ago I checked through just a few headlines in *Newsweek*'s stacks: Litton Industries agrees to reimburse government $82 million for overcharging on a defense contract. Northrop Corporation found guilty of rigging bids on the stealth bomber to the tune of over $60 million. Grumman agrees to pay U.S. government $20 million to avert criminal charges for defrauding the U.S. Navy. Curtiss-Wright pays $17.5 million to settle charges that top executives fraudulently overcharged the Navy. Gencorp and Alliant Techsystems pay $12 million for conspiring to cheat the military on the sale of cluster bombs. United Technologies pays $150 million for improperly billing the government for military helicopters. Eight suppliers were accused of providing substandard hardware to the U.S. military, defective stuff such as ammo and aviation parts used by our warriors in combat.

According to the General Accounting Office, CEOs of giant defense contractors are some of the highest-paid executives in the U.S.A. Here's a recap of the latest available figures: James Mellor of General Dynamics ($11.3 million); Bernard Schwartz of Loral ($4.6 million); Dennis Picard of Raytheon ($2.5 million); Daniel Tellep of Lockheed Martin ($2.4 million); Harry Stonecipher of McDonnell Douglas ($1.6 million); Krent Kresa of Northrop Grumman ($1.6 million). By one report, Schwartz was expecting to get an $18 million bonus for arranging the sale of Loral's defense division to Lockheed Martin, a merger that would mean thousands of layoffs.

The military brass at the Pentagon make up the third leg of the Iron Triangle. Sadly, many three- and four-star generals put service over country and forget their oath of office: to defend America, not their service's budget. These Perfumed Princes fight for their individual service's budget harder than they fight any external enemy. I learned while working in the Pentagon that the U.S. Air Force was a bigger enemy to the U.S. Army than the North Vietnamese we were fighting at the time or the Soviets we were preparing to fight. Almost every three- and four-star is not a war fighter but a super arms salesman, whose job it is to schmooze the lawmakers and hype their service's needs.

The most recent schmooze champs come from the Air Force, madly selling the need for those four hundred new F-22 fighters as well as a fleet of $350-million-a-copy C-17 cargo planes. If we acquired only a dozen C-17s for combat assaults, the balance of the air fleet could be modernized with Boeing 747 cargo aircraft at half the cost. And we do not need all four hundred of those new fighters. The Navy is hustling for new carriers and submarines, and fighting hard to stop the fleet from being trimmed. The Army is spinning the need for a multibillion-dollar digital battlefield and fleets of new helicopters and tanks even though the tank itself will soon be obsolete. And the Marine Corps is beating the drums for that costly lemon, the V-22 Osprey.

Each service jealously guards its overlapping missions and competes with the other services for more, more, more, just the way poker players always want more chips. Their interservice rivalry blinds them. They can't admit that their forces are too large for today and will be inadequate and too expensive for tomorrow. Despite all the we-can-bomb-them-back-to-the-Stone-Age bluster, no single service can win a war by itself. Wars take a team effort despite the hype put out by the U.S. Air Force about how it won Desert Storm.

In this era of huge national debt with our bonehead politicians eagerly assuming the role of Global Cop regardless of the cost, there is no room for duplication, overlap, or waste. But we are wallowing in redundancy. To pick just one example, we have two ground forces—the U.S. Army and the U.S. Marine Corps. Both are in sharp competition, especially for post–Cold-War 911 missions in places like Somalia and Haiti, the only games in town. Both forces do the same thing: break things and kill people. The Marine Corps, which is in theory a primarily light, hit-and-pull-out amphibious force, has not hit any beaches for real since the 1950 Inchon invasion during the Korean War. After that, Marine divisions fought as Army divisions for the next two years in Korea, the decade of the Vietnam War, and again during Desert Storm. Now the Marine Corps, like the Army, has developed a heavy-tank capability. Meanwhile, the Army, while whistling the Marine Corps hymn, is buying ships to replicate the U.S. Marine Corps's floating reserve strategic mission. It sounds like something from *M*A*S*H*, except that it's real and we're paying for it with scarce tax dollars.

President Clinton hasn't just joined the Military Industrial Congressional Complex, he's become its main cheerleader. One month before the 1994 State of the Union speech, President Clinton gave the Pentagon another $11 billion, then told Congress, "We must not cut defense spending any further." Even that was not enough. After the Republicans ripped his knickers in the off-year congressional elections for neglecting defense, he breathlessly flung another $25 billion into the Pentagon larder. That was supposed to square away the "readiness gap." Now we have a readiness gap. Remember the bomber gap? Remember the missile gap? The readiness gap is the same type of con. Ready for what? Mexico to invade? A Canadian blitzkrieg attack? As things now stand, President Clinton is spending more on defense than Richard Nixon proposed at the height of the Cold War.

Instead of whipping our forces into realistic shape for post–Cold-War missions, President Bill Clinton caves in to every Perfumed Prince and defense contractor who wails Gimme, Gimme, Gimme. Here's why. He's politically vulnerable because not only did he not serve in Vietnam, but he got caught lying about his draft status. Then he ushered in his administration with his misguided gays-in-the-military initiative, which turned off just about everyone who has ever worn a uniform. Scrounging votes in Connecticut, he gave the kiss of life to the Sea Wolf after George Bush cut off its air supply. The B-2 bomber was dead on arrival in California. Then President Clinton, snapping after the state's fifty-four electoral votes, did a double shuffle to keep it alive. First, he announced he would not fund the B-2 in the 1997 budget. Then he waffled, going ahead with the $496 million allocated for 1996. He did this, even though the chiefs of the Army, Navy, Marine Corps, and Air Force have said no more B-2s. My guess is that the B-2, like the B-1 resuscitated by Ronald Reagan, will be born again. And of course the President isn't the only politician playing this game. Anywhere in America where there's a defense factory, you find a politician and a deal. This mess is bipartisan. Campaigning in California, Bob Dole sounded like Clinton—

though in slow motion—about the B-2. The bottom line: You pay $30 billion, they get to hustle twenty-five thousand California votes.

Ike knew all about this insidious cabal. Just before leaving office, when he coined the phrase the Military Industrial Congressional Complex, his advisers convinced him he should drop "congressional" because it would rub too many porkers the wrong way. Before he turned the White House over to JFK, he said, "In the councils of governments we must guard against the acquisition of unwarranted influence, whether sought or unsought, by the military industrial complex. The potential for the disastrous rise of misplaced power exists and will persist. We must never let the weight of this combination endanger our liberties or democratic process. We should take nothing for granted. Only an alert and knowledgeable citizenry can couple the proper meshing of the huge industrial and military machine of defense with our peaceful methods and goals, so that security and liberty may prosper together."

Because most of us slept through the speech, we now have pretty much four of everything. To start with, we have four air forces, one each for the U.S. Navy, the U.S. Army, the U.S. Marine Corps, and, of course, the U.S. Air Force itself. All do long-range "deep strike" operations and all do close air support missions. Would you believe we also have: four legal corps; four intelligence commands; four personnel centers; four chaplain branches; and four information/data commands. We also have: four training systems from boot to flight training right up to charm schools for new generals; four supply systems and four research and development commands. Hey, there are even four different color socks and four different hats!

Imagine GM or Ford having four sets of everything. The stockholders would revolt. CEO heads would roll. But at the Pentagon piggy bank these are the rules: Don't worry about the taxpayers; bigger is better; always preserve pork fat while cutting military muscle. Whenever you need more money, simply scream, "The bogeyman is coming." Believe me, the Pentagon's cash addiction is America's real enemy, the enemy within.

Each service is spending millions of dollars duplicating research and development work. Today, the Army, the Navy, and the Air Force are all developing defense systems to deal with missiles. The same is true on close air support. The Army has the Apache helicopter; the U.S. Marine Corps has the F-18C, the Harrier jet, and Super Cobra gunship; the U.S. Navy has the F-18D; and the U.S. Air Force has the F-15E and F-16. Of course, they are slowly phasing out the best one, the A-10 Warthog, because it's apparently not sexy or costly enough. Only the A-10, the Apache, and Cobra choppers reliably put iron down on the close air support target. The others—all fast burners—can't stay over the targets long enough, fly too fast, and are too thin-skinned. Using a fast burner aircraft to provide combat air support is like trying to throw a golf ball from a car going 500 miles per hour into a coffee mug from a freeway overpass.

Our waste would run a small nation. We are now paying $5.5 billion a year for forty-two Guard combat brigades. Hanging on to them is like General Motors keeping a buggy division. I brought this up one day with Mike Stone, as we were sitting in his Pentagon office, not long after Desert Storm. The downsizers were all over him and he was griping to me about not having enough defense dollars to keep the Army fighting fit.

"You know," I said, "you've got to get rid of the Army National Guard's combat brigades and divisions."

"Hack," he said, "it would be easier to get rid of God."

The National Guard is entrenched, totally bulletproof. Every politician on the Hill fights to keep it, because the Guard is in every district and state. It's been that way since before the days of Abe Lincoln. The Joint Chiefs of Staff say they only need fifteen National Guard combat brigades. President Clinton insists for political reasons on keeping forty-two. Those twenty-seven unneeded brigades cost $3 billion a year. So we're keeping the Guard on the porkroll even though Desert Storm proved once again that the Guard's combat brigades were a waste of good money. War is now come as you are, so there is no longer time for the National Guard to get ready. Today's Army and Marine combat units have to be as ready as a jet fighter sitting on a carrier's catapult, always prepared to launch.

We are even being ripped off on what we pay defense contractors for research and development. Here's how it works: The taxpayer pays billions to develop a new weapons system. The contractor then sells the system overseas and his profits skyrocket, because he hasn't had to pay the full bill for R and D. There have been attempts to control this scam, but the porkers have resisted them. The 1996 Defense Authorization Bill, going their way, slid right into the trough.

Blubber, bureaucracy, and duplication do not give the United States a stronger force or the edge in a fight. In fact, they produce a weaker military force, because during lean times no individual service has enough to do the job and eventually all become as hollow and limp as a used straw. To get an idea of our current girth, just check out the Pentagon in Washington D.C., ten square miles of posturing and stealing surrounded by reality. In 1945, at the end of

World War II when the U.S. military had 13 million men and women in uniform, there were 25,000 people in the Pentagon. Today, when our military is down to 1.5 million there are 26,000 people trying to look busy in that same five-cornered, concrete bunker and thousands more big and little bureaucrats squirreled away in office buildings all over town. Amazing. More clerks than trigger pullers. Doesn't make much sense. Clerks do not put holes in enemy soldiers.

At present, we have only 192,000 trigger pullers out of 732,000 people on active duty in the Army and Marines. That is like leaving 75 percent of the cops in the police station to shuffle papers, rather than out on the beat fighting crime. You can find the same blubber at every headquarters in the military. Down in the trenches, where warriors fight and die, we never have enough people.

Looking back, as long as I've been around, backstabbing and interservice rivalry have been a running sore. It amazes me we won World War II with so much throat cutting going on between the admirals and generals. Back in the 1940s, the Army and Navy were constantly fighting and undercutting each other. The Navy wanted priority for the Pacific and the Army wanted priority for Europe. Neither would release critical resources—aircraft, landing craft, ships, and other war-fighting gear—to the other, and in some tactical situations they would not even support each other. Vietnam was even worse. Naval Air had its own set of targets and the U.S. Air Force had theirs. The Marines and Army each had their separate war, and neither would coordinate their efforts. It was as if the four services were fighting four different wars. In the Delta, the Army was pretending it was the Marines. In the north, fighting big land battles, the Marines were acting more like the Army. Meanwhile, almost 60,000 young men were KIA and 300,000 more were ground up as WIA, partly because there was no unity of command, a vital principle of war, or concentration of effort (the principle of mass) or even one single game plan (the principle of objective).

We, a superpower, lost our first war in U.S. history to a third world army that never had one airplane over our battle positions, one sub or ship attacking our convoys, or one SCUD missile sputtering overhead; yet they kicked our ass. In 1980, Jimmy Carter's hostage rescue mission failed in the Iranian desert because all four services jumped into the act to justify their budgets. Everything failed. A U.S. Navy chopper flown by U.S. Marine Corps pilots crashed into a U.S. Air Force C-130 aircraft killing U.S. Army troopers. The operation was a humiliating disaster. The invasion of Grenada in 1983 was just as bad. It was a Keystone Kop comedy. The Navy bombed the Army. The Marines, with only one infantry battalion, outperformed the Army's eight parachute battalions. SEALs drowned. Green Berets were killed in badly planned missions. Things got so bad an officer on the beach in a firefight had to use a credit card to call Fort Bragg, North Carolina, the fastest way to get the U.S. Navy off Grenada to adjust its gunfire, since the Army and Navy radios couldn't communicate with each other. That was almost forty years after World War II, where this kind of stupidity happened all the time. The same thing happened during the rehearsals for the 1994 invasion of Haiti. These disasters all occurred because we had four different squads playing under four different coaches, all trying to stick it to each other. Joint service operations have been anything but joint.

And when they do try to play together—watch out. Not long ago the Pentagon decided to conduct an antiterrorist exercise on the island of Guam. The ostensible idea was to see what would happen if terrorists tried to capture a nuclear weapon. Their real purpose was to justify maintaining a seldom-used air base on the island.

Here's what happened. U.S. Marines posing as terrorists set out to see if they could steal a nuke coming to Guam aboard a plane from South Korea. The "bomb" was supposed to be a dummy. But unknown to most of the participants in the exercise, the plane arrived carrying live weapons along with the dummies. Then the aircraft was parked on a sector of a runway over an underground fuel dump.

During the mock raid, the Marine "terrorists" tossed a concussion grenade that rolled up against the leg of a USAF security guard. In this case, the soldier happened to be a young woman, Airman Laurie "Ranger" Lucas. The grenade blew off her foot and killed her. The Marines had been ordered to use concussion grenades since they were running an exercise anyway, the idea was they might as well take out a broken gate in order to replace it at taxpayers, expense from the training exercise budget. What about Ranger Lucas? Training accident, you know.

If the Marines had lobbed that grenade a mite harder, the exercise would have created more devastation than the Oklahoma City bombing. The grenade would have ignited over the fuel dump, torching off 220,000 gallons of aviation fuel stored under the plane with the nukes. The explosion would have been conventional, not nuclear, but there's no doubt it would have spewed radioactive debris into the winds and across the Pacific. Oh, the colonel would have gotten not only his new gate but a rebuilt airfield—and a new asshole.

A very close call.

CHAPTER 13

CHANGE— OR BELLY UP: WHITEFISH, 1996

Now that you've gotten inside what's gone down from Desert Storm to Bosnia, you can see that our military machine is sputtering like a worn-out tank. Any mechanic can see it's only hitting on two cylinders while using a hell of a lot of expensive gas. If a doctor looked at our armed forces, he'd see a bloated patient, lacking coordination and into advanced denial. A business management expert would see redundancy, inefficiency, and obsolescence; skewed priorities; a corrupt personnel setup; and a lousy accounting system. Our military is a sick institution, and if we don't do something about it, the Republic, the very source of our freedom, is going to die.

It is always a bitch to cause change. Woodrow Wilson said, "If you want to make enemies, try to change something." The first thing that needs to be changed is a lot of minds. For many in our country, U.S. military spending has become a sacred cow. A cow that has been well milked. This is the prevailing mind-set: We must not challenge the idea of national defense. It has become like social security and Medicare—entitlements few lawmakers have the guts to challenge or cut.

Those who have a vested interest in the system—or are soaring on a hawk trip—say, "You gotta be a Commie or a loony tune to want to cut defense spending." This is the prevailing view of a lot of wild-eyed Congress members, most generals, all weapons makers, and one hell of a lot of uninformed citizens. They have all been conditioned to believe and fear that the Nazis, the Russians, the bogeymen, are coming. When they hear me pushing for military reform, these people say, "Only a traitor or a fool would want to tamper with the defense apparatus of the U.S.A."

My answer is, the hell with that. Change is urgently required. We've got to say, "What's the smart way to defend America without going broke, without collapsing like the Greeks and Romans, the Spanish, the French, the Dutch, the British, and the Soviets?"

In arguing for peace through strength, Ronald Reagan had it half right. But all he did was throw money at the problems. He solved nothing. I shudder when I look at how our 1995 tax dollars were spent: 50 cents for entitlements, 16 cents for interest on the debt, 17 cents for defense. That leaves us only 17 cents on the dollar to run the nation—for education, health, highways, transportation, and everything else. If entitlements, interest, and defense spending are not brought under control, even that 17 cents is going to shrink like a cheap T-shirt. In a few cases we are headed in a positive direction, such as closing redundant bases—we've already cut over a hundred of them—for an enormous savings. But we need to apply the same nonpartisan approach to every aspect of military spending, from tent pegs to satellites, from squads to divisions. Everything has to be challenged. We must constantly ask ourselves, "Is this particular expenditure necessary?" We need to say, "Look, is there a smarter way of defending America?"

I think there is.

The first step is to clean up the military's top leadership. We must find leaders who will put country and soldiers first, not their individual service and their career, war fighters in the great tradition of Matt Ridgway, Chesty Nimitz, Jimmy Doolittle, and Vic "the Brute" Krulak. To paraphrase JFK, we need leaders who will ask not what their service can do for them, but what they—and their service—can do for their country. We haven't lost all the good guys, studs like General Hugh Shelton and Admiral Snuffy Smith. And as much as it pains me to say it, President Clinton deserves full marks for putting three top warriors who care about people and want to do the right thing in charge of the Air Force, the Army, and the Marine Corps: General Ronald Fogleman, General Dennis Reimer, and General Charles Krulak, son of the Brute. But these good appointments haven't changed the behavior of the Perfumed Princes entrenched throughout the system. The culture is all-pervasive and the Perfumed Princes are in position to outscheme and outlast these few good men. When a problem is systemic, reform must also be systemic. What we need is a total overhaul, a task that could take ten years.

We have to get back to the bare bone basics, like a sergeant before going on patrol. That sergeant doesn't get into an esoteric exercise. He asks his captain, "What's the mission? How many men am I going to have? What's the enemy situation?" Then he scopes out the weather and terrain and moves out. The first thing America has to do is identify its real enemies—today, next year, and on into the middle of the twenty-first century. We can no longer afford to inflate our enemies list just to justify our weapons list.

Sure, Cuba, Libya, Iran, Iraq, North Korea, and Syria are out there growling today. Down the track, China and Russia could be far bigger and more dangerous fire breathers. Let's consider for a moment the kind of threat they present. With North Korea, Iran, and Iraq, we could be looking at smaller versions of Desert Storm. With the rest of the little guys it's terrorism, bombs, chemical and biological weapons. With China and Russia the danger is total war: ICBMs, massed armies, and mass destruction right here on Main Street U.S.A. as well as the enemy's home ground. That is a worst-case scenario, of course. None of it may happen. But it's what we have to plan and spend for intelligently. Right now we are not gearing our efforts to genuine threats but to the overwhelming momentum of the Military Industrial Congressional Complex. We have to plan for two quite different kinds of war: low-tech and high-tech. Low-tech fighting of the kind we saw in Somalia hasn't changed much since the boys took up throwing rocks in the Stone Age to decide who got to be chief and which tribe would sit on the top of the mountain. Low-tech is man pitted against man—small-scale, deadly, and with significant political repercussions. High-tech is laser against laser, long-distance war with satellites and digital battlefields, the mighty computer chip driving whiz-bang weapons only now in their infancy.

Right now, our high-tech and low-tech capabilities are out of sync. We are behaving and spending as if we are already living and fighting in a *Star Wars* galaxy. But you've seen that Mogadishu, Port-au-Prince, or Tuzla aren't exactly Jedi warrior stuff. For the immediate future, given the end of the superpower face-off and the nature of our new, fragmented world, we are a lot more likely to find ourselves in a shitload of low-tech Mogadishu-style fights than high-tech shootouts. But our obsession with high tech, the search for the ultimate wonder weapon, has kept us from striking the right low-tech–high-tech balance.

We have to have a high-tech force that is ready to defend our skies and to fight futuristic over-the-horizon wars. Contrary to the general impression, Desert Storm came nowhere near what lies ahead. The irony is that even in high tech, our thinking is behind the power curve. We have a lot of good weapons systems with fifteen or twenty years shelf life left in them. We are replacing them early with stuff that's supposed to be whiz-bang, but is at best only next year's model. For example, we are replacing the F-15 fighter, one of the best ever made, with the F-22, when we should be thinking about a whole new family of missiles controlled by satellites and computer chips to take over the Wild Blue Yonder. We need to be looking even farther over the horizon. The weapons we have now, with updates, should last us easily into the 2010–2020 time frame. They are good enough to protect us against anybody out there right now. We should hold the line with these weapons while we invent and test a new generation of genuine future-shock hardware capable of convincing any Nasty, big or small, that if he slaps leather we will shut off his lights.

In the meantime, we have to recognize that our most frequent fights are going to be low-tech and then put a far greater priority on getting our warriors the right stuff. We have to spend more money to provide them a new family of small arms and lightweight, reliable communications gear. They need a better mine detector that can sniff out plastic mines. Our research and development people should devote the same attention they give to stealth technology to better personal gear in the form of body armor as well as improved detection and protection from nuclear-biological-chemical attack. Not as in Bosnia, where the supply system, crashing at the last minute, couldn't put body armor and winter gear on our warriors, throw a bridge across the Sava River in less than a week without making it a rat fuck, or get combat forces to the right place at the right time.

We should put people with hands-on experience, not just whiz-bang engineers and salesmen, in charge of weapons development. As things stand now, the whole process of getting war toys is staffed out to people who stand to make bigger bucks the more we spend, people with little fiefdoms to defend, people who have never been shot at, people who have worn white lab coats all their lives. Among the people who set the priorities for new weapons, we should also have a lot more experienced warriors. And we should give them veto power over projects like MILSTAR and the early Bradley that turn into dollar-sucking monsters.

The truth is, we need to change and reform the entire U.S. armed forces. To do the job, we need a task force of the best brains in America. Here are the sorts of ideas I think we should be looking into. The most important, long past due, is to consolidate all our fighting forces into one unified service.

- We could merge the Army and the Marine Corps, giving the new outfit its own air arm including strategic bombers, and eliminate the Air Force.
- We could put the Navy in charge of all strategic missiles. The Navy would keep its traditional role and its own air arm, as well. In the short term, this would mean that swabbies would be manning missile silos on land. Long-range, and as quickly as possible, the missiles would be moved to subs at sea.
- We could form a new Strategic Mobility Command, taking the planes from the Air Force and the cargo ships from the Navy, and tasking it with all our air lift and sea lift needs.
- We could reconfigure the Pentagon, eliminating the separate service chiefs and the civilian secretaries of the Army, Navy, and Air Force in favor of a combined Defense Force headquarters, run by a civilian Secretary of Defense.
- We could eliminate the current evaluation-report system and the zero-defects mentality that produce highly inflated evaluations for Perfumed Princes who avoid all risks while destroying original thinkers and truth tellers. The existing system only encourages lying and officers who are afraid to step up to the plate. We would be better off with a simple report that asked, "Would you want to see your son serve under this guy in combat?"
- We could get rid of the Pentagon's command assignment system, its promotion boards, and its insistence that everyone be a jack-of-all-trades: romping, stomping combat leader, clever manager, brilliant staff officer, and West Point-caliber instructor. To destroy ticket punching and get warriors where we need them, we could return

the choice of battalion and brigade leaders to division commanders; squadrons and groups to wing commanders; and ships to fleet commanders—and establish a professional Command Corps.

- We could, in light of the Boorda suicide, set up a better fail-safe system to make sure that our four stars in all the services are emotionally stable and able to stand up to the enormous stress and psychological pressures they must face every day.
- We could merge the National Guard and the Reserves into one streamlined organization. To cut waste and sleaze, the new outfit would be under federal, not state, control, where the politics of pork is even worse than in the Beltway.
- We could merge the duplicate, non-war-fighting functions of the services—intelligence, medical, legal, acquisitions, research and development, logistics, training, chaplain, and support—so that we have one, not four, outfits for each task. We could also consolidate all the service academies into a single American Defense Academy.
- We could set up a Weapons System Closing Commission to operate like the Base Closing Commission. No more than 5 percent of the commissioners to come from defense contractors—and none of those could vote on any weapon they themselves make.
- We could transform the Federal Acquisition Regulations (FAR), isolating the final decision on defense contracts from politicians and generals. We could make it impossible for Bill Clinton to promise Sea Wolfs to Connecticut and Bob Dole to promise B-2s to California. Those decisions could be in the hands of independent boards of review composed of people who are barred from ever working for defense contractors.
- We could control congressional porkers swilling at the trough by making sure that no more than one fourth of the members of the Armed Services/National Security Committees or the Defense and Military Construction Subcommittees of the House and Senate Appropriations Committees represent states or districts with major military installations, military contracts, or large numbers of civilians working on defense contracts. No member of

those bodies should accept any Political Action Committee contributions from any company that has received more than $1 million in defense contracts or that earns more than half its total revenue from the Defense Department. The terms of senators and representatives chairing those committees and subcommittees should be limited to no more than four years. The fun and profit would go out of serving, but we would sharpen our focus on the tip of the nation's spear and wind up with a less costly, more effective defense.

- We could nail shut the MICC's revolving door by banning anyone who serves on a military-related committee of Congress or who serves in a flag-rank position in the military from working for any defense contractor for at least five years after leaving the job. That goes for the senior staff of both institutions. Period. No exceptions.
- We could restore the draft in the form of universal national service for all young men and women, who could choose between military or civilian assignments. This would save money and restore a sense of civic duty and other basic American values to our youth while keeping the military in better tune with democracy.

These ideas are not as extreme as they may sound. Senator Barry Goldwater, an Air Force Reserve major general who knew the armed forces from the inside, fought hard to reform the military. In 1986, it looked for a time like help was on the way when Congress passed the Goldwater-Nichols Defense Reform Act. The legislation made the Chairman of the Joint Chiefs of Staff the head man of our military. The goal was to no longer have the four service chiefs bickering like fish salesmen on a hot day. This act increased attention to joint operations—the coordination of all the combat power of all the services in pursuit of a single mission, the winning of wars.

The nature of high-tech warfare, such as we saw in Desert Storm, requires total integration, total unity of effort. In other words, one coach to ensure that everyone runs down the field in the same direction and uses only one game plan. This single act gave the United States quick and decisive victories with minimum casualties during the Panama invasion and Desert Storm. The recent operation in Haiti again showed unity in action. Army choppers flying off Navy carriers. All services worked very well together. As a result, they got the job done quickly and with minimum fuss.

But the 1986 Goldwater-Nichols Act was just the first wobbly step toward military reform. In 1995, Congress established a bottom-up commission headed by Harvard's John White to study the roles and missions of each service with the goal of streamlining the military machine and preparing it for the twenty-first century. The commission looked into twenty-five areas, from equipment acquisition to procurement to supply management to war fighting. Unlike the Goldwater-Nichols Act, this commission accomplished zilch. The commission merely rubber-stamped the status quo, perhaps because its brain trust was composed largely of retired generals and admirals who had created the mess in the first place. Reformers 0—MICC 10. White, who accomplished considerably more for himself than for his country, followed that time-tested principle: If you can't beat 'em, join 'em. He became the number two man in the Pentagon.

The idea of a bottom-up review is brilliant, but it can only work when the people conducting it are the sort of straight shooters you've met in this book: Jim Burton, Bill Carpenter, Dave Evans, Dave Hunt, Sandy Mangold, Jim Morrison, Jim Mukoyama, Mike Wyly. They would bring integrity, vision, and moral courage to the job. They would not be bought, bent, or intimidated by the MICC. They would slay that evil sucker.

The military will never volunteer for this trip. The Perfumed Princes won't reform themselves. That would be like expecting the Mafia to share crime intelligence with the FBI. They have had it too good for too long. You cannot expect the hangman to burn the rope. If nothing is done, economics alone will force change, but it will be the wrong kind. The point is soon coming when Congress will have to say to the Pentagon, "We don't have $300 billion to give you. We have only $200 billion." But what will happen then? The ticket punchers will preserve the flagpoles, the headquarters, the staff cars for the brass; but there will no longer be enough warriors in the foxholes, and flagpoles don't shoot cannonballs.

We have to wake up. Paradoxically, the larger society is driven by well-meaning but misplaced idealism that constantly gets us stuck in the wrong fights. We keep writing moral checks we can't cover. We have seen them bounce in Vietnam and Somalia and we will probably see "Returned for Insufficient Funds" in Haiti and Bosnia. We can't just jump into every fight around the globe, no matter how hard television tugs at our heartstrings. We have to balance compassion with realism and ask ourselves before each mission, Is our national security endangered? Is this operation really necessary? Are there things we better fix at home first? As things stand now, we don't ask these questions until the flag-draped coffins turn up at Dover and Travis Air Force bases. We have to think harder about consequences before we act, before the honor guards are firing their last salutes, before we get more white headstones and mothers like Gail Joyce wondering whether their young warriors have died in vain.

The same idealism has led us into a wrongheaded form of political correctness that now threatens to tear our armed forces apart at the very moment it has been going through the trauma of downsizing. The purpose of our military is to defend a democracy, not to be a democracy. To forget this is to invite disaster. Desert Storm and Somalia, Haiti and Bosnia, show beyond any doubt that we've got an enormous job ahead of us to make sure that our armed forces are combat-ready to whip anyone who wants to destroy us. Under the circumstances, it is astounding to me that the first item on President Clinton's political agenda after his inauguration was gays in the military followed by his energetic efforts to put women into combat positions. Beyond that, the Army right now is getting ready to cut twenty thousand people; so we are going to see one or two more divisions disappear. At the same moment, maddeningly, we are keeping more than 16,000 people with asthma, heart conditions, and other medical problems that make them nondeployable. If they were discharged, there would be no need to lose those divisions. This PC-think has to stop. If our society continues to put its bleeding heart ahead of military muscles already stretched to the limit, we're going to end up knocking our ownselves out.

It may be mission impossible to cure everything wrong with America at once, but cleaning up the military is the first step in shoveling out the barn. If the military isn't as clean as an M-16 rifle before firing, then there's no way America can be put right, because it is going to jam. A "new look" military would give our nation a strong nuclear deterrence, a high-tech force ready for the twenty-first century and a low-tech force to protect our national security interests and to fire brigade hot spots. Now is the time to strike, while no serious enemy is breathing down our necks. It will take about ten years to make the transition. Israel did it when they formed the Israeli Defense Force, surrounded at the time 200 to 1. So it's not an impossible mission. But, as Ike warned, citizens must get involved. Change begins when people finally get angry enough to say, "Enough is enough," and demand reform.

Change can save a lot of money, too, right at the moment we need it most. History shows that uncontrolled defense budgets are a terminal illness for empires and superpowers. I believe that the savings from this kind of reform, in today's dollars, would run $100 billion a year. Eventually we could save $150 billion each year even as we are creating a much more effective military machine. According to the Center for Defense Information, a reform-minded think tank headed by John J. Shanahan, a retired vice admiral, each billion dollars spent on military procurement produces 25,000 jobs. If spent in the civilian sector, the same billion would create 30,000 jobs in mass transport, or 36,000 in housing or 41,000 in education or 47,000 in health care. As the *Defense Monitor*, the think tank's journal, has put it: "The irony is that continued military spending to support military related jobs is forcing budget cuts for superior job creating civilian activities." So reform could produce more jobs and more meaningful jobs. At the same time, maybe we should just give some of the savings back to the people from whom it was taken—the taxpayers.

I have been around soldiers and wars long enough to see the Death Wheel turn a lot of times. I have seen things get splattered over and over again, always the same patterns, the same mistakes. We never remember or learn from the past. But unlike so many of our top brass, fortunately—or unfortunately—I don't suffer from CRS.

Perhaps my memory's so good because as a teenager I saw so many fine young men wasted because of impostors masquerading as combat leaders and slick, shallow politicians who got off on the strongest aphrodisiac of them all—POWER—all of them pretending to be altruistic leaders.

Hackworths live a long time. Several of my forebears made it to the over-100 mark and a lot of them were raising hell and drinking good whiskey well into their nineties. So be warned, all you Perfumed Princes and Propaganda Poets, all you slick political porkers and weapons makers with your hands in the till. I intend to keep sniffing around like an old coyote, chewing on the Military Industrial Congressional Complex and calling 'em as I see 'em.

I intend to continue to tell it like it is to my fellow citizens with the hope that one day they will become so damn mad they'll stomp out the bad guys and retake charge of this great but sinking republic.

Since I'm no longer able to defend America by swinging my sword with the young studs, I will continue picking my targets and honing my pen into the ultimate bayonet. Hopefully the pen *will* prove to be mightier than the sword.

Meanwhile, as the troops say: Keep ten, watch out for mines—and stay up on the radio.

AFTERWORD

Hazardous Duty is the third leg of a trilogy. I wrote the first installment, *Vietnam Primer*, thirty years ago so grunts and their skippers in the 'Nam could learn from the brave warriors who went before them and not walk through the same minefields. At the end of the eighties, I added *About Face* to spike the lie that we won all the battles only to lose in Vietnam because of politicians, press and peaceniks. I also wanted to pass on the lessons I had learned from a lifetime in combat. The mission in *Hazardous Duty* was to tell the story of the main American military engagements since the Berlin Wall came tumbling down and to suggest ideas for reforming our military machine so we'll be ready for the fights coming our way in the twenty-first century as surely as the sun rises in the east.

Since *Hazardous Duty* came out, I've received thousands of letters from soldiers saying "Iron on the target." A master sergeant in the Army wrote to say that in his eighteen years of service, he'd "seen a lot of political crap. But today it's far worse. People get promoted because their uniform is perfect, but their job performance sucks." From within the Air Force, a major said he was worried that upcoming defense cuts would come from "training, ammunition, individual equipment and quality-of-life budgets for our soldiers and units, before they come out of the big ticket, high roller programs." And a lieutenant colonel from the U.S. Marines cut me a new set of orders for the future in just three words: "Don't let up."

Scores of readers taken with the book's message—that bad senior leadership, outmoded thinking, and political correctness are killing our armed forces—have told me they sent copies to the President, the Secretary of Defense and members of Congress asking them to read it. So far, the Commander-in-Chief hasn't called to say, "Hack, you're full of it! What the hell you been smoking, boy?" let alone, "What did you mean? Come by and talk to me and my staff about those radical changes you want to see in the military. I think it would be healthy to find out the 'why' behind what you're saying."

Despite the support from the warriors who are down at the bottom, I haven't heard a word from anyone in power who could start the critical changes needed to stop the waste, convert blubber to muscle, and provide the right kind of leadership for the killing fields that lie ahead of us. The powers in residence are not eager to talk, let alone reform. I'll never forget the warning of a U.S. Navy commander who wrote me to say, "Right now our military is antiquated, broken, not prepared for future wars, and political corruption is so entrenched in the upper echelons of all services that it will probably take another Pearl Harbor to wake up the American people."

After *Hazardous Duty* was published in hardback I talked to audiences all across the United States. People would nod their head and say, "I agree. We gotta do something."

But that's where it ends. No action.

We have become a nation of zombies zoned out on apathy. It's nuts. Most people want to know the truth; but once they find out, they retreat from action. It's hard to make change. It requires thinking. Getting off our butts.

One day after I'd given a talk in Florida, an elderly lady walked up and said to me, "So the Pentagon is a mess. So it's wasting billions of our dollars. So it's putting young men and women in harms way. What are *you* doing about it?"

I felt stunned.

"Madam, that's exactly why I wrote *Hazardous Duty*. To wake up the people."

"Book, shmook," she said. "People steeple. You talk such nonsense. Books don't wake up people, organizations do."

She was absolutely right. Over the past ten years, in *About Face*, *Hazardous Duty*, and "Defending America," my nationally syndicated newspaper column—and in so many radio and television talk interviews and talk shows I've lost count—I have tried to inform Americans about the shame of our Perfumed Princes in the military and the fat cat defense industry.

Now, I'm worried that I may only have been preaching to the choir. What that little old lady wanted to tell me was, 'Hey, Hack, you've got to fight fire with fire. Get organized."

Following her advice—and in response to all the others who keep asking me what they can do—I've decided to set up an outfit called SOLDIERS FOR THE TRUTH. The purpose for this lash-up will be to expose inefficiency, incompetence and outright corruption within the Pentagon and the defense industry.

And we need you to make it work.

I am hoping this paperback edition will reach an even larger audience of people willing to scream, "I've had enough. I want change. Now."

If what you are seeing makes you sick, join up. Write to: PO Box 430, Whitefish, Montana, 59937, or send E-mail to: teagles@hackworth.com.

If we, the people, don't save our country by getting the right defense for the right price, who will? The Pentagon? The politicians? The defense industry? Forget it. For more than fifty years they have misused our warriors and weakened the national defense by putting pork above principle.

It's not going to change unless *you* take charge and make them do the right thing.

Any volunteers?

—Hack

ACKNOWLEDGMENTS

I couldn't have told this story without the help of John Boyd, Jim Burton, Bill Carpenter, Roger Charles, David Evans, Ernie Fitzgerald, Dave Hunt, Larry Joyce, Sandy Mangold, Jim Morrison, Jim Mukoyama, Charles Murphy, Chuck Spinney, Jim Stevenson, Mike Suessmann, Mike Wyly, and all those other hundreds of honest soldiers who live by the adage that loyalty is not to the boss or the institution, but to the truth. I am particularly grateful to those who can't be individually named for their courageous assistance because to do so would get their gold watches broken along with their necks.

Maynard Parker at *Newsweek* had the vision and nerve to put me back on hazardous duty along the fronts of the New World Disorder.

Henry Morrison, agent supreme, shepherded this book safely through Manhattan's Valley of the Shadow of Death, ably assisted by Joan Gurgold.

Will Schwalbe at William Morrow and Company, Inc., was a

publisher with guts, Jim Wade, an editor with a fine eye and ear for the military, and Zachary Schisgal, a sensitive coordinator.

Heidi Duncan, Liza Hamm, Rita Hanson, Chris London, and Charlot Schwartz kept the copy moving under pressure above and beyond the call of duty.

Tim Grattan, best friend and partner in crime, offered solid advice and the shelter of Grouse Mountain Lodge in Whitefish, and our pals at Hawk's Cay Hotel provided emergency commo and logistic support.

Lucille Beachy Mathews kept the fort at Fire Support Base Hoboken after Eilhys, Tom, and I crossed the line of departure.

Ben and David Hackworth selflessly sacrificed time with their dad while he closed in on a writer's scariest enemy—the deadline.

GLOSSARY

ABRAMS The official name of the latest United States combat tank designated the M1, including the entire series of tanks, the latest model of which is the M1A1. It mounts a 120-millimeter main gun and weighs 70 tons. Its cross-country cruising speed is 30 miles per hour.

AK-47 The standard Russian-made assault rifle. It has been reproduced and sold worldwide, especially to former clients of the old Soviet Union but also to other groups and countries, some of them inimical to the United States. The AK-47 was the standard infantry weapon issued to the North Vietnamese Army and the Viet Cong during the nine years of U.S. involvement in Southeast Asia, 1964–1973.

APC "Armored personnel carrier," a generic term used to denote several types of armored vehicles designed to carry and protect infantry in combat.

A-10 A U.S. Air Force attack aircraft, nicknamed the Warthog because of its ungamely appearance. Its ability to cruise at relatively slow speed, identify targets, and sustain damage without interference to mission accomplishment, combined to make it the most effective tank-killing aircraft in the Gulf War.

AWOL "Absent without leave," a term no longer in official usage, having now been replaced by UA, which stands for "unauthorized absence." AWOL continues among military personnel as a slang word to connote that someone is *not* where he is supposed to be.

Beast Barracks At West Point, the United States Military Academy, freshmen, called "plebes," report during the summer before the academic year begins. During this first summer of their military lives, they undergo training akin to enlisted "boot camp," with a strong emphasis on discipline. The summer training period is known as Beast Barracks.

B-52 The B-52 Stratofortress bomber has been in service since the Vietnam War and was employed extensively in the Gulf War. A multimission intercontinental heavy bomber, it flies at 50,000 feet and can carry 70,000 pounds of ordnance, nuclear or conventional.

B-1 U.S. Air Force's strategic bomber. It carries a payload of 61,000 kilograms, has a range of 4,580 kilometers, and can reach speeds of Mach 1.25.

Bradley Fighting Vehicle — The Bradley is the U.S. Army's latest armored infantry fighting vehicle. Each vehicle can carry part of an infantry squad and is armed with a 25-millimeter cannon, mounted in a turret.

Bradley platoon — A U.S. Army mechanized infantry platoon, whose squads are mounted in Bradley Fighting Vehicles, is often referred to as a Bradley platoon.

B-2 — The B-2 stealth bomber is programmed to be the U.S. Air Force's next top-of-the-line long-range strategic bomber aircraft.

CG — Abbreviation for "Commanding General."

C1 and C2 readiness reports — All combat units in the U.S. armed forces are required to submit periodic reports, based upon which their combat readiness status is rated from C1, the highest attainable state, incrementally down to C5, which is "unready."

C-141 — U.S. Air Force heavy-lift cargo aircraft.

C-130 — U.S. Air Force medium-lift cargo aircraft.

DefCon 1 — The highest of several "Defense Conditions" established by the Department of Defense and mandating specified preparations for mobilization. DefCon 1 is set when war is imminent.

Dish — Any of the command, control, and communications centers that coordinate aviation operations, so called because of the dish-shaped radar transmitter/receiver. Also, GI slang for Mogadishu in Somalia.

DMZ — "Demilitarized zone." DMZs were established between North and South Korea and North and South Vietnam in order to establish buffer areas in hopes of reducing friction between the conflicting ideologies of the neighboring belligerents.

E-RING Within the Pentagon, offices are divided among five concentric five-sided "rings," lettered alphabetically, the E-ring being the outermost opening to an interior courtyard and the A-ring, the innermost. Because senior officers' offices and headquarters of the more prestigious entities face the courtyard, the E-ring has become symbolic of an inner sanctum of military hierarchy.

F-15 The Eagle, a high-performance, supersonic, all-weather fighter. It carries radar-guided and infrared homing air-to-air missiles and 20-millimeter gun in "air superiority" role; and guided and unguided air-to-ground weapons in air-to-surface role. It has a 9,960-mile radius and flies at supersonic speeds.

.50 CAL A machine gun that fires bullets that measure .50 caliber in size; otherwise known as a "heavy machine gun."

FORSCOM U.S. Forces Command. The U.S. Army command that governs the manning of combat units, worldwide.

FRAG GRENADE Fragmentation grenade: That is the official nomenclature of the infantryman's baseball-size bombshell, commonly called a "hand grenade."

F-22 The stealth fighter aircraft, designed, but yet to become operational.

G-DAY The day designated to begin the U.S. ground attack in the Gulf War. Any letter may be used in a given attack plan to designate the day of the attack. D-day is the most common designator. The calendar date may fluctuate without invalidating the plan. The famous World War II Normandy plan, for instance, was written to detail preparations, etc. that had to be in place a precise period in advance of "D-day." D-day was not set for

June 6 until well after the plan was written and not until acceptable conditions regarding imponderables such as the weather and enemy were known. In the Gulf War, there were two separate major U.S. attack D-days, January 17, when the air attack began, and G-day, February 24, when the ground attack began.

Gore-Tex Commercial material used for outdoor field clothing; it insulates, keeping the soldier warm, yet "breathes" to preclude excessive perspiring and to enhance comfort.

Hail Mary The name assigned in the Gulf War to the overall tactical scheme of holding on the right while going in deep on the left. It is named for the classic football play, which, if diagrammed, would appear the same; that is the form of a priest with one hand raised, giving a blessing.

Hanoi Hilton The nickname given the building that housed U.S. prisoners of war in Hanoi, North Vietnam, during the Vietnam War.

Hawks "Black Hawk," UH-60 assault helicopters. The Black Hawk is presently the U.S. Army's primary assault helicopter, able to transport a combat-equipped infantry squad of eleven soldiers or an external cargo load of up to 8,000 pounds.

Humvee This vehicle has taken the place of the jeep on the American battlefield. "Humvee" is a pronounceable substitute for the official letter designation of the vehicle: HMMWV, which stands for "high-mobility, multipurpose wheeled vehicle." It is a four-wheel drive built on a 1¼-ton payload chassis. Variants include the cargo/troop carrier, communications system carrier, ambulance, TOW missile carrier, and light artillery prime mover.

JIB Joint Information Bureau. See definition below.

JOINT INFORMATION BUREAU An organization within General Norman Schwarzkopf's Central Command, established to control information revealed to the news media.

KEVLAR VESTS Kevlar is a synthetic material used in earlier versions of protective vests issued to soldiers to stop shrapnel and indirect hits from small arms. It is much heavier than state-of-the-art body armor. It is commonly called the "Ranger vest."

KIA "Killed in action." A standard abbreviation used in military administration.

K-1 A tank conceptualized, designed, developed, and manufactured in South Korea; especially suited for mobility in complex, broken, hilly terrain.

K-POT GI slang for the Kevlar helmet.

MAIN SUPPLY ROUTE Any route designated in logistic planning to become a main thoroughfare for transportation of supplies and equipment from rear-area depots to front-line combat troops. Commonly abbreviated MSR.

MARK 19 A 40-millimeter machine gun designed for mounting on combat vehicles, including the U.S. Marine "assault amphibian vehicle" (AAV).

MASH "Mobile army surgical hospital," the Army's standard organization, equipment, and mobile facilities that combine to create a field hospital suitable for combat conditions.

MAVERICK MISSILE A U.S. air-to-surface missile fired from the A-6, A-10, AV-8B, F-16, F-4G, and F/A-18 aircraft. Mavericks are equipped with either a television guidance system or a passive imaging infrared seeker, which gives the missile a "fire and forget" capability. It has a range of fifteen miles and can destroy point targets.

MH-6 Little Bird A small U.S. Army helicopter designed for reconnaissance and command and control.

MiG-21 An aging Russian-made tactical fighter aircraft, named the Fishbed, it has a max speed of Mach 2.0, a radius of 600 kilometers, and carries a payload of 600 kilograms of ordnance.

Milan missile A French-made missile that can be carried by a single man and fired from the shoulder, or mounted on and fired from a vehicle. It is ejected from a launch tube by a booster and propelled by a dual-thrust solid-propellant sustainer rocket motor and guided by a wire. The missile was originally developed to protect ground troops from helicopters; however, it has proven to be extremely effective as a tank killer and a bunker buster. A thermal imaging system gives the missile a night-firing capability.

M1A1A The Abrams tank (see definition above), the M1A1A being the latest improved model. The major improvements over the M-1 that were introduced in 1981 are the 120-millimeter main gun that replaces the M-1's 105-millimeter, and enhanced survivability and fire control.

M-60 The standard U.S. infantry light machine gun, 7.62-millimeter, in use today and since before the Vietnam War.

MSR "Main Supply Route." See definition above.

NCO "Noncommissioned officer," a term designating enlisted personnel qualified for leadership roles including corporals up to sergeants major.

Nimitz-class carrier The Nimitz-class carrier is the Navy's nuclear-powered "supercarrier," designated CVN. These ships are 1,092 feet long, displacing 96,300 tons, and embarking ninety combat aircraft each. They are manned by 6,300 sailors and travel at speeds in excess of 30 knots. The U.S.S. *Eisenhower* (CVN 69) and *Theodore Roosevelt* (CVN 71) were Nimitz-class carriers employed in the Gulf War.

PAC "Political action committee." A group with its own political agenda, usually endeavoring to glean money from or for a constituency.

PAO "Public affairs officer," a military officer whose duties are to brief the press and control information allowed to flow to the press.

Pentagon Follies A pejorative term to describe the daily briefings given in the Pentagon or other senior headquarters for the consumption of the press and/or the "top brass."

Phase line A tactical "control measure" characteristic of World War I-type tactics. A phase line is drawn on the map in front of a moving unit, typically perpendicular to its direction of march. When the unit reaches the phase line, it is required to report its position; sometimes it is also required to halt until a certain time or until orders are received for continued movement. Thus, it serves the function of "keeping the line straight," not a significant or even desirable requirement in modern, fast-moving, fluid, and decisive combat. Heavy use of phase lines in tactics tends to result in an army moving at the pace of the slowest instead of endeavoring to keep up with the fastest. This makes for indecision, and often high casualties while allowing the enemy the luxury of time.

RANGER BODY ARMOR The state-of-the-art bulletproof vest, lightweight and effective.

REMF "Rear Echelon Motherfucker," he who weathers the storm indoors, content to let others face the elements.

REPUBLICAN GUARD Saddam Hussein's elite troops, presumed the pride of the Iraqi Army in the Gulf War.

ROK Republic of Korea, or South Korea. The ROKs were the "good guys" during the Korean War.

RPG "Rocket-propelled grenade," usually applied to Russian- and Chinese-made weapons, which are light, fired from an infantryman's shoulder, and capable of penetrating limited armor and busting bunkers.

RTO Tactical "radiotelephone operator"; an infantryman with a radio strapped to his back.

SAM "Surface-to-air missile." It shoots down aircraft from the ground. SAMs include: missiles that are light, man-packed, and fired from the shoulder; heavier missiles that are vehicle mounted; and others that are fired from fixed installations.

SAM-6 MISSILE A Russian-made, vehicle-mounted surface-to-air missile designed for rapid reaction against low-flying aircraft. It has a slant range of 30–60 kilometers and travels at Mach 2.8. Name: "Gainful."

SAW "Squad automatic weapon"; a lightweight automatic weapon with a bipod that has become the centerpiece of today's U.S. infantry squad.

SCUD A Russian-built surface-to-surface missile that weighs over 13,000 pounds and is usually launched from a truck. It has a range of 280 kilometers and is powered by a storable liquid propellant motor.

SEALs Elite Navy warriors whose duty is to go ashore in advance of an amphibious landing, perform reconnaissance, and destroy beach obstacles. The intense training and extraordinary physical fitness required to perform this mission make these sailors among the toughest U.S. warriors in service today. They are used for special missions on land as well as the seas.

Sea Wolf The latest class of U.S. Navy nuclear-powered attack submarine (SSN).

VII Corps Seventh U.S. Army Corps. One of the major subordinate commands under General Norman Schwarzkopf's Central Command during Operation Desert Storm, VII Corps was commanded by Lieutenant General Frederick Franks. His command included: 1st Armored Division, 3rd Armored Division, 1st Cavalry Division, 2nd Armored Cavalry Regiment, 1st Infantry Division, and United Kingdom's 1st Armoured Division. During the attack, VII Corps swept into Iraq immediately west of Kuwait with final objectives along Kuwait's northern border. Other major commands were XVIII Airborne Corps, which moved along VII Corps' left flank, Joint Forces Command North on VII Corps' right flank, and First U.S. Marine Expeditionary Force and Joint Forces Command East, on the right.

Sheridan A U.S. Army tank, lightly armored, designed more for mobility than firepower or armor protection.

Siegfried Line A defensive line in Germany established by Adolf Hitler in conjunction with Field Marshal Gerd von Rundstedt in the fall of 1944 as the Allies closed on the German border after the Normandy invasion. "Siegfried Line" has become figurative of a line presumed to be impassable, intended to stop an enemy at a given point, and is often connected with ultimate failure.

SITREP "Situation Report." A report routinely required by higher headquarters of subordinate units, either periodically, or upon the occurrence of a noteworthy event.

SMART MISSILES Ordnance that is guided to the target by some force other than gravity, such as infrared heat seeking, radar, lasers, wire, or some other means.

SPECIAL OPS The range of military operations beyond conventional warfare, including counterterrorism, counterguerrilla, demolitions, deep insertions of small independent units, and clandestine intelligence.

STEALTH Aircraft designed to defy radar and other means of detection by virtue of their design, metallurgy, electronics, and other countermeasures.

SUN TZU A Chinese warrior-philosopher who lived approximately 500 B.C. and whose writings have been compiled into a classic book entitled *The Art of War.* His lessons are in essence timeless and continue to provide valuable study material for modern professional warriors.

TACTICAL OPERATIONS CENTER For ground combat units, battalion level and higher, the tactical operations center serves as a command and communications hub. Communications are maintained with ground maneuver elements, higher headquarters, artillery, air, and other supporting arms. Watch officers keep situation maps up-to-date, including maps that reflect friendly operations and the enemy situation. It is often the site of command briefings.

T-55 An obsolete Russian-built tank surviving from the 1950s—no contest for any U.S. or

NATO tank now in operation. It mounts a 100-millimeter main gun.

TOW — TOW stands for "*t*ube-launched, *o*ptically tracked, *w*ire-guided." The standard U.S. infantry antitank weapon, the TOW missile has a range of 3,750 meters and is said to be able to penetrate the armor of any tank in the world today. While the high-explosive shaped-charge warhead weighs only 8 pounds, the system in its entirety weighs 175 pounds and is usually transported by vehicle, though it can be broken down into four units for carriage by infantry. It can be fired from its vehicle mounts (usually from a Humvee), or it can be dismounted for firing.

TRADOC — "Training and Doctrine Command," the U.S. Army command responsible for writing current and up-to-date fighting doctrine and establishing Armywide guidance to train to it.

T-72 — The Russian-built tanks most prevalent among third world former Soviet clients, the T-72 mounts a 125-millimeter main gun and can attain road speeds of 50 miles per hour.

UH-60 — "Black Hawk," UH-60 assault helicopters. The Black Hawk is presently the U.S. Army's primary assault helicopter, able to transport a combat-equipped infantry squad of eleven soldiers or an external cargo load of up to 8,000 pounds.

WARTHOGS — A fond nickname for the A-10.

WIA — "Wounded in action." A standard abbreviation used in military administration.

WOLFHOUNDS — "Wolfhounds" is the nickname assigned to the 27th Infantry Regiment, the unit the author fought in during the Korean War as an enlisted man and then a lieutenant after his battlefield commission.

Index